ScaLAPACK Users Guide

SOFTWARE • ENVIRONMENTS • TOOLS

The series includes handbooks and software guides, as well as monographs
on practical implementation of computational methods, environments, and tools.
The focus is on making recent developments available in a practical format
to researchers and other users of these methods and tools.

ScaLAPACK Users' Guide

L. S. Blackford ▪ J. Choi ▪ A. Cleary ▪ E. D'Azevedo
J. Demmel ▪ I. Dhillon ▪ J. Dongarra ▪ S. Hammarling
G. Henry ▪ A. Petitet ▪ K. Stanley ▪ D. Walker ▪ R. C. Whaley

Society for Industrial and Applied Mathematics

Philadelphia

1997

Library of Congress Catalog Card Number: 97-68164

ISBN 0-89871-400-1 (paperback edition)
ISBN 0-89871-401-X (CD-ROM)
ISBN 0-89871-397-8 (set)

The royalties from the sales of this book are being placed in a fund to help students attend SIAM meetings and other SIAM-related activities. This fund is administered by SIAM and qualified individuals are encouraged to write directly to SIAM for guidelines.

Dedication

This work is dedicated to the pioneers of high-performance computing who blazed a trail, set standards, and made our job easier.

Acknowledgment

We give credit to and thank all of the LAPACK authors for allowing us to reuse large portions of the LAPACK Users' Guide in creating this users guide for ScaLAPACK.

Authors' Affiliations:

L. S. Blackford (formerly L. S. Ostrouchov)
University of Tennessee, Knoxville

J. Choi
Soongsil University, Korea

A. Cleary
Lawrence Livermore National Laboratory

E. D'Azevedo
Oak Ridge National Laboratory

J. Demmel
University of California, Berkeley

I. Dhillon
University of California, Berkeley

J. Dongarra
*University of Tennessee, Knoxville, and
Oak Ridge National Laboratory*

S. Hammarling
Numerical Algorithms Group Ltd.

G. Henry
Intel Corporation

A. Petitet
University of Tennessee, Knoxville

K. Stanley
University of California, Berkeley

D. Walker
University of Wales, Cardiff

R. C. Whaley
University of Tennessee, Knoxville

Contents

List of Figures

List of Tables

Preface

Following the initial release of LAPACK and the emerging importance of distributed memory computing, work began on adapting LAPACK to distributed-memory architectures. Since porting software efficiently from one distributed-memory architecture to another is a challenging task, this work is an effort to establish standards for library development in the varied world of distributed-memory computing.

ScaLAPACK is an acronym for Scalable Linear Algebra PACKage, or Scalable LAPACK. As in LAPACK, the ScaLAPACK routines are based on block-partitioned algorithms in order to minimize the frequency of data movement between different levels of the memory hierarchy. (For distributed-memory machines, the memory hierarchy includes the off-processor memory of other processors, in addition to the hierarchy of registers, cache, and local memory on each processor.) The fundamental building block of the ScaLAPACK library is a distributed-memory version of the Level 1, 2, and 3 BLAS, called the PBLAS (Parallel BLAS). The PBLAS are in turn built on the BLAS for computation on single nodes and on a set of Basic Linear Algebra Communication Subprograms (BLACS) for communication tasks that arise frequently in parallel linear algebra computations. For optimal performance, it is necessary, first, that the BLAS be implemented efficiently on the target machine, and second, that an efficient version of the BLACS be available.

Versions of the BLACS exist for both MPI and PVM, as well as versions for the Intel series (NX), IBM SP series (MPL), and Thinking Machines CM-5 (CMMD). A vendor-optimized version of the BLACS is available for the Cray T3 series. Thus, ScaLAPACK is portable on any computer or network of computers that supports MPI or PVM (as well as the aforementioned native message-passing protocols).

Most of the ScaLAPACK code is written in standard Fortran 77; the PBLAS and the BLACS are written in C, but with Fortran 77 interfaces.

The first ScaLAPACK software was written in 1989–1990, and the appearance of the code has undergone many changes since then in our pursuit to resemble and enable code reuse from LAPACK.

The first public release (version 1.0) of ScaLAPACK occurred on February 28, 1995, and subsequent releases occurred in 1996.

The ScaLAPACK library is only one facet of the "ScaLAPACK Project," which is a collaborative effort involving several institutions:

- Oak Ridge National Laboratory

- Rice University

- University of California, Berkeley

- University of California, Los Angeles

- University of Illinois, Champaign-Urbana

- University of Tennessee, Knoxville

and comprises four components:

- dense and band matrix software (ScaLAPACK)

- large sparse eigenvalue software (P_ARPACK)

- sparse direct systems software (CAPSS)

- preconditioners for large sparse iterative solvers (ParPre)

For further information on any of the related ScaLAPACK projects, please refer to the scalapack index on *netlib*:

```
http://www.netlib.org/scalapack/index.html
```

This users guide describes version 1.5 of the dense and band matrix software package (ScaLAPACK).

The University of Tennessee, Knoxville, provided the routines for the solution of dense, band, and tridiagonal linear systems of equations, condition estimation and iterative refinement, for LU and Cholesky factorization, matrix inversion, full-rank linear least squares problems, orthogonal and generalized orthogonal factorizations, orthogonal transformation routines, reductions to upper Hessenberg, bidiagonal and tridiagonal form, and reduction of a symmetric-definite generalized eigenproblem to standard form. And finally, the BLACS, the PBLAS, and the HPF wrappers were also written at the University of Tennessee, Knoxville.

The University of California, Berkeley, provided the routines for the symmetric and generalized symmetric eigenproblem and the singular value decomposition.

Greg Henry at Intel Corporation provided the routines for the nonsymmetric eigenproblem.

Oak Ridge National Laboratory provided the out-of-core linear solvers for LU, Cholesky, and QR factorizations.

ScaLAPACK has been incorporated into several commercial packages, including the NAG Parallel Library, IBM Parallel ESSL, and Cray LIBSCI, and is being integrated into the VNI IMSL Numerical Library, as well as software libraries for Fujitsu, Hewlett-Packard/Convex, Hitachi, and NEC. Additional information can be found on the respective Web pages:

```
http://www.nag.co.uk:80/numeric/FM.html
http://www.rs6000.ibm.com/software/sp_products/esslpara.html
http://www.cray.com/PUBLIC/product-info/sw/PE/LibSci.html
http://www.sgi.com/Products/hardware/Power/ch_complib.html
http://www.vni.com/products/imsl/index.html
```

A number of technical reports have been written during the development of ScaLAPACK and published as LAPACK Working Notes by the University of Tennessee. Refer to the following URL for a complete set of working notes:

```
http://www.netlib.org/lapack/lawns/index.html
```

Many of these reports subsequently appeared as journal articles. The Bibliography gives the most recent published reference.

As the distributed-memory version of LAPACK, ScaLAPACK has drawn heavily on the software and documentation standards set by LAPACK. The test and timing software for the Level 2 and 3 BLAS was used as a model for the PBLAS test and timing software, and the ScaLAPACK test suite was patterned after the LAPACK test suite. Because of the large amount of software, all BLACS, PBLAS, and ScaLAPACK routines are maintained in basefiles whereby the codes can be re-extracted as needed. Final formatting of the software was done using Toolpack/1 [105].

We have tried to be consistent with our documentation and coding style throughout ScaLAPACK in the hope that it will serve as a model for other distributed-memory software development efforts. ScaLAPACK has been designed as a source of building blocks for larger parallel applications.

The development of ScaLAPACK was supported in part by National Science Foundation Grant ASC-9005933; by the Defense Advanced Research Projects Agency under contract DAAH04-95-1-0077, administered by the Army Research Office; by the Division of Mathematical, Information, and Computational Sciences, of the U.S. Department of Energy, under Contract DE-AC05-96OR22464; and by the National Science Foundation Science and Technology Center Cooperative Agreement CCR-8809615.

The performance results presented in this book were obtained using computer resources at various sites:

- Cray T3E, located at Lawrence Berkeley National Laboratory, National Energy Research Scientific Computing Center (NERSC), supported by the Director, Office of Computational and Technology Research, Division of Mathematical, Information, and Computational Sciences of the U.S. Department of Energy under contract number 76SF00098.

- IBM SP-2, located at the Cornell Theory Center, which receives major funding from the National Science Foundation (NSF) and New York State, with additional support from the Defense Advanced Research Projects Agency (DARPA), the National Center for Research Resources at the National Institutes of Health (NIH), IBM Corporation, and other members of the center's Corporate Partnership Program.

- Intel MP Paragon XPS/35, located at Intel Corporation, Portland, Oregon.

- Intel ASCI Option Red Supercomputer Technology located in Beaverton, Oregon.

- Network of Sun Ultra Enterprise 2 (Model 2170s) workstations, located in the Department of Computer Science at the University of Tennessee, funded by National Science Foundation Grant CDA-9529459, the Center of Excellence – Science Alliance, UT Networking Services, and the UT Computer Science Department.

- Network of Sun UltraSPARC-1 workstations, located in the Department of Computer Science at the University of California, Berkeley supported by DARPA Grant F30602-95-C-0014, NSF Grants CCR-9257974 and PFF-CCR-9253705, as well as California MICRO Grants. Corporate sponsors are: the AT&T Foundation, Digital Equipment Corporation, Exabyte Corporation, Hewlett-Packard Company, Informix Software Inc, Intel Corporation, International Business Machines, Internet Archive, Microsoft Corporation, Mitsubishi Electric Research Laboratories, Myricom Inc, Siemens Corporation, Sun Microsystems, Synoptics Corporation, Tandem Corporation, and TIBCO Inc.

The cover of this book was designed by Andy Cleary at Lawrence Livermore National Laboratory.

We acknowledge with gratitude the support that we have received from the following organizations, and the help of individual members of their staff: Cornell Theory Center, Cray Research, a Silicon Graphics Company, IBM (Parallel ESSL Development and Research), Lawrence Berkeley National Laboratory, National Energy Research Scientific Computing Center (NERSC), Maui High Performance Computer Center, Minnesota Supercomputing Center, NAG Ltd., and Oak Ridge National Laboratory Center for Computational Sciences (CCS).

We also thank the many, many people who have contributed code, criticism, ideas and encouragement. We especially acknowledge the contributions of Mark Adams, Peter Arbenz, Scott Betts, Shirley Browne, Henri Casanova, Soumen Chakrabarti, Mishi Derakhshan, Frederic Desprez, Brett Ellis, Ray Fellers, Markus Hegland, Nick Higham, Adolfy Hoisie, Velvel Kahan, Xiaoye Li, Bill Magro, Osni Marques, Paul McMahan, Caroline Papadopoulos, Beresford Parlett, Loic Prylli, Yves Robert, Howard Robinson, Tom Rowan, Shilpa Singhal, Françoise Tisseur, Bernard Tourancheau, Anne Trefethen, Robert van de Geijn, and Andrey Zege.

We express appreciation to all those who helped in the preparation of this work, in particular to Gail Pieper for her tireless efforts in proofreading the draft and improving the quality of the presentation.

Finally, we thank all the test sites that received several test releases of the ScaLAPACK software and that ran an extensive series of test programs for us.

The royalties from the sales of this book are being placed in a fund to help students attend SIAM meetings and other SIAM-related activities. This fund is administered by SIAM and qualified individuals are encouraged to write directly to SIAM for guidelines.

Suggestions for Reading

This users guide is divided into two parts. **Part I: Guide** contains chapters and appendices providing a thorough explanation of the design and functionality of the ScaLAPACK library. These chapters should be read in the order in which they are presented. **Part II: Specifications of Routines** is a reference manual of the leading comments of each routine in alphabetical order by routine name. A **Bibliography** is also provided, as well as two indexes– **Index by Keyword** and **Index by Routine Name**.

This book assumes a basic knowledge of distributed-memory parallel programming, and is written for an audience of both novice and expert users. Users intimately familiar with specific concepts discussed in this book may choose to not read certain chapters or sections within this book. Some of the chapters can be regarded as stand-alone and read independently. Novice users are directed to focus their attention on special introductory chapters, sections, and example programs, as detailed below.

All users are encouraged to frequently refer to the **List of Notation** and the **Glossary**. The first time notation from the glossary appears in the text, it will be *italicized*. If the user is unfamiliar with any of the concepts defined, a number of books provide background information in parallel programming [6, 33, 65, 66, 67, 70, 75, 87, 92, 95, 99, 112].

- **Chapter 1: Essentials** provides a brief overview of the components of the library, downloading instructions, and details of support for the package. Users who are familiar with the design of the BLAS and LAPACK and acquainted with the existing Web pages may wish to skip this chapter.

- **Chapter 2: Getting Started with ScaLAPACK** presents the basic requirements to enable users to call ScaLAPACK software, together with a very simple example program. Users who are well versed in using ScaLAPACK software may choose to skip this chapter.

- **Chapter 3: Contents of ScaLAPACK** outlines the functionality provided by the package. This is a stand-alone chapter and important for both expert and novice users.

- **Chapter 4: Data Distributions and Software Conventions** discusses process grid layout, contexts, block and block-cyclic data distributions, and documentation and software conventions. This chapter is essential reading for any user who is not familiar with data distributions, array descriptors, and the calling sequences of ScaLAPACK routines.

- **Chapter 5: Performance of ScaLAPACK** provides guidelines to achieve high performance by using ScaLAPACK and presents performance results for a subset of the ScaLAPACK routines on a variety of distributed-memory MIMD computers and networks of workstations. This is a stand-alone chapter.

- **Chapter 6: Accuracy and Stability** discusses the accuracy and stability of the algorithms used in ScaLAPACK, as well as issues of heterogeneous computing. This chapter provides varying degrees of detail, catering to novice as well as expert users.

- **Chapter 7: Troubleshooting** provides a set of installation and application debugging hints for first-time ScaLAPACK users.

- **Appendix A** provides a list of routine names for all driver, computational, and auxiliary routines in ScaLAPACK, as well as the matrix redistribution/copy routines.

- **Appendix B** provides a brief tutorial on how to convert programs using the BLAS to the PBLAS and LAPACK to ScaLAPACK.

- **Appendix C** provides two additional example programs. **Section C.1** contains a more memory-efficient and practical example program, which reads a matrix from a file, distributes this matrix to the process grid, calls the desired ScaLAPACK routine, and writes the solution matrix to a file. **Section C.2** provides a brief description of the HPF interface to ScaLAPACK, as well as an example program.

- **Appendix D** contains Quick Reference Guides for ScaLAPACK, the PBLAS, and the BLACS.

- **Part II: Specifications of Routines** is a reference manual of the leading comments from the source code of all driver and computational routines. This manual can be read selectively as needed.

List of Notation

1D	One-dimensional (as in 1D data distribution)
$1/t_v$	Bandwidth (or throughput) for the network
2D	Two-dimensional (as in 2D data distribution)
$C_f N^3$	Total number of floating-point operations
$C_m N/NB$	Total number of messages
CSRC_	Entry in DESC_ indicating the process column over which the first column of the array is distributed
CTXT_	Entry in DESC_ indicating the BLACS context associated with the global array
$C_v N^2/\sqrt{P}$	Total number of data items communicated
DESC_	Array descriptor for a global array
$DESCA$	Array descriptor for global array A
$DESCB$	Array descriptor for global array B
DLEN_	Length of the descriptor DESC_
DTYPE_	First entry of the descriptor DESC_, identifying the descriptor type
$E()$	Estimated parallel efficiency
F_{MM}	Time per floating point operation in matrix-matrix multiply
Gflop/s	Gigaflops (10^9 floating point operations) per second
IA	Global row index in the global array A indicating the first row of $sub(A)$
$ICTXT$	BLACS context associated with a process grid
INFO	Output integer argument of driver and computational routines indicating the success or failure of the routine
JA	Global column index in the global array A indicating the first column of $sub(A)$
$lcm(P_r, P_c)$	Least common multiple of (P_r, P_c)
LLD_	Entry in DESC_ indicating the local leading dimension of the local array
LLD_A	Local leading dimension of the local array A
LLD_B	Local leading dimension of the local array B
$LOC_c(\text{K}\_)$	Number of columns that a process receives if $K\_$ columns of a matrix are distributed over c columns of its process row.
$LOC_r(\text{K}\_)$	Number of rows that a process would receive if $K\_$ rows

	of a matrix are distributed over r rows of its process column.
M	Global number of rows of the distributed submatrix $sub(A)$
M_	Entry of DESC_ indicating the number of rows in the global array
MB	Global row block size for partitioning the global matrix
MB_	Entry in DESC_ indicating the block size used to distribute the rows of the global array
MB/s	Megabyte per second
Mflop/s	Megaflops (10^6 floating point operations) per second
$MYCOL$	The calling process's column coordinate in the process grid.
$MYROW$	The calling process's row coordinate in the process grid
N	Global number of columns of the distributed submatrix $sub(A)$
N_	Entry of DESC_ indicating the number of columns in the global array
NB	Global column block size for partitioning the global matrix
NB_	Entry in DESC_ indicating the block size used to distribute the columns of the global array
$NBRHS$	Global column block size for the solution matrix
$NPCOL$	Number of process columns in the process grid (equivalent to P_c)
$NPROCS$	Total number of processes in the process grid (equivalent to P)
$NPROW$	Number of process rows in the process grid (equivalent to P_r)
$NRHS$	Global number of columns in the global solution matrix B
NUMROC	TOOLS routine used to calculate the number of rows or columns in a local array
P	Total number of processes in the process grid, i.e., $P_r \times P_c$
P_c	Number of process columns in the process grid
P_r	Number of process rows in the process grid
RSRC_	Entry in DESC_ indicating the process row over which the first row of the array is distributed
$sub(A)$	Distributed submatrix $A(IA : IA + M - 1, JA : JA + N - 1)$
$T()$	Estimated parallel execution time
$T_{\text{seq}}()$	Estimated serial execution time
t_f	Time per floating-point operation (typically F_{mm})
t_m	Time per message (latency)
t_N	Time required to solve a problem of size N on P processors
$t_{N/2}$	Time required to solve a problem of size $N/2$ on P processors
t_v	Time per data item communicated

Part I

Guide

Chapter 1

Essentials

1.1 ScaLAPACK

ScaLAPACK is a library of high-performance linear algebra routines for distributed-memory message-passing MIMD computers and networks of workstations supporting PVM [68] and/or MPI [64, 110]. It is a continuation of the LAPACK [3] project, which designed and produced analogous software for workstations, vector supercomputers, and shared-memory parallel computers. Both libraries contain routines for solving systems of linear equations, least squares problems, and eigenvalue problems. The goals of both projects are efficiency (to run as fast as possible), scalability (as the problem size and number of processors grow), reliability (including error bounds), portability (across all important parallel machines), flexibility (so users can construct new routines from well-designed parts), and ease of use (by making the interface to LAPACK and ScaLAPACK look as similar as possible). Many of these goals, particularly portability, are aided by developing and promoting *standards*, especially for low-level communication and computation routines. We have been successful in attaining these goals, limiting most machine dependencies to two standard libraries called the BLAS, or Basic Linear Algebra Subprograms [57, 59, 74, 93], and BLACS, or Basic Linear Algebra Communication Subprograms [50, 54]. LAPACK will run on any machine where the BLAS are available, and ScaLAPACK will run on any machine where both the BLAS and the BLACS are available.

The library is currently written in Fortran 77 (with the exception of a few symmetric eigenproblem auxiliary routines written in C to exploit IEEE arithmetic) in a Single Program Multiple Data (SPMD) style using explicit message passing for interprocessor communication. The name ScaLAPACK is an acronym for Scalable Linear Algebra PACKage, or Scalable LAPACK.

1.2 Structure and Functionality

ScaLAPACK can solve systems of linear equations, linear least squares problems, eigenvalue problems, and singular value problems. ScaLAPACK can also handle many associated computations such as matrix factorizations or estimating condition numbers.

Like LAPACK, the ScaLAPACK routines are based on block-partitioned algorithms in order to minimize the frequency of data movement between different levels of the memory hierarchy. The fundamental building blocks of the ScaLAPACK library are distributed-memory versions of the Level 1, Level 2, and Level 3 BLAS, called the Parallel BLAS or PBLAS [26, 104], and a set of Basic Linear Algebra Communication Subprograms (BLACS) [54] for communication tasks that arise frequently in parallel linear algebra computations. In the ScaLAPACK routines, the majority of interprocessor communication occurs within the PBLAS, so the source code of the top software layer of ScaLAPACK looks similar to that of LAPACK.

ScaLAPACK contains **driver routines** for solving standard types of problems, **computational routines** to perform a distinct computational task, and **auxiliary routines** to perform a certain subtask or common low-level computation. Each driver routine typically calls a sequence of computational routines. Taken as a whole, the computational routines can perform a wider range of tasks than are covered by the driver routines. Many of the auxiliary routines may be of use to numerical analysts or software developers, so we have documented the Fortran source for these routines with the same level of detail used for the ScaLAPACK computational routines and driver routines.

Dense and band matrices are provided for, but not general sparse matrices. Similar functionality is provided for real and complex matrices. See Chapter 3 for a complete summary of the contents.

Not all the facilities of LAPACK are covered by Release 1.5 of ScaLAPACK.

1.3 Software Components

Figure 1.1 describes the ScaLAPACK software hierarchy. The components below the line, labeled **Local**, are called on a single processor, with arguments stored on single processors only. The components above the line, labeled **Global**, are synchronous parallel routines, whose arguments include matrices and vectors distributed across multiple processors. We describe each component in turn.

1.3.1 LAPACK

As mentioned before, **LAPACK**, or Linear Algebra PACKage [3], is a collection of routines for solving linear systems, least squares problems, eigenproblems, and singular problems. High performance is attained by using algorithms that do most of their work in calls to the BLAS, with an emphasis on matrix-matrix multiplication. Each routine has one or more *performance tuning parameters*, such as the sizes of the blocks operated on by the BLAS. These parameters are machine dependent and are obtained from a table defined when the package is installed and referenced at runtime.

The LAPACK routines are written as a single thread of execution. LAPACK can accommodate shared-memory machines, provided parallel BLAS are available (in other words, the only parallelism is implicit in calls to BLAS). Extensive performance results for LAPACK can be found in the LAPACK Users' Guide [3].

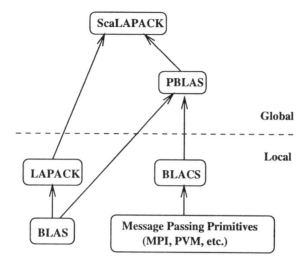

Figure 1.1: ScaLAPACK software hierarchy

1.3.2 BLAS

The **BLAS** (Basic Linear Algebra Subprograms) [57, 59, 93] include subroutines for common linear algebra computations such as dot-products, matrix-vector multiplication, and matrix-matrix multiplication. As is well known, using matrix-matrix operations (in particular, matrix multiplication) tuned for a particular architecture can mask the effects of the memory hierarchy (cache misses, TLB misses, etc.) and permit floating-point operations to be performed near peak speed of the machine. An important aim of the BLAS is to provide a portability layer for computation.

1.3.3 PBLAS

To simplify the design of ScaLAPACK, and because the BLAS have proven to be useful tools outside LAPACK, we chose to build a parallel set of BLAS, called PBLAS [26, 104], which perform message-passing and whose interface is as similar to the BLAS as possible. This decision has permitted the ScaLAPACK code to be quite similar, and sometimes nearly identical, to the analogous LAPACK code.

We hope that the PBLAS will provide a distributed memory standard, just as the BLAS have provided a shared memory standard. This would simplify and encourage the development of high performance and portable parallel numerical software, as well as providing manufacturers with a small set of routines to be optimized. Further details of the PBLAS can be found in [26], [104], and Appendix D.2.

1.3.4 BLACS

The **BLACS** (Basic Linear Algebra Communication Subprograms) [50, 54] are a message-passing library designed for linear algebra. The computational model consists of a one- or two-dimensional *process grid*, where each process stores pieces of the matrices and vectors. The BLACS include synchronous send/receive routines to communicate a matrix or submatrix from one process to another, to broadcast submatrices to many processes, or to compute global reductions (sums, maxima and minima). There are also routines to construct, change, or query the process grid. Since several ScaLAPACK algorithms require broadcasts or reductions among different subsets of processes, the BLACS permit a process to be a member of several overlapping or disjoint process grids, each one labeled by a **context**. Some message-passing systems, such as MPI [64, 110], also include this context concept; MPI calls this a **communicator**. The BLACS provide facilities for safe inter-operation of system contexts and BLACS contexts. Further details of the BLACS can be found in [54]. An important aim of the BLACS is to provide a portable, linear algebra specific layer for communication.

1.4   Efficiency and Portability

ScaLAPACK is designed to give high efficiency on MIMD distributed memory concurrent supercomputers, such as the Intel Paragon, IBM SP series, and the Cray T3 series. In addition, the software is designed so that it can be used with clusters of workstations through a networked environment and with a heterogeneous computing environment via PVM or MPI. Indeed, ScaLAPACK can run on any machine that supports either PVM or MPI. See Chapter 5 for some examples of the performance achieved by ScaLAPACK routines.

The ScaLAPACK strategy for combining efficiency with portability is to construct the software so that as much as possible of the computation is performed by calls to the Parallel Basic Linear Algebra Subprograms (PBLAS). The PBLAS [26, 104] perform global computation by relying on the Basic Linear Algebra Subprograms (BLAS) [93, 59, 57] for local computation and the Basic Linear Algebra Communication Subprograms (BLACS) [54, 113] for communication.

The efficiency of ScaLAPACK software depends on the use of block-partitioned algorithms and on efficient implementations of the BLAS and the BLACS being provided by computer vendors (and others) for their machines. Thus, the BLAS and the BLACS form a low-level interface between ScaLAPACK software and different machine architectures. Above this level, all of the ScaLAPACK software is portable.

The BLAS, PBLAS, and the BLACS are not, strictly speaking, part of ScaLAPACK. C code for the PBLAS is included in the ScaLAPACK distribution. Since the performance of the package depends upon the BLAS and the BLACS being implemented efficiently, we have not included this software with the ScaLAPACK distribution. A machine-specific implementation of the BLAS and the BLACS should be used. If a machine-optimized version of the BLAS is not available, a Fortran 77 reference implementation of the BLAS is available from *netlib* (see section 1.5). This code constitutes the "model implementation" [58, 56]. The model implementation of the BLAS is not expected to perform as well as a specially tuned implementation on most high-performance computers — on some machines it may give *much* worse performance — but it allows users to run ScaLAPACK codes on machines that do not offer any other implementation of the BLAS.

If a vendor-optimized version of the BLACS is not available for a specific architecture, efficiently ported versions of the BLACS are available on *netlib*. Currently, the BLACS have been efficiently ported on machine-specific message-passing libraries such as the IBM (MPL) and Intel (NX) message-passing libraries, as well as more generic interfaces such as PVM and MPI. The BLACS overhead has been shown to be negligible in [54]. Refer to the URL for the *blacs* directory on *netlib* for more details:

```
http://www.netlib.org/blacs/index.html
```

1.5 Availability

The complete ScaLAPACK package is freely available on *netlib* [60, 22, 23] and can be obtained via the World Wide Web or anonymous ftp.

The ScaLAPACK homepage can be accessed on the World Wide Web via the URL address:

```
http://www.netlib.org/scalapack/index.html
```

Prebuilt ScaLAPACK and BLACS libraries are available on *netlib* for a variety of architectures. Refer to the following URLs:

```
http://www.netlib.org/scalapack/archives/index.html
http://www.netlib.org/blacs/archives/index.html
```

At the time of this writing, the e-mail addresses for *netlib* [22, 23] are

```
netlib@www.netlib.org
netlib@research.bell-labs.com
```

Both repositories provide electronic mail and anonymous ftp service (the `netlib@www.netlib.org` site is available via anonymous ftp to ftp.netlib.org). The URL for *netlib* is `http://www.netlib.org/`.

The following sites are mirror repositories:

Tennessee, U.S.A.	`http://www.netlib.org/`
New Jersey, U.S.A.	`http://netlib.bell-labs.com/`
Bergen, Norway	`http://www.netlib.no/`
Kent, UK	`http://www.hensa.ac.uk/ftp/mirrors/netlib/master/`
Germany	`ftp://ftp.zib.de/netlib/`
Taiwan	`ftp://ftp.nchc.gov.tw/netlib/`
Japan	`http://phase.etl.go.jp/netlib/`

General information about ScaLAPACK (and the PBLAS) can be obtained by contacting any of the URLs listed above. If additional information is desired, feel free to contact the authors at `scalapack@cs.utk.edu`.

The complete ScaLAPACK package, including test code and timing programs in four different data types, constitutes some 500,000 lines of Fortran and C source and comments.

1.6   Commercial Use

LAPACK and ScaLAPACK are freely available software packages provided on the World Wide Web via *netlib,* anonymous ftp, and http access. Thus they can be included in commercial packages (and have been). We ask only that proper credit be given to the authors.

Like all software, these packages are copyrighted. They are not trademarked; however, if modifications are made that affect the interface, functionality, or accuracy of the resulting software, the name of the routine should be changed. Any modification to our software should be noted in the modifier's documentation.

We will gladly answer questions regarding our software. If modifications are made to the software, however, it is the responsibility of the individuals/company who modified the routine to provide support.

1.7   Installation

To ease the installation process, prebuilt ScaLAPACK libraries are available on *netlib* for a variety of architectures.

```
http://www.netlib.org/scalapack/archives/
```

Included with each prebuilt library archive is the make include file `SLmake.inc` detailing the compiler options, and so on, used to compile the library. If a prebuilt library is not available for the specific architecture, the user will need to download the source code from *netlib*

```
http://www.netlib.org/scalapack/scalapack.tar.gz
```

and build the library as instructed in the ScaLAPACK Installation Guide [24]. Sample `SLmake.inc` files for various architectures are included in the distribution tar file and will require only limited modifications to customize for a specific architecture.

A comprehensive ScaLAPACK Installation Guide (LAPACK Working Note 93) [24] is distributed with the complete package and contains descriptions of the testing programs, as well as detailed installation instructions.

A BLAS library and BLACS library must have been installed or be available on the architecture on which the user is planning to run ScaLAPACK. Users who plan to run ScaLAPACK on top of PVM [68] or MPI [64, 110] must also have PVM and/or MPI available.

If a vendor-optimized version of the BLAS is not available, one can obtain a Fortran77 reference implementation from the *blas* directory on *netlib*. If a BLACS library is not available, prebuilt BLACS libraries are available in the *blacs/archives* directory on *netlib* for a variety of architecture and message-passing library combinations. Otherwise, BLACS implementations for the Intel series, IBM SP series, PVM, and MPI are available from the *blacs* directory on *netlib*. PVM is available from the *pvm3* directory on *netlib*, and a reference implementation of MPI is also available. Refer to the following URLs:

```
http://www.netlib.org/blas/index.html
http://www.netlib.org/blacs/index.html
http://www.netlib.org/blacs/archives/
http://www.netlib.org/pvm3/index.html
http://www.netlib.org/mpi/index.html
```

Comprehensive test suites for the BLAS, BLACS, and PVM are provided on *netlib*, and it is highly recommended that these test suites be run to ensure proper installation of the packages.

If the user will be using PVM, it is important to note that only PVM version 3.3 or later is supported with the BLACS [113, 52]. Because of major changes in PVM and the resulting changes required in the BLACS, earlier versions of PVM are not supported. User who have a previous release of PVM must obtain version 3.3 or later to run the PVM BLACS and thus ScaLAPACK.

1.8 Documentation

This users guide provides an informal introduction to the design of the package, a detailed description of its contents, and a reference manual for the leading comments of the source code. A brief discussion of the contents of each chapter, as well as guidance for novice and expert users, can be found in the **Suggestions for Reading** at the beginning of this book. A **List of Notation** and **Glossary** are also provided.

On-line manpages (troff files) for ScaLAPACK routines, as well as for LAPACK and the BLAS, are available on *netlib*. These files are automatically generated at the time of each release. For more information, see the `manpages.tar.gz` entry on the *scalapack* index on *netlib*. A comprehensive Installation Guide for ScaLAPACK [24] is also available; refer to section 1.7 for further details.

Using a World Wide Web browser such as Netscape, one can access the ScaLAPACK homepage via the URL:

```
http://www.netlib.org/scalapack/index.html
```

This homepage contains hyperlinks for additional documentation as well as the ability to view individual ScaLAPACK driver and computational routines.

1.9 Support

ScaLAPACK has been thoroughly tested before release, on many different types of computers and configurations. The ScaLAPACK project supports the package in the sense that reports of errors or poor performance will gain immediate attention from the developers. Refer to section 7 for a list of questions asked when the user submits a bug report. Such reports — and also descriptions of interesting applications and other comments — should be sent to

> ScaLAPACK Project
> c/o J. J. Dongarra
> Computer Science Department
> University of Tennessee
> Knoxville, TN 37996-1301
> USA
> E-mail: `scalapack@cs.utk.edu`

1.10 Errata

A list of known problems, bugs, and compiler errors for ScaLAPACK and the PBLAS, as well as an errata list for this guide, is maintained on *netlib*. For a copy of this report, refer to the URL

> `http://www.netlib.org/scalapack/errata.scalapack`

Similarly, an errata file for the BLACS can be obtained by the request:

> `http://www.netlib.org/blacs/errata.blacs`

A ScaLAPACK FAQ (Frequently Asked Questions) file is also maintained via the URL

> `http://www.netlib.org/scalapack/faq.html`

1.11 Related Projects

As mentioned in the Preface, the ScaLAPACK library discussed in this Users' Guide is only one facet of the ScaLAPACK project. A variety of other software is also available in the *scalapack* directory on *netlib*.

> `http://www.netlib.org/scalapack/index.html`

P_ARPACK (Parallel ARPACK) is an extension of the ARPACK software package used for solving large-scale eigenvalue problems on distributed-memory parallel architectures. The message-passing layers currently supported are BLACS and MPI. Serial ARPACK must be retrieved and installed prior to installing P_ARPACK. All core ARPACK routines are available in single-precision real, double-precision real, single-precision complex, and double-precision complex. An extensive set of driver routines is available for ARPACK, and a subset of these is available for parallel computation with P_ARPACK. These may be used as templates that are easily modified to construct a problem specific parallel interface to P_ARPACK.

CAPSS is a fully parallel package to solve a sparse linear system of the form $Ax = b$ on a message passing multiprocessor; the matrix A is assumed to be symmetric positive definite and associated with a mesh in two or three dimensions. This version has been tested on the Intel Paragon and makes possible efficient parallel solution for several right-hand-side vectors.

ParPre is a package of parallel preconditioners for general sparse matrices. It includes classical point/block relaxation methods, generalized block SSOR preconditioners (this includes ILU), and domain decomposition methods (additive and multiplicative Schwarz, Schur complement). The communication protocol is MPI, and low-level routines from the PETSc [109] library are used, but installing the complete PETSc library is not necessary.

Prototype codes are provided for out-of-core solvers [55] for LU, Cholesky, and QR, the matrix sign function for eigenproblems [14, 13, 12], an HPF interface to a subset of ScaLAPACK routines, and SuperLU [96, 39, 41].

```
http://www.netlib.org/scalapack/prototype/
```

These software contributions are classified as *prototype codes* because they are still under development and their calling sequences may change. They have been tested only on a limited number of architectures and have not been rigorously tested on all of the architectures to which the ScaLA-PACK library is portable.

Refer to Appendix C.2 for a brief description of the HPF interface to ScaLAPACK, as well as an example program.

1.12   Contents of the CD-ROM

Each Users' Guide includes a CD-ROM containing

- the HTML version of the ScaLAPACK Users' Guide,

- the source code for the ScaLAPACK, PBLAS, BLACS, and LAPACK packages, including testing and timing programs,

- prebuilt ScaLAPACK, BLACS, and LAPACK libraries for a variety of architectures,

- example programs, and

- the full set of LAPACK Working Notes in postscript and pdf format.

Instructions for reading and traversing the directory structure on the CD-ROM are provided in the booklet packaged with the CD-ROM. A *readme* file is provided in each directory on the CD-ROM. The directory structure on the CD-ROM mimics the *scalapack*, *blacs*, and *lapack* directory contents on *netlib*.

Chapter 2

Getting Started with ScaLAPACK

This chapter provides the background information to enable users to call ScaLAPACK software, together with a simple example program. The chapter begins by presenting a set of instructions to execute a ScaLAPACK example program, followed by the source code of Example Program #1. A detailed explanation of the example program is given, and finally the necessary steps to call a ScaLAPACK routine are described by referencing the example program.

For an explanation of the terminology used within this chapter, please refer to the **List of Notation** and/or **Glossary** at the beginning of this book.

2.1   How to Run an Example Program Using MPI

This section presents the instructions for installing ScaLAPACK and running a simple example program in parallel. The example assumes that the underlying system is a Sun Solaris system; the problem is run on one physical processor, using six processes that do message passing. The example uses MPI as the message-passing layer. The version of MPI used in this example is MPICH (version 1.0.13), and we assume the user has this version installed. MPICH is a freely available, portable implementation of MPI. If MPICH is not installed, refer to http://www.netlib.org/mpi/. If this is run on a different architecture, the user will have to make a number of changes. In particular, the prebuilt libraries will have to be changed. If prebuilt libraries do not exist for the specific architecture, the user will need to download the source (http://www.netlib.org/scalapack/scalapack.tar.gz) and build them.

To use ScaLAPACK for the first time (on a network of workstations using MPI), one should

1. Make a directory for this testing.

   ```
   mkdir SCALAPACK
   cd SCALAPACK
   ```

2. Download the ScaLAPACK example program (about 7 KB) into directory SCALAPACK.
   http://www.netlib.org/scalapack/examples/example1.f

3. Download the prebuilt ScaLAPACK library (about 3MB) for the specific architecture into directory SCALAPACK (e.g., SUN4SOL2) and uncompress.
   http://www.netlib.org/scalapack/archives/scalapack_SUN4SOL2.tar.gz

   ```
   gunzip scalapack_SUN4SOL2.tar.gz
   tar xvf scalapack_SUN4SOL2.tar
   rm scalapack_SUN4SOL2.tar
   ```

   (Note that this tar file contains the library archive and the SLmake.inc used to build the library. Details of compiler flags, etc. can be found in this make include file.)

4. Download the prebuilt BLACS library (about 60 KB) for the architecture (e.g., SUN4SOL2) and message-passing layer (e.g., MPICH), and uncompress into directory SCALAPACK.
   http://www.netlib.org/blacs/archives/blacs_MPI-SUN4SOL2-0.tar.gz

   ```
   gunzip blacs_MPI-SUN4SOL2-0.tar.gz
   tar xvf blacs_MPI-SUN4SOL2-0.tar
   rm blacs_MPI-SUN4SOL2-0.tar
   ```

   (Note that this tar file contains the library archive(s) and the Bmake.inc used to build the library. Details of compiler flags, etc. can be found in this make include file.)

5. Find the optimized BLAS library on the specific architecture.

   If not available, download reference implementation (about 1 MB) into directory SCALA-PACK/BLAS, compile, and build the library.

   ```
   mkdir BLAS
   cd BLAS
   ```

   Download http://www.netlib.org/blas/blas.shar.

   ```
   sh blas.shar
   f77 -O -f -c *.f
   ar cr ../blas_SUN4SOL2.a *.o
   cd ..
   ```

   (Note that this reference implementation of the BLAS will not deliver high performance.)

6. Compile and link to prebuilt libraries.

   ```
   sun4sol2> f77 -f -o example1 example1.f scalapack_SUN4SOL2.a \
             blacsF77init_MPI-SUN4SOL2-0.a blacs_MPI-SUN4SOL2-0.a \
             blacsF77init_MPI-SUN4SOL2-0.a blas_SUN4SOL2.a \
             $MPI_ROOT/lib/solaris/ch_p4/libmpi.a -lnsl -lsocket
   example1.f:
    MAIN example1:
         matinit:
   ```

Note that the `-lnsl` `-lsocket` libraries are machine specific to Solaris. Refer to the `SLmake.inc` for details. MPICH can be found in the directory $MPI_ROOT. On our system we did

```
sun4sol2> setenv MPI_ROOT /src/icl/MPI/mpich
```

7. Run the ScaLAPACK example program.

To run an MPI program with MPICH, one will need to add $MPI_ROOT/bin to the path. On our system we did

```
sun4sol2> set path = ($path $MPI_ROOT/bin)
```

To run the example:

```
sun4sol2> mpirun -np 6 example1
```

The example runs on six processes and prints out a statement that "the solution is correct" or "the solution is incorrect".

2.2 Source Code for Example Program #1

This program is also available in the scalapack directory on netlib (http://www.netlib.org/scalapack/examples/example1.f).

```
      PROGRAM EXAMPLE1
*
*     Example Program solving Ax=b via ScaLAPACK routine PDGESV
*
*     .. Parameters ..
      INTEGER            DLEN_, IA, JA, IB, JB, M, N, MB, NB, RSRC,
     $                   CSRC, MXLLDA, MXLLDB, NRHS, NBRHS, NOUT,
     $                   MXLOCR, MXLOCC, MXRHSC
      PARAMETER          ( DLEN_ = 9, IA = 1, JA = 1, IB = 1, JB = 1,
     $                   M = 9, N = 9, MB = 2, NB = 2, RSRC = 0,
     $                   CSRC = 0, MXLLDA = 5, MXLLDB = 5, NRHS = 1,
     $                   NBRHS = 1, NOUT = 6, MXLOCR = 5, MXLOCC = 4,
     $                   MXRHSC = 1 )
      DOUBLE PRECISION   ONE
      PARAMETER          ( ONE = 1.0D+0 )
*     ..
*     .. Local Scalars ..
      INTEGER            ICTXT, INFO, MYCOL, MYROW, NPCOL, NPROW
      DOUBLE PRECISION   ANORM, BNORM, EPS, RESID, XNORM
*     ..
*     .. Local Arrays ..
      INTEGER            DESCA( DLEN_ ), DESCB( DLEN_ ),
     $                   IPIV( MXLOCR+NB )
      DOUBLE PRECISION   A( MXLLDA, MXLOCC ), AO( MXLLDA, MXLOCC ),
     $                   B( MXLLDB, MXRHSC ), BO( MXLLDB, MXRHSC ),
     $                   WORK( MXLOCR )
*     ..
*     .. External Functions ..
```

```
      DOUBLE PRECISION   PDLAMCH, PDLANGE
      EXTERNAL           PDLAMCH, PDLANGE
*     ..
*     .. External Subroutines ..
      EXTERNAL           BLACS_EXIT, BLACS_GRIDEXIT, BLACS_GRIDINFO,
     $                   DESCINIT, MATINIT, PDGEMM, PDGESV, PDLACPY,
     $                   SL_INIT
*     ..
*     .. Intrinsic Functions ..
      INTRINSIC          DBLE
*     ..
*     .. Data statements ..
      DATA               NPROW / 2 / , NPCOL / 3 /
*     ..
*     .. Executable Statements ..
*
*     INITIALIZE THE PROCESS GRID
*
      CALL SL_INIT( ICTXT, NPROW, NPCOL )
      CALL BLACS_GRIDINFO( ICTXT, NPROW, NPCOL, MYROW, MYCOL )
*
*     If I'm not in the process grid, go to the end of the program
*
      IF( MYROW.EQ.-1 )
     $   GO TO 10
*
*     DISTRIBUTE THE MATRIX ON THE PROCESS GRID
*     Initialize the array descriptors for the matrices A and B
*
      CALL DESCINIT( DESCA, M, N, MB, NB, RSRC, CSRC, ICTXT, MXLLDA,
     $               INFO )
      CALL DESCINIT( DESCB, N, NRHS, NB, NBRHS, RSRC, CSRC, ICTXT,
     $               MXLLDB, INFO )
*
*     Generate matrices A and B and distribute to the process grid
*
      CALL MATINIT( A, DESCA, B, DESCB )
*
*     Make a copy of A and B for checking purposes
*
      CALL PDLACPY( 'All', N, N, A, 1, 1, DESCA, A0, 1, 1, DESCA )
      CALL PDLACPY( 'All', N, NRHS, B, 1, 1, DESCB, B0, 1, 1, DESCB )
*
*     CALL THE SCALAPACK ROUTINE
*     Solve the linear system A * X = B
*
      CALL PDGESV( N, NRHS, A, IA, JA, DESCA, IPIV, B, IB, JB, DESCB,
     $             INFO )
*
      IF( MYROW.EQ.0 .AND. MYCOL.EQ.0 ) THEN
         WRITE( NOUT, FMT = 9999 )
         WRITE( NOUT, FMT = 9998 )M, N, NB
         WRITE( NOUT, FMT = 9997 )NPROW*NPCOL, NPROW, NPCOL
         WRITE( NOUT, FMT = 9996 )INFO
      END IF
*
*     Compute residual ||A * X  - B|| / ( ||X|| * ||A|| * eps * N )
*
      EPS = PDLAMCH( ICTXT, 'Epsilon' )
      ANORM = PDLANGE( 'I', N, N, A, 1, 1, DESCA, WORK )
      BNORM = PDLANGE( 'I', N, NRHS, B, 1, 1, DESCB, WORK )
      CALL PDGEMM( 'N', 'N', N, NRHS, N, ONE, A0, 1, 1, DESCA, B, 1, 1,
     $             DESCB, -ONE, B0, 1, 1, DESCB )
      XNORM = PDLANGE( 'I', N, NRHS, B0, 1, 1, DESCB, WORK )
```

```fortran
      RESID = XNORM / ( ANORM*BNORM*EPS*DBLE( N ) )
*
      IF( MYROW.EQ.0 .AND. MYCOL.EQ.0 ) THEN
         IF( RESID.LT.10.0D+0 ) THEN
            WRITE( NOUT, FMT = 9995 )
            WRITE( NOUT, FMT = 9993 )RESID
         ELSE
            WRITE( NOUT, FMT = 9994 )
            WRITE( NOUT, FMT = 9993 )RESID
         END IF
      END IF
*
*     RELEASE THE PROCESS GRID
*     Free the BLACS context
*
      CALL BLACS_GRIDEXIT( ICTXT )
   10 CONTINUE
*
*     Exit the BLACS
*
      CALL BLACS_EXIT( 0 )
*
 9999 FORMAT( / 'ScaLAPACK Example Program #1 -- May 1, 1997' )
 9998 FORMAT( / 'Solving Ax=b where A is a ', I3, ' by ', I3,
     $          ' matrix with a block size of ', I3 )
 9997 FORMAT( 'Running on ', I3, ' processes, where the process grid',
     $          ' is ', I3, ' by ', I3 )
 9996 FORMAT( / 'INFO code returned by PDGESV = ', I3 )
 9995 FORMAT( /
     $   'According to the normalized residual the solution is correct.'
     $       )
 9994 FORMAT( /
     $ 'According to the normalized residual the solution is incorrect.'
     $       )
 9993 FORMAT( / '||A*x - b|| / ( ||x||*||A||*eps*N ) = ', 1P, E16.8 )
      STOP
      END


      SUBROUTINE MATINIT( AA, DESCA, B, DESCB )
*
*     MATINIT generates and distributes matrices A and B (depicted in
*     figures 2.5 and 2.6) to a 2 x 3 process grid
*
*     .. Array Arguments ..
      INTEGER            DESCA( * ), DESCB( * )
      DOUBLE PRECISION   AA( * ), B( * )
*     ..
*     .. Parameters ..
      INTEGER            CTXT_, LLD_
      PARAMETER          ( CTXT_ = 2, LLD_ = 9 )
*     ..
*     .. Local Scalars ..
      INTEGER            ICTXT, MXLLDA, MYCOL, MYROW, NPCOL, NPROW
      DOUBLE PRECISION   A, C, K, L, P, S
*     ..
*     .. External Subroutines ..
      EXTERNAL           BLACS_GRIDINFO
*     ..
*     .. Executable Statements ..
*
      ICTXT = DESCA( CTXT_ )
      CALL BLACS_GRIDINFO( ICTXT, NPROW, NPCOL, MYROW, MYCOL )
*
      S = 19.0D0
```

```
      C = 3.0D0
      A = 1.0D0
      L = 12.0D0
      P = 16.0D0
      K = 11.0D0
*
      MXLLDA = DESCA( LLD_ )
*
      IF( MYROW.EQ.0 .AND. MYCOL.EQ.0 ) THEN
         AA( 1 ) = S
         AA( 2 ) = -S
         AA( 3 ) = -S
         AA( 4 ) = -S
         AA( 5 ) = -S
         AA( 1+MXLLDA ) = C
         AA( 2+MXLLDA ) = C
         AA( 3+MXLLDA ) = -C
         AA( 4+MXLLDA ) = -C
         AA( 5+MXLLDA ) = -C
         AA( 1+2*MXLLDA ) = A
         AA( 2+2*MXLLDA ) = A
         AA( 3+2*MXLLDA ) = A
         AA( 4+2*MXLLDA ) = A
         AA( 5+2*MXLLDA ) = -A
         AA( 1+3*MXLLDA ) = C
         AA( 2+3*MXLLDA ) = C
         AA( 3+3*MXLLDA ) = C
         AA( 4+3*MXLLDA ) = C
         AA( 5+3*MXLLDA ) = -C
         B( 1 ) = 0.0D0
         B( 2 ) = 0.0D0
         B( 3 ) = 0.0D0
         B( 4 ) = 0.0D0
         B( 5 ) = 0.0D0
      ELSE IF( MYROW.EQ.0 .AND. MYCOL.EQ.1 ) THEN
         AA( 1 ) = A
         AA( 2 ) = A
         AA( 3 ) = -A
         AA( 4 ) = -A
         AA( 5 ) = -A
         AA( 1+MXLLDA ) = L
         AA( 2+MXLLDA ) = L
         AA( 3+MXLLDA ) = -L
         AA( 4+MXLLDA ) = -L
         AA( 5+MXLLDA ) = -L
         AA( 1+2*MXLLDA ) = K
         AA( 2+2*MXLLDA ) = K
         AA( 3+2*MXLLDA ) = K
         AA( 4+2*MXLLDA ) = K
         AA( 5+2*MXLLDA ) = K
      ELSE IF( MYROW.EQ.0 .AND. MYCOL.EQ.2 ) THEN
         AA( 1 ) = A
         AA( 2 ) = A
         AA( 3 ) = A
         AA( 4 ) = -A
         AA( 5 ) = -A
         AA( 1+MXLLDA ) = P
         AA( 2+MXLLDA ) = P
         AA( 3+MXLLDA ) = P
         AA( 4+MXLLDA ) = P
         AA( 5+MXLLDA ) = -P
      ELSE IF( MYROW.EQ.1 .AND. MYCOL.EQ.0 ) THEN
         AA( 1 ) = -S
         AA( 2 ) = -S
```

```
         AA( 3 ) = -S
         AA( 4 ) = -S
         AA( 1+MXLLDA ) = -C
         AA( 2+MXLLDA ) = -C
         AA( 3+MXLLDA ) = -C
         AA( 4+MXLLDA ) = C
         AA( 1+2*MXLLDA ) = A
         AA( 2+2*MXLLDA ) = A
         AA( 3+2*MXLLDA ) = A
         AA( 4+2*MXLLDA ) = -A
         AA( 1+3*MXLLDA ) = C
         AA( 2+3*MXLLDA ) = C
         AA( 3+3*MXLLDA ) = C
         AA( 4+3*MXLLDA ) = C
         B( 1 ) = 1.0D0
         B( 2 ) = 0.0D0
         B( 3 ) = 0.0D0
         B( 4 ) = 0.0D0
      ELSE IF( MYROW.EQ.1 .AND. MYCOL.EQ.1 ) THEN
         AA( 1 ) = A
         AA( 2 ) = -A
         AA( 3 ) = -A
         AA( 4 ) = -A
         AA( 1+MXLLDA ) = L
         AA( 2+MXLLDA ) = L
         AA( 3+MXLLDA ) = -L
         AA( 4+MXLLDA ) = -L
         AA( 1+2*MXLLDA ) = K
         AA( 2+2*MXLLDA ) = K
         AA( 3+2*MXLLDA ) = K
         AA( 4+2*MXLLDA ) = K
      ELSE IF( MYROW.EQ.1 .AND. MYCOL.EQ.2 ) THEN
         AA( 1 ) = A
         AA( 2 ) = A
         AA( 3 ) = -A
         AA( 4 ) = -A
         AA( 1+MXLLDA ) = P
         AA( 2+MXLLDA ) = P
         AA( 3+MXLLDA ) = -P
         AA( 4+MXLLDA ) = -P
      END IF
      RETURN
      END
```

2.3 Details of Example Program #1

This example program demonstrates the basic requirements to call a ScaLAPACK routine — initializing the process grid, assigning the matrix to the processes, calling the ScaLAPACK routine, and releasing the process grid. For further details on each of these steps, please refer to section 2.4.

This example program solves the 9×9 system of linear equations given by

$$\begin{pmatrix} 19 & 3 & 1 & 12 & 1 & 16 & 1 & 3 & 11 \\ -19 & 3 & 1 & 12 & 1 & 16 & 1 & 3 & 11 \\ -19 & -3 & 1 & 12 & 1 & 16 & 1 & 3 & 11 \\ -19 & -3 & -1 & 12 & 1 & 16 & 1 & 3 & 11 \\ -19 & -3 & -1 & -12 & 1 & 16 & 1 & 3 & 11 \\ -19 & -3 & -1 & -12 & -1 & 16 & 1 & 3 & 11 \\ -19 & -3 & -1 & -12 & -1 & -16 & 1 & 3 & 11 \\ -19 & -3 & -1 & -12 & -1 & -16 & -1 & 3 & 11 \\ -19 & -3 & -1 & -12 & -1 & -16 & -1 & -3 & 11 \end{pmatrix} \begin{pmatrix} x_1 \\ x_2 \\ x_3 \\ x_4 \\ x_5 \\ x_6 \\ x_7 \\ x_8 \\ x_9 \end{pmatrix} = \begin{pmatrix} 0 \\ 0 \\ 1 \\ 0 \\ 0 \\ 0 \\ 0 \\ 0 \\ 0 \end{pmatrix}$$

using the ScaLAPACK driver routine PDGESV. The ScaLAPACK routine PDGESV solves a system of linear equations $A * X = B$, where the coefficient matrix (denoted by A) and the right-hand-side matrix (denoted by B) are real, general distributed matrices. The coefficient matrix A is distributed as depicted below, and for simplicity, we shall solve the system for one right-hand side ($NRHS = 1$); that is, the matrix B is a vector. The third element of the matrix B is equal to 1, and all other elements are equal to 0. After solving this system of equations, the solution vector X is given by

$$\begin{pmatrix} x_1 \\ x_2 \\ x_3 \\ x_4 \\ x_5 \\ x_6 \\ x_7 \\ x_8 \\ x_9 \end{pmatrix} = \begin{pmatrix} 0 \\ \frac{-1}{6} \\ \frac{1}{2} \\ 0 \\ 0 \\ 0 \\ \frac{-1}{2} \\ \frac{1}{6} \\ 0 \end{pmatrix}.$$

Let us assume that the matrix A is partitioned and distributed as denoted in figure 4.6; that is, we have chosen the row and column block sizes as $MB = NB = 2$, and the matrix is distributed on a 2×3 process grid ($P_r = 2, P_c = 3$). The partitioning and distribution of our example matrix A is represented in figures 2.1 and 2.2, where, to aid visualization, we use the notation $s = 19$, $c = 3$, $a = 1$, $l = 12$, $p = 16$, and $k = 11$.

s	c	a	l	a	p	a	c	k
-s	c	a	l	a	p	a	c	k
-s	-c	a	l	a	p	a	c	k
-s	-c	-a	l	a	p	a	c	k
-s	-c	-a	-l	a	p	a	c	k
-s	-c	-a	-l	-a	p	a	c	k
-s	-c	-a	-l	-a	-p	a	c	k
-s	-c	-a	-l	-a	-p	-a	c	k
-s	-c	-a	-l	-a	-p	-a	-c	k

9 x 9 matrix partitioned in 2 x 2 blocks

Figure 2.1: Partitioning of global matrix A ($s = 19; c = 3; a = 1; l = 12; p = 16; k = 11$)

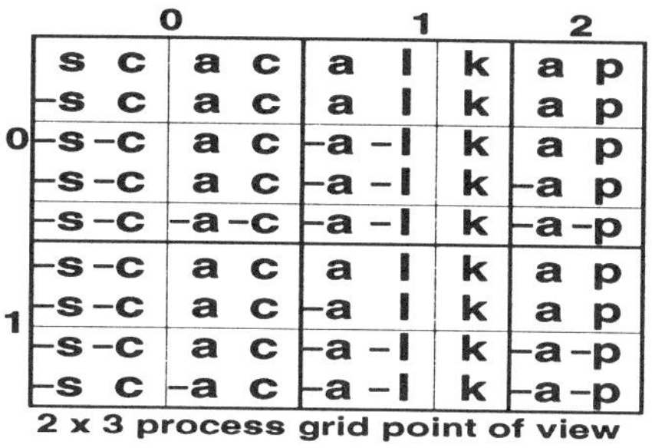

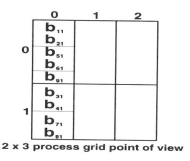

Figure 2.2: Mapping of matrix A onto process grid ($P_r = 2$,$P_c = 3$). For example, note that process $(0,0)$ contains a local array of size $A(5,4)$.

The partitioning and distribution of our example matrix B are demonstrated in figure 2.3. Note

Figure 2.3: Mapping of matrix B onto process grid ($P_r = 2$,$P_c = 3$)

that matrix B is distributed only in column 0 of the process grid. All other columns in the process grid possess an empty local portion of the matrix B.

On exit from PDGESV, process $(0,0)$ contains (in the global view) the global vector X and (in the local view) the local array B given by

$$
\begin{pmatrix} x_1 \\ x_2 \\ x_5 \\ x_6 \\ x_9 \end{pmatrix} \quad \begin{pmatrix} b_1 \\ b_2 \\ b_3 \\ b_4 \\ b_5 \end{pmatrix} = \begin{pmatrix} 0 \\ \frac{-1}{6} \\ 0 \\ 0 \\ 0 \end{pmatrix},
$$

and process $(1,0)$ contains (in the global view) the global vector X and (in the local view) local array B given by

$$
\begin{pmatrix} x_3 \\ x_4 \\ x_7 \\ x_8 \end{pmatrix} \quad \begin{pmatrix} b_1 \\ b_2 \\ b_3 \\ b_4 \end{pmatrix} = \begin{pmatrix} \frac{1}{2} \\ 0 \\ \frac{-1}{2} \\ \frac{1}{6} \end{pmatrix}.
$$

The normalized residual check

$$
\frac{\|A * x - b\|}{(\|x\| * \|A\| * eps * N)}
$$

is performed on the solution to verify the accuracy of the results.

For more information on the BLACS routines called in this program, please refer to section 2.4, Appendix D.3, [54], and the BLACS homepage (`http://www.netlib.org/blacs/index.html`). Further details of the matrix distribution and storage scheme can be found in section 2.3.2, figure 4.6, and table 4.8. Complete details on matrix distribution can be found in Chapter 4 and details of the array descriptors can be found in section 4.3.2. For a more flexible and memory efficient example program, please refer to Appendix C.1.

2.3.1   Simplifying Assumptions Used in Example Program

Several simplifying assumptions and/or restrictions have been made in this example program in order to present the most basic example for the user:

- We have chosen a small block size, $MB = NB = 2$; however, this should not be regarded as a typical choice of block size in a user's application. For best performance, a choice of $MB = NB = 32$ or $MB = NB = 64$ is more suitable. Refer to Chapter 5 for further details.

- A simplistic subroutine MATINIT is used to assign matrices A and B to the process grid. Note that this subroutine hardcodes the local arrays on each process and does not perform communication. It is not a ScaLAPACK routine and is provided only for the purposes of this example program.

- We assume $RSRC = CSRC = 0$, and thus both matrices A and B are distributed across the process grid starting with process $(0, 0)$. In general, however, any process in the current process grid can be assigned to receive the first element of the distributed matrix.

- We have set the local leading dimension of local array A and the local leading dimension of local array B to be the same over all process rows in the process grid. The variable $MXLLDA$ is equal to the maximum local leading dimension for array A (denoted LLD_A) over all process rows. Likewise, variable $MXLLDB$ is the maximum local leading dimension for array B (denoted LLD_B) over all process rows. In general, however, the local leading dimension of the local array can differ from process to process in the process grid.

- The system is solved by using the entire matrix A, as opposed to a submatrix of A, so the global indices, denoted by IA, JA, IB, and JB, into the matrix are equal to 1. Refer to figure 4.7 in section 4.3.5 for more information on the representation of global addressing into a distributed submatrix.

2.3.2   Notation Used in Example Program

The following is a list of notational variables and definitions specific to Example Program #1. A complete **List of Notation** can be found at the beginning of this book.

Variable	Definition
$CSRC$	(global) Process column over which the first column of the matrix is distributed.
$DESCA$	(global and local) Array descriptor for matrix A.
$DESCB$	(global and local) Array descriptor for matrix B.
$ICTXT$	(global) BLACS context associated with a process grid.
M	(global) Number of rows in the global matrix A.
MB	(global) Row block size for the matrix A.
$MXLLDA$	(global) Maximum local leading dimension of the array A.
$MXLLDB$	(global) Maximum local leading dimension of the array B.
$MXLOCC$	(global) Maximum number of columns of the matrix A owned by any process column.
$MXLOCR$	(global) Maximum number of rows of the matrix A owned by any process row.
$MXRHSC$	(global) Maximum number of columns of the matrix B owned by any process column.
$MYCOL$	(local) Calling process's column coordinate in the process grid.
$MYROW$	(local) Calling process's row coordinate in the process grid.
N	(global) Number of columns in the global matrix A, and the number of rows of the global solution matrix B.
NB	(global) Column block size for the matrix A, and the row block size for the matrix B.
$NBRHS$	(global) Column block size for the global solution matrix B.
$NPCOL$	(global) Number of columns in the process grid.
$NPROW$	(global) Number of rows in the process grid.
$NRHS$	(global) Number of columns in the global solution matrix B.
$RSRC$	(global) Process row over which the first row of the matrix is distributed.

2.3.3 Output of Example Program #1 Using MPI

When this example program is executed on a Sun Solaris architecture using MPICH (version 1.0.13) and the MPI BLACS, the following output[1] is received:

```
sun4sol2> f77 -o example1 example1.f scalapack_SUN4SOL2.a \
        blacsF77init_MPI-SUN4SOL2-0.a blacs_MPI-SUN4SOL2-0.a \
        blacsF77init_MPI-SUN4SOL2-0.a blas_SUN4SOL2.a \
        $MPI_ROOT/lib/solaris/ch_p4/libmpi.a -lnsl -lsocket
example1.f:
 MAIN example1:
        matinit:
sun4sol2> mpirun -np 6 example1
```

---

[1]This example program computes the relative machine precision which causes, on some systems, the IEEE floating-point exception flags to be raised. This may result in the printing of a warning message. This is normal.

```
ScaLAPACK Example Program #1 -- May 1, 1997

Solving Ax=b where A is a    9 by    9 matrix with a block size of    2
Running on    6 processes, where the process grid is    2 by    3

INFO code returned by PDGESV =    0

According to the normalized residual the solution is correct.

||A*x - b|| / ( ||x||*||A||*eps*N ) =    0.00000000E+00
```

2.3.4   Output of Example Program #1 Using PVM

When this example program is executed on a SUN4 architecture using PVM (version 3.3.11) and the PVM BLACS, the following output[2] is received:

```
sun4> f77 -o example1 example1.f scalapack_SUN4.a \
      blacs_PVM-SUN4-0.a blas_SUN4.a $PVM_ROOT/lib/SUN4/libpvm3.a
example1.f:
 MAIN example1:
        matinit:
sun4> cp example1 $HOME/pvm3/bin/SUN4/
sun4> cd $HOME/pvm3/bin/SUN4/
sun4> pvm
pvm> quit

pvmd still running.
sun4> example1
File 'blacs_setup.dat' not found.  Spawning processes to current
configuration.
Enter the name of the executable to run: example1
Spawning 5 more copies of example1
Spawning process 'example1' to host sun4
[t40003] BEGIN
Spawning process 'example1' to host sun4
[t40004] BEGIN
Spawning process 'example1' to host sun4
[t40005] BEGIN
Spawning process 'example1' to host sun4
[t40006] BEGIN
Spawning process 'example1' to host sun4
[t40007] BEGIN
```

---

[2]This example program computes the relative machine precision which causes, on some systems, the IEEE floating-point exception flags to be raised. This may result in the printing of a warning message. This is normal.

```
ScaLAPACK Example Program #1 -- May 1, 1997

Solving Ax=b where A is a   9 by   9 matrix with a block size of   2
Running on   6 processes, where the process grid is   2 by   3

INFO code returned by PDGESV =   0

According to the normalized residual the solution is correct.

||A*x - b|| / ( ||x||*||A||*eps*N ) =   0.00000000E+00
sun4> pvm
pvmd already running.
pvm> halt
sun4>
```

2.4 Four Basic Steps Required to Call a ScaLAPACK Routine

Four basic steps are required to call a ScaLAPACK routine.

1. Initialize the process grid

2. Distribute the matrix on the process grid

3. Call ScaLAPACK routine

4. Release the process grid

Each of these steps is detailed below. The example program in section 2.3 illustrates these basic requirements. Refer to section 2.3.2 for an explanation of notational variables.

For more information on the BLACS routines called in this program, and more specifically their calling sequences, please refer to Appendix D.3, [54], and the BLACS homepage (http://www.netlib.org/blacs/index.html). Further details of the matrix distribution and storage scheme can be found in Chapter 4 and section 4.3.2.

2.4.1 Initialize the Process Grid

A call to the ScaLAPACK TOOLS routine SL_INIT initializes the process grid. This routine initializes a $P_r \times P_c$ (denoted $NPROW \times NPCOL$ in the source code) process grid by using a row-major ordering of the processes, and obtains a default system *context*. For more information on contexts, refer to section 4.1.2 or [54].

The user can then query the process grid to identify each process's coordinates ($MYROW, MYCOL$)) via a call to BLACS_GRIDINFO.

A typical code fragment (as obtained from the example program in section 2.3) to accomplish this task would be the following:

```
CALL SL_INIT( ICTXT, NPROW, NPCOL )
CALL BLACS_GRIDINFO( ICTXT, NPROW, NPCOL, MYROW, MYCOL )
```

where details of the calling sequence for SL_INIT are provided below. Detailed descriptions of calling sequences for each BLACS routine can be found in [54] and Appendix D.3. *Note that underlined arguments in the calling sequence denote output arguments.*

- SL_INIT(<u>ICTXT</u>, NPROW, NPCOL)

ICTXT (global output) INTEGER
　　　　ICTXT specifies the BLACS context identifying the created process grid.

NPROW (global input) INTEGER
　　　　NPROW specifies the number of process rows in the process grid to be created.

NPCOL (global input) INTEGER
　　　　NPCOL specifies the number of process columns in the process grid to be created.

For a description of these variables names, please refer to the example program notation in section 2.3 and [54].

2.4.2   Distribute the Matrix on the Process Grid

All global matrices must be distributed on the process grid prior to the invocation of a ScaLAPACK routine. It is the user's responsibility to perform this data distribution. For further information on the appropriate data distribution, please refer to Chapter 4.

Each global matrix that is to be distributed across the process grid must be assigned an *array descriptor*. Details of the entries in the array descriptor can be found in section 4.3.3. This array descriptor is most easily initialized with a call to a ScaLAPACK TOOLS routine called DESCINIT and *must* be set prior to the invocation of a ScaLAPACK routine.

As an example, the array descriptors for the matrices in figures 2.2 and 2.3 are assigned with the following code excerpt from the example program in section 2.3.

```
CALL DESCINIT( DESCA, M, N, MB, NB, RSRC, CSRC, ICTXT, MXLLDA,
$              INFO )
CALL DESCINIT( DESCB, N, NRHS, NB, NBRHS, RSRC, CSRC, ICTXT,
$              MXLLDB, INFO )
```

These two calls to DESCINIT are equivalent to the assignment statements:

```
      DESCA( 1 ) = 1
      DESCA( 2 ) = ICTXT
      DESCA( 3 ) = M
      DESCA( 4 ) = N
      DESCA( 5 ) = MB
      DESCA( 6 ) = NB
      DESCA( 7 ) = RSRC
      DESCA( 8 ) = CSRC
      DESCA( 9 ) = MXLLDA
*
      DESCB( 1 ) = 1
      DESCB( 2 ) = ICTXT
      DESCB( 3 ) = N
      DESCB( 4 ) = NRHS
      DESCB( 5 ) = NB
      DESCB( 6 ) = NBRHS
      DESCB( 7 ) = RSRC
      DESCB( 8 ) = CSRC
      DESCB( 9 ) = MXLLDB
```

Details of the entries in the array descriptor can be found in section 4.3.3.

A simplistic mapping of the global matrix in figure 2.2 to a process grid is accomplished in the example program in section 2.3 via a call to the subroutine MATINIT. Please note that the routine MATINIT is not a ScaLAPACK routine and is used in this example program for demonstrative purposes only.

Appendix C.1 provides a more detailed example program, which reads a matrix from a file, distributes it onto a process grid, and then writes the solution to a file.

2.4.3   Call the ScaLAPACK Routine

All ScaLAPACK routines assume that the data has been distributed on the process grid prior to the invocation of the routine. Detailed descriptions of the appropriate calling sequences for each of the ScaLAPACK routines can be found in the leading comments of the source code or in Part II of this users guide. The required data distribution for the ScaLAPACK routine, as well as the amount of input error checking to be performed, is described in Chapter 4. For debugging hints, the user should refer to Chapter 7.

2.4.4   Release the Process Grid

After the desired computation on a process grid has been completed, it is advisable to release the process grid via a call to BLACS_GRIDEXIT. When all computations have been completed, the program is exited with a call to BLACS_EXIT.

A typical code fragment to accomplish these steps would be

```
CALL BLACS_GRIDEXIT( ICTXT )
CALL BLACS_EXIT( 0 )
```

A detailed explanation of the BLACS calling sequences can be found in Appendix D and [54].

Chapter 3

Contents of ScaLAPACK

3.1 Structure of ScaLAPACK

3.1.1 Levels of Routines

The routines in ScaLAPACK are classified into three broad categories:

- **Driver** routines, each of which solves a complete problem, for example, solving a system of linear equations or computing the eigenvalues of a real symmetric matrix. Users are recommended to use a driver routine if one meets their requirements. Driver routines are described in section 3.2, and a complete list of routine names can be found in Appendix A.1. Global and local input error-checking are performed where possible for these routines.

- **Computational** routines, each of which performs a distinct computational task, for example an *LU* factorization or the reduction of a real symmetric matrix to tridiagonal form. Each driver routine calls a sequence of computational routines. Users (especially software developers) may need to call computational routines directly to perform tasks, or sequences of tasks, that cannot conveniently be performed by the driver routines. Computational routines are described in section 3.3 and a complete list of routine names can be found in Appendix A.1. Global and local input error-checking are performed for these routines.

- **Auxiliary** routines, which in turn can be classified as follows:

  - routines that perform subtasks of block-partitioned algorithms — in particular, routines that implement unblocked versions of the algorithms; and

  - routines that perform some commonly required low-level computations, for example, scaling a matrix, computing a matrix-norm, or generating an elementary Householder matrix; some of these may be of interest to numerical analysts or software developers and could be considered for future additions to the PBLAS.

In general, no input error-checking is performed in the auxiliary routines. The exception to this rule is for the auxiliary routines that are Level 2 equivalents of computational routines (e.g., PxGETF2, PxGEQR2, PxORMR2, PxORM2R). For these routines, local input error-checking is performed.

Both driver routines and computational routines are fully described in this users guide, but not the auxiliary routines. A list of the auxiliary routines, with brief descriptions of their functions, is given in Appendix A.2. LAPACK auxiliary routines are also used whenever possible for local computation. Refer to the LAPACK Users' Guide [3] for details.

The PBLAS, BLAS, BLACS, and LAPACK are strictly-speaking not part of the ScaLAPACK routines. However, the ScaLAPACK routines make frequent calls to these packages.

ScaLAPACK also provides two matrix redistribution/copy routines for each data type [107, 49, 106]. These routines provide a truly general copy from any block cyclically distributed (sub)matrix to any other block cyclically distributed (sub)matrix. These routines are the only ones in the entire ScaLAPACK library which provide *inter-context* operations. Because of the generality of these routines, they may be used for many operations not usually associated with copy routines. For instance, they may be used to a take a matrix on one process and distribute it across a process grid, or the reverse. If a supercomputer is grouped into a virtual parallel machine with a workstation, for instance, this routine can be used to move the matrix from the workstation to the supercomputer and back. In ScaLAPACK, these routines are called to copy matrices from a two-dimensional process grid to a one-dimensional process grid. They can be used to redistribute matrices so that distributions providing maximal performance can be used by various component libraries, as well. For further details on these routines, refer to Appendix A.3.

3.1.2 Data Types and Precision

ScaLAPACK provides the same range of functionality for **real** and **complex** data, with a few exceptions. The complex Hermitian eigensolver (PCHEEV) and complex singular value decomposition (PCGESVD) are still under development. They may be available in a future release of ScaLAPACK.

Matching routines for real and complex data have been coded to maintain a close correspondence between the two, wherever possible. However, there are cases where the corresponding complex version calling sequence has more arguments than the real version.

All routines in ScaLAPACK are provided in both **single-** and **double**-precision versions.

Double-precision routines for complex matrices require the nonstandard Fortran 77 data type COMPLEX*16, which is available on most machines where double precision computation is usual.

3.1.3 Naming Scheme

Each subroutine name in ScaLAPACK, which has an LAPACK equivalent, is simply the LAPACK name prepended by a P. Thus, we have relaxed (violated) the Fortran 77 standard by allowing subroutine names to be greater than 6-characters in length and allowing an underscore _ in the names of certain TOOLS routines.

All driver and computational routines have names of the form **PXYYZZZ**, where for some driver routines the seventh character is blank.

The second letter, **X**, indicates the data type as follows:

S   REAL
D   DOUBLE PRECISION
C   COMPLEX
Z   COMPLEX*16 or DOUBLE COMPLEX

When we wish to refer to a ScaLAPACK routine generically, regardless of data type, we replace the second letter by "x". Thus PxGESV refers to any or all of the routines PSGESV, PCGESV, PDGESV, and PZGESV.

The next two letters, **YY**, indicate the type of matrix (or of the most significant matrix). Most of these two-letter codes apply to both real and complex matrices; a few apply specifically to one or the other, as indicated in table 3.1.

Table 3.1: Matrix types in the ScaLAPACK naming scheme

DB	general band (diagonally dominant-like)
DT	general tridiagonal (diagonally dominant-like)
GB	general band
GE	general (i.e., unsymmetric, in some cases rectangular)
GG	general matrices, generalized problem (i.e., a pair of general matrices)
HE	(complex) Hermitian
OR	(real) orthogonal
PB	symmetric or Hermitian positive definite band
PO	symmetric or Hermitian positive definite
PT	symmetric or Hermitian positive definite tridiagonal
ST	(real) symmetric tridiagonal
SY	symmetric
TR	triangular (or in some cases quasi-triangular)
TZ	trapezoidal
UN	(complex) unitary

A **diagonally dominant-like** matrix is one for which it is known *a priori* that pivoting for stability is NOT required in the *LU* factorization of the matrix. Diagonally dominant matrices themselves are examples of diagonally dominant-like matrices.

When we wish to refer to a class of routines that performs the same function on different types of matrices, we replace the second, third, and fourth letters by "xyy". Thus, PxyySVX refers to all the expert driver routines for systems of linear equations that are listed in table 3.2.

The last three letters **ZZZ** indicate the computation performed. Their meanings will be explained in section 3.3. For example, PSGEBRD is a single-precision routine that performs a bidiagonal reduction (BRD) of a real general matrix.

The names of auxiliary routines follow a similar scheme except that the third and fourth characters YY are usually LA (for example, PSLASCL or PCLARFG). There are two kinds of exception. Auxiliary routines that implement an unblocked version of a block-partitioned algorithm have similar names to the routines that perform the block-partitioned algorithm, with the seventh character being "2" (for example, PSGETF2 is the unblocked version of PSGETRF). A few routines that may be regarded as extensions to the BLAS are named similar to the BLAS naming schemes (for example, PCMAX1, PSCSUM1).

3.2 Driver Routines

This section describes the driver routines in ScaLAPACK. Further details on the terminology and the numerical operations they perform are given in section 3.3, which describes the computational routines. If the parallel algorithm or implementation differs significantly from the serial LAPACK equivalent, this fact will be noted and the user directed to consult the appropriate LAPACK working note.

3.2.1 Linear Equations

Two types of driver routines are provided for solving systems of linear equations:

- a **simple** driver (name ending -SV), which solves the system $AX = B$ by factorizing A and overwriting B with the solution X;

- an **expert** driver (name ending -SVX), which can also perform some or all of the following functions (some of them optionally):

  - solve $A^T X = B$ or $A^H X = B$ (unless A is symmetric or Hermitian);
  - estimate the condition number of A, check for near-singularity, and check for pivot growth;
  - refine the solution and compute forward and backward error bounds;
  - equilibrate the system if A is poorly scaled.

  The expert driver requires roughly twice as much storage as the simple driver in order to perform these extra functions.

Both types of driver routines can handle multiple right-hand sides (the columns of B).

Different driver routines are provided to take advantage of special properties or storage schemes of the matrix A, as shown in table 3.2.

These driver routines cover all the functionality of the computational routines for linear systems, except matrix inversion. It is seldom necessary to compute the inverse of a matrix explicitly, and such computation is certainly not recommended as a means of solving linear systems.

At present, only simple drivers (name ending -SV) are provided for systems involving band and tridiagonal matrices. It is important to note that in the banded and tridiagonal factorizations (PxDBTRF, PxDTTRF, PxGBTRF, PxPBTRF, and PxPTTRF) used within these drivers, the resulting factorization is *not* the same factorization as returned from LAPACK. Additional permutations are performed on the matrix for the sake of parallelism. Further details of the algorithmic implementations can be found in [32].

Table 3.2: Driver routines for linear equations

Type of matrix and storage scheme	Operation	Single precision		Double precision	
		Real	Complex	Real	Complex
general (partial pivoting)	simple driver	PSGESV	PCGESV	PDGESV	PZGESV
	expert driver	PSGESVX	PCGESVX	PDGESVX	PZGESVX
general band (partial pivoting)	simple driver	PSGBSV	PCGBSV	PDGBSV	PZGBSV
general band (no pivoting)	simple driver	PSDBSV	PCDBSV	PDDBSV	PZDBSV
general tridiagonal (no pivoting)	simple driver	PSDTSV	PCDTSV	PDDTSV	PZDTSV
symmetric/Hermitian positive definite	simple driver	PSPOSV	PCPOSV	PDPOSV	PZPOSV
	expert driver	PSPOSVX	PCPOSVX	PDPOSVX	PZPOSVX
symmetric/Hermitian positive definite band	simple driver	PSPBSV	PCPBSV	PDPBSV	PZPBSV
symmetric/Hermitian positive definite tridiagonal	simple driver	PSPTSV	PCPTSV	PDPTSV	PZPTSV

3.2.2 Linear Least Squares Problems

The **linear least squares (LLS)** problem is:

$$\operatorname*{minimize}_{x} \|b - Ax\|_2, \tag{3.1}$$

where A is an m-by-n matrix, b is a given m element vector and x is the n element solution vector.

In the most usual case, $m \geq n$ and $\operatorname{rank}(A) = n$. In this case the solution to problem (3.1) is unique. The problem is also referred to as finding a **least squares solution** to an **overdetermined** system of linear equations.

When $m < n$ and $\operatorname{rank}(A) = m$, there are an infinite number of solutions x that exactly satisfy $b - Ax = 0$. In this case it is often useful to find the unique solution x that minimizes $\|x\|_2$, and the problem is referred to as finding a **minimum norm solution** to an **underdetermined** system of linear equations.

The driver routine PxGELS solves problem (3.1) on the assumption that $\text{rank}(A) = \min(m, n)$ — in other words, A has **full rank** — finding a least squares solution of an overdetermined system when $m > n$, and a minimum norm solution of an underdetermined system when $m < n$. PxGELS uses a QR or LQ factorization of A and also allows A to be replaced by A^T in the statement of the problem (or by A^H if A is complex).

In the general case when we may have $\text{rank}(A) < \min(m, n)$ — in other words, A may be **rank-deficient** — we seek the **minimum norm least squares** solution x that minimizes both $\|x\|_2$ and $\|b - Ax\|_2$.

The LLS driver routines are listed in table 3.3.

All routines allow several right-hand-side vectors b and corresponding solutions x to be handled in a single call, storing these vectors as columns of matrices B and X, respectively. Note, however, that equation 3.1 is solved for each right-hand-side vector independently; this is *not* the same as finding a matrix X that minimizes $\|B - AX\|_2$.

Table 3.3: Driver routines for linear least squares problems

Operation	Single Precision		Double Precision	
	Real	Complex	Real	Complex
solve LLS using QR or LQ factorization	PSGELS	PCGELS	PDGELS	PZGELS

3.2.3   Standard Eigenvalue and Singular Value Problems

3.2.3.1   Symmetric Eigenproblems

The **symmetric eigenvalue problem (SEP)** is to find the **eigenvalues**, λ, and corresponding **eigenvectors**, $z \neq 0$, such that

$$Az = \lambda z, \quad A = A^T, \text{ where } A \text{ is real.}$$

For the **Hermitian eigenvalue problem** we have

$$Az = \lambda z, \quad A = A^H.$$

For both problems the eigenvalues λ are real.

When all eigenvalues and eigenvectors have been computed, we write

$$A = Z\Lambda Z^T,$$

where Λ is a diagonal matrix whose diagonal elements are the eigenvalues, and Z is an orthogonal (or unitary) matrix whose columns are the eigenvectors. This is the classical **spectral factorization** of A.

Two types of driver routines are provided for symmetric or Hermitian eigenproblems:

- a **simple** driver (name ending -EV), which computes all the eigenvalues and (optionally) the eigenvectors of a symmetric or Hermitian matrix A;

- an **expert** driver (name ending -EVX), which can compute either all or a selected subset of the eigenvalues, and (optionally) the corresponding eigenvectors.

The driver routines are shown in table 3.4. Currently the only simple drivers provided are PSSYEV and PDSYEV.

3.2.3.2   Singular Value Decomposition

The **singular value decomposition (SVD)** of an m-by-n matrix A is given by

$$A = U\Sigma V^T, \quad (A = U\Sigma V^H \quad \text{in the complex case}),$$

where U and V are orthogonal (unitary) and Σ is an m-by-n diagonal matrix with real diagonal elements, σ_i, such that

$$\sigma_1 \geq \sigma_2 \geq \ldots \sigma_{\min(m,n)} \geq 0.$$

The σ_i are the **singular values** of A and the first $\min(m,n)$ columns of U and V are the **left** and **right singular vectors** of A.

The singular values and singular vectors satisfy

$$Av_i = \sigma_i u_i \quad \text{and} \quad A^T u_i = \sigma_i v_i \quad (\text{or} \quad A^H u_i = \sigma_i v_i),$$

where u_i and v_i are the ith columns of U and V, respectively.

A single driver routine, PxGESVD, computes the "economy size" or "thin" singular value decomposition of a general nonsymmetric matrix (see table 3.4). Thus, if A is m-by-n with $m > n$, then only the first n columns of U are computed and Σ is an n-by-n matrix. For a detailed discussion of the "thin" singular value decomposition, refer to [71, p. 72].

Currently, only PSGESVD and PDGESVD are provided.

Table 3.4: Driver routines for standard eigenvalue and singular value problems

Type of Problem	Function	Single Precision		Double Precision	
		Real	Complex	Real	Complex
SEP	simple driver	PSSYEV		PDSYEV	
	expert driver	PSSYEVX	PCHEEVX	PDSYEVX	PZHEEVX
SVD	singular values/vectors	PSGESVD		PDGESVD	

3.2.4   Generalized Symmetric Definite Eigenproblems (GSEP)

An expert driver is provided to compute all the eigenvalues and (optionally) the eigenvectors of the following types of problems:

1. $Az = \lambda Bz$

2. $ABz = \lambda z$

3. $BAz = \lambda z$

where A and B are symmetric or Hermitian and B is positive definite. For all these problems the eigenvalues λ are real. When A and B are symmetric, the matrices Z of computed eigenvectors satisfy $Z^T A Z = \Lambda$ (problem types 1 and 3) or $Z^{-1} A Z^{-T} = I$ (problem type 2), where Λ is a diagonal matrix with the eigenvalues on the diagonal. Z also satisfies $Z^T B Z = I$ (problem types 1 and 2) or $Z^T B^{-1} Z = I$ (problem type 3). When A and B are Hermitian, the matrices Z of computed eigenvectors satisfy $Z^H A Z = \Lambda$ (problem types 1 and 3) or $Z^{-1} A Z^{-H} = I$ (problem type 2), where Λ is a diagonal matrix with the eigenvalues on the diagonal. Z also satisfies $Z^H B Z = I$ (problem types 1 and 2) or $Z^H B^{-1} Z = I$ (problem type 3).

The routine is listed in table 3.5.

Table 3.5: Driver routine for the generalized symmetric definite eigenvalue problems

Type of problem	Function	Single precision		Double precision	
		Real	Complex	Real	Complex
GSEP	expert driver	PSSYGVX	PCHEGVX	PDSYGVX	PZHEGVX

3.3   Computational Routines

As previously stated, if the parallel algorithm or implementation differs significantly from the serial LAPACK equivalent, this fact will be noted and the user directed to consult the appropriate LAPACK Working Note.

3.3.1   Linear Equations

We use the standard notation for a system of simultaneous linear equations:

$$Ax = b, \tag{3.2}$$

where A is the **coefficient matrix**, b is the **right-hand side**, and x is the **solution**. In (3.2) A is assumed to be a square matrix of order n, but some of the individual routines allow A to be rectangular. If there are several right-hand sides we write

$$AX = B \tag{3.3}$$

where the columns of B are the individual right-hand sides, and the columns of X are the corresponding solutions. The basic task is to compute X, given A and B.

If A is upper or lower triangular, (3.2) can be solved by a straightforward process of backward or forward substitution. Otherwise, the solution is obtained after first factorizing A as a product of triangular matrices (and possibly also a diagonal matrix or permutation matrix).

The form of the factorization depends on the properties of the matrix A. ScaLAPACK provides routines for the following types of matrices, based on the stated factorizations:

- **general** matrices (LU factorization with partial pivoting):

$$A = PLU,$$

  where P is a permutation matrix, L is lower triangular with unit diagonal elements (lower trapezoidal if $m > n$), and U is upper triangular (upper trapezoidal if $m < n$).

- **symmetric and Hermitian positive definite** matrices (Cholesky factorization):

$$A = U^T U \ \text{ or } \ A = LL^T (\text{in the symmetric case})$$

$$A = U^H U \ \text{ or } \ A = LL^H (\text{in the Hermitian case}),$$

  where U is an upper triangular matrix and L is lower triangular.

- **general band** matrices (LU factorization with partial pivoting):

  If A is m-by-n with bwl subdiagonals and bwu superdiagonals, the factorization is

$$A = PLUQ,$$

  where P and Q are permutation matrices and L and U are banded lower and upper triangular matrices, respectively.

- **general diagonally dominant-like band** matrices including **general tridiagonal** matrices (LU factorization without pivoting):

  A **diagonally dominant-like** matrix is one for which it is known *a priori* that pivoting for stability is NOT required in the LU factorization of the matrix. Diagonally dominant matrices themselves are examples of diagonally dominant-like matrices.

  If A is m-by-n with bwl subdiagonals and bwu superdiagonals, the factorization is

$$A = PLUP^T,$$

  where P is a permutation matrix and L and U are banded lower and upper triangular matrices respectively.

- **symmetric and Hermitian positive definite band** matrices (Cholesky factorization):

$$A = PU^T U P^T \ \text{ or } \ A = PLL^T P^T \text{(in the symmetric case)}$$

$$A = PU^H U P^T \ \text{ or } \ A = PLL^H P^T \text{(in the Hermitian case)},$$

where P is a permutation matrix and U and L are banded upper and lower triangular matrices, respectively.

- **symmetric and Hermitian positive definite tridiagonal** matrices (LDL^T factorization):

$$A = PU^T D U P^T \ \text{ or } \ A = PLDL^T P^T \text{(in the symmetric case)}$$

$$A = PU^H D U P^T \ \text{ or } \ A = PLDL^H P^T \text{(in the Hermitian case)},$$

where P is a permutation matrix and U and L are bidiagonal upper and lower triangular matrices respectively.

Note: In the banded and tridiagonal factorizations (PxDBTRF, PxDTTRF, PxGBTRF, PxPB-TRF, and PxPTTRF), the resulting factorization is *not* the same factorization as returned from LAPACK. Additional permutations are performed on the matrix for the sake of parallelism. Further details of the algorithmic implementations can be found in [32].

The factorization for a general diagonally dominant-like tridiagonal matrix is like that for a general diagonally dominant-like band matrix with $bwl = 1$ and $bwu = 1$. Band matrices use the band storage scheme described in section 4.4.3.

While the primary use of a matrix factorization is to solve a system of equations, other related tasks are provided as well. Wherever possible, ScaLAPACK provides routines to perform each of these tasks for each type of matrix and storage scheme (see table 3.6). The following list relates the tasks to the last three characters of the name of the corresponding computational routine:

Pxy*y***TRF**: factorize (obviously not needed for triangular matrices);

Pxy*y***TRS**: use the factorization (or the matrix A itself if it is triangular) to solve (3.3) by forward or backward substitution;

Pxy*y***CON**: estimate the reciprocal of the condition number $\kappa(A) = \|A\|.\|A^{-1}\|$; Higham's modification [81] of Hager's method [72] is used to estimate $\|A^{-1}\|$ (not provided for band or tridiagonal matrices);

Pxy*y***RFS**: compute bounds on the error in the computed solution (returned by the PxyyTRS routine), and refine the solution to reduce the backward error (see below) (not provided for band or tridiagonal matrices);

Pxy*y***TRI**: use the factorization (or the matrix A itself if it is triangular) to compute A^{-1} (not provided for band matrices, because the inverse does not in general preserve bandedness);

PxyyEQU: compute scaling factors to equilibrate A (not provided for band, tridiagonal, or triangular matrices). These routines do not actually scale the matrices: auxiliary routines PxLAQyy may be used for that purpose — see the code of the driver routines PxyySVX for sample usage.

Note that some of the above routines depend on the output of others:

PxyyTRF: may work on an equilibrated matrix produced by PxyyEQU and PxLAQyy, if yy is one of {GE, PO};

PxyyTRS: requires the factorization returned by PxyyTRF;

PxyyCON: requires the norm of the original matrix A and the factorization returned by PxyyTRF;

PxyyRFS: requires the original matrices A and B, the factorization returned by PxyyTRF, and the solution X returned by PxyyTRS;

PxyyTRI: requires the factorization returned by PxyyTRF.

The RFS ("refine solution") routines perform iterative refinement and compute backward and forward error bounds for the solution. Iterative refinement is done in the same precision as the input data. In particular, the residual is *not* computed with extra precision, as has been traditionally done. The benefit of this procedure is discussed in section 6.5.

Table 3.6: Computational routines for linear equations

Matrix Type and Storage	Operation	Single Precision		Double Precision	
		Real	Complex	Real	Complex
general partial pivoting	factorize	PSGETRF	PCGETRF	PDGETRF	PZGETRF
	solve using fact.	PSGETRS	PCGETRS	PDGETRS	PZGETRS
	est. cond. number	PSGECON	PCGECON	PDGECON	PZGECON
	error bds. for soln.	PSGERFS	PCGERFS	PDGERFS	PZGERFS
	invert using fact.	PSGETRI	PCGETRI	PDGETRI	PZGETRI
	equilibrate	PSGEEQU	PCGEEQU	PDGEEQU	PZGEEQU
general band partial pivoting	factorize	PSGBTRF	PCGBTRF	PDGBTRF	PZGBTRF
	solve using fact.	PSGBTRS	PCGBTRS	PDGBTRS	PZGBTRS
general band no pivoting	factorize	PSDBTRF	PCDBTRF	PDDBTRF	PZDBTRF
	solve using fact.	PSDBTRS	PCDBTRS	PDDBTRS	PZDBTRS
general tridiagonal no pivoting	factorize	PSDTTRF	PCDTTRF	PDDTTRF	PZDTTRF
	solve using fact.	PSDTTRS	PCDTTRS	PDDTTRS	PZDTTRS
symmetric/ Hermitian positive definite	factorize	PSPOTRF	PCPOTRF	PDPOTRF	PZPOTRF
	solve using fact.	PSPOTRS	PCPOTRS	PDPOTRS	PZPOTRS
	est. cond. number	PSPOCON	PCPOCON	PDPOCON	PZPOCON
	error bds. for soln.	PSPORFS	PCPORFS	PDPORFS	PZPORFS
	invert using fact.	PSPOTRI	PCPOTRI	PDPOTRI	PZPOTRI
	equilibrate	PSPOEQU	PCPOEQU	PDPOEQU	PZPOEQU
symmetric/ Hermitian positive definite band	factorize	PSPBTRF	PCPBTRF	PDPBTRF	PZPBTRF
	solve using fact.	PSPBTRS	PCPBTRS	PDPBTRS	PZPBTRS
symmetric/ Hermitian positive definite tridiagonal	factorize	PSPTTRF	PCPTTRF	PDPTTRF	PZPTTRF
	solve using fact.	PSPTTRS	PCPTTRS	PDPTTRS	PZPTTRS
triangular	solve	PSTRTRS	PCTRTRS	PDTRTRS	PZTRTRS
	est. cond. number	PSTRCON	PCTRCON	PDTRCON	PZTRCON
	error bds. for soln.	PSTRRFS	PCTRRFS	PDTRRFS	PZTRRFS
	invert	PSTRTRI	PCTRTRI	PDTRTRI	PZTRTRI

3.3.2 Orthogonal Factorizations and Linear Least Squares Problems

ScaLAPACK provides a number of routines for factorizing a general rectangular m-by-n matrix A, as the product of an **orthogonal** matrix (**unitary** if complex) and a **triangular** (or possibly trapezoidal) matrix.

A real matrix Q is **orthogonal** if $Q^T Q = I$; a complex matrix Q is **unitary** if $Q^H Q = I$. Orthogonal or unitary matrices have the important property that they leave the two-norm of a vector invariant:

$$\|x\|_2 = \|Qx\|_2, \quad \text{if } Q \text{ is orthogonal or unitary.}$$

As a result, they help to maintain numerical stability because they do not amplify rounding errors.

Orthogonal factorizations are used in the solution of linear least squares problems. They may also be used to perform preliminary steps in the solution of eigenvalue or singular value problems.

Table 3.7 lists all routines provided by ScaLAPACK to perform orthogonal factorizations and the generation or pre- or post-multiplication of the matrix Q for each matrix type and storage scheme.

3.3.2.1 QR Factorization

The most common, and best known, of the factorizations is the **QR factorization** given by

$$A = Q \left(\begin{array}{c} R \\ 0 \end{array} \right), \quad \text{if } m \geq n,$$

where R is an n-by-n upper triangular matrix and Q is an m-by-m orthogonal (or unitary) matrix. If A is of full rank n, then R is nonsingular. It is sometimes convenient to write the factorization as

$$A = \left(\begin{array}{cc} Q_1 & Q_2 \end{array} \right) \left(\begin{array}{c} R \\ 0 \end{array} \right),$$

which reduces to

$$A = Q_1 R,$$

where Q_1 consists of the first n columns of Q, and Q_2 the remaining $m - n$ columns.

If $m < n$, R is trapezoidal, and the factorization can be written

$$A = Q \left(\begin{array}{cc} R_1 & R_2 \end{array} \right), \quad \text{if } m < n,$$

where R_1 is upper triangular and R_2 is rectangular.

The routine PxGEQRF computes the QR factorization. The matrix Q is not formed explicitly, but is represented as a product of elementary reflectors, as described in section 3.4. Users need not be aware of the details of this representation, because associated routines are provided to work with Q: PxORGQR (or PxUNGQR in the complex case) can generate all or part of Q, while PxORMQR (or PxUNMQR) can pre- or post-multiply a given matrix by Q or Q^T (Q^H if complex).

The QR factorization can be used to solve the linear least squares problem (3.1) when $m \geq n$ and A is of full rank, since

$$\|b - Ax\|_2 = \|Q^T b - Q^T Ax\|_2 = \left\|\begin{array}{c} c_1 - Rx \\ c_2 \end{array}\right\|_2, \quad \text{where } c \equiv \left(\begin{array}{c} c_1 \\ c_2 \end{array}\right) = \left(\begin{array}{c} Q_1^T b \\ Q_2^T b \end{array}\right) = Q^T b;$$

c can be computed by PxORMQR (or PxUNMQR), and c_1 consists of its first n elements. Then x is the solution of the upper triangular system

$$Rx = c_1,$$

which can be computed by PxTRTRS. The residual vector r is given by

$$r = b - Ax = Q\left(\begin{array}{c} 0 \\ c_2 \end{array}\right)$$

and may be computed using PxORMQR (or PxUNMQR). The residual sum of squares $\|r\|_2^2$ may be computed without forming r explicitly, since

$$\|r\|_2 = \|b - Ax\|_2 = \|c_2\|_2.$$

3.3.2.2 *LQ* Factorization

The **LQ factorization** is given by

$$A = \left(\begin{array}{cc} L & 0 \end{array}\right) Q = \left(\begin{array}{cc} L & 0 \end{array}\right) \left(\begin{array}{c} Q_1 \\ Q_2 \end{array}\right) = LQ_1, \quad \text{if } m \leq n,$$

where L is m-by-m lower triangular, Q is n-by-n orthogonal (or unitary), Q_1 consists of the first m rows of Q, and Q_2 consists of the remaining $n - m$ rows.

This factorization is computed by the routine PxGELQF, and again Q is represented as a product of elementary reflectors; PxORGLQ (or PxUNGLQ in the complex case) can generate all or part of Q, and PxORMLQ (or PxUNMLQ) can pre- or post-multiply a given matrix by Q or Q^T (Q^H if Q is complex).

The LQ factorization of A is essentially the same as the QR factorization of A^T (A^H if A is complex), since

$$A = \left(\begin{array}{cc} L & 0 \end{array}\right) Q \quad \Longleftrightarrow \quad A^T = Q^T \left(\begin{array}{c} L^T \\ 0 \end{array}\right).$$

The LQ factorization may be used to find a minimum norm solution of an underdetermined system of linear equations $Ax = b$, where A is m-by-n with $m < n$ and has rank m. The solution is given by

$$x = Q^T \left(\begin{array}{c} L^{-1}b \\ 0 \end{array}\right)$$

and may be computed by calls to PxTRTRS and PxORMLQ.

3.3.2.3 QR Factorization with Column Pivoting

To solve a linear least squares problem (3.1) when A is not of full rank, or the rank of A is in doubt, we can perform either a QR factorization with column pivoting or a singular value decomposition (see subsection 3.3.6).

The **QR factorization with column pivoting** is given by

$$A = Q \begin{pmatrix} R \\ 0 \end{pmatrix} P^T, \quad m \geq n,$$

where Q and R are as before and P is a permutation matrix, chosen (in general) so that

$$|r_{11}| \geq |r_{22}| \geq \ldots \geq |r_{nn}|$$

and moreover, for each k,

$$|r_{kk}| \geq \|R_{k:j,j}\|_2 \quad \text{for } j = k+1, \ldots, n.$$

In exact arithmetic, if $\text{rank}(A) = k$, then the whole of the submatrix R_{22} in rows and columns $k+1$ to n would be zero. In numerical computation, the aim must be to determine an index k such that the leading submatrix R_{11} in the first k rows and columns is well conditioned and R_{22} is negligible:

$$R = \begin{pmatrix} R_{11} & R_{12} \\ 0 & R_{22} \end{pmatrix} \simeq \begin{pmatrix} R_{11} & R_{12} \\ 0 & 0 \end{pmatrix}.$$

Then k is the effective rank of A. See Golub and Van Loan [71] for a further discussion of numerical rank determination.

The so-called basic solution to the linear least squares problem (3.1) can be obtained from this factorization as

$$x = P \begin{pmatrix} R_{11}^{-1} \hat{c}_1 \\ 0 \end{pmatrix},$$

where $\hat{c}_1$ consists of just the first k elements of $c = Q^T b$.

The routine PxGEQPF computes the QR factorization with column pivoting but does not attempt to determine the rank of A. The matrix Q is represented in exactly the same way as after a call of PxGEQRF, and so the routines PxORGQR and PxORMQR can be used to work with Q (PxUNGQR and PxUNMQR if Q is complex).

3.3.2.4 Complete Orthogonal Factorization

The QR factorization with column pivoting does not enable us to compute a *minimum norm* solution to a rank-deficient linear least squares problem unless $R_{12} = 0$. However, by applying further orthogonal (or unitary) transformations from the right to the upper trapezoidal matrix $\begin{pmatrix} R_{11} & R_{12} \end{pmatrix}$, using the routine PxTZRZF, R_{12} can be eliminated:

$$\begin{pmatrix} R_{11} & R_{12} \end{pmatrix} Z = \begin{pmatrix} T_{11} & 0 \end{pmatrix}.$$

This gives the **complete orthogonal factorization**

$$AP = Q \begin{pmatrix} T_{11} & 0 \\ 0 & 0 \end{pmatrix} Z^T$$

from which the minimum norm solution can be obtained as

$$x = PZ \begin{pmatrix} T_{11}^{-1} \hat{c}_1 \\ 0 \end{pmatrix}.$$

The matrix Z is not formed explicitly but is represented as a product of elementary reflectors, as described in section 3.4. Users need not be aware of the details of this representation, because associated routines are provided to work with Z: PxORMRZ (or PxUNMRZ) can pre- or post-multiply a given matrix by Z or Z^T (Z^H if complex).

3.3.2.5   Other Factorizations

The QL and RQ factorizations are given by

$$A = Q \begin{pmatrix} 0 \\ L \end{pmatrix}, \quad \text{if } m \geq n,$$

and

$$A = \begin{pmatrix} 0 & R \end{pmatrix} Q, \quad \text{if } m \leq n.$$

These factorizations are computed by PxGEQLF and PxGERQF, respectively; they are less commonly used than either the QR or LQ factorizations described above, but have applications in, for example, the computation of generalized QR factorizations [5].

All the factorization routines discussed here (except PxTZRZF) allow arbitrary m and n, so that in some cases the matrices R or L are trapezoidal rather than triangular. A routine that performs pivoting is provided only for the QR factorization.

3.3.3   Generalized Orthogonal Factorizations

3.3.3.1   Generalized QR Factorization

The **generalized QR (GQR)** factorization of an n-by-m matrix A and an n-by-p matrix B is given by the pair of factorizations

$$A = QR \quad \text{and} \quad B = QTZ,$$

where Q and Z are respectively n-by-n and p-by-p orthogonal matrices (or unitary matrices if A and B are complex). R has the form

$$R = \begin{matrix} m \\ n-m \end{matrix} \begin{pmatrix} \overset{m}{R_{11}} \\ 0 \end{pmatrix}, \quad \text{if } n \geq m,$$

Table 3.7: Computational routines for orthogonal factorizations

Type of Fact. and Matrix	Operation	Single Precision		Double Precision	
		Real	Complex	Real	Complex
QR, general	fact. with pivoting	PSGEQPF	PCGEQPF	PDGEQPF	PZGEQPF
	fact., no pivoting	PSGEQRF	PCGEQRF	PDGEQRF	PZGEQRF
	generate Q	PSORGQR	PCUNGQR	PDORGQR	PZUNGQR
	multiply by Q	PSORMQR	PCUNMQR	PDORMQR	PZUNMQR
LQ, general	fact., no pivoting	PSGELQF	PCGELQF	PDGELQF	PZGELQF
	generate Q	PSORGLQ	PCUNGLQ	PDORGLQ	PZUNGLQ
	multiply by Q	PSORMLQ	PCUNMLQ	PDORMLQ	PZUNMLQ
QL, general	fact., no pivoting	PSGEQLF	PCGEQLF	PDGEQLF	PZGEQLF
	generate Q	PSORGQL	PCUNGQL	PDORGQL	PZUNGQL
	multiply by Q	PSORMQL	PCUNMQL	PDORMQL	PZUNMQL
RQ, general	fact., no pivoting	PSGERQF	PCGERQF	PDGERQF	PZGERQF
	generate Q	PSORGRQ	PCUNGRQ	PDORGRQ	PZUNGRQ
	multiply by Q	PSORMRQ	PCUNMRQ	PDORMRQ	PZUNMRQ
RZ, trapezoidal	fact., no pivoting	PSTZRZF	PCTZRZF	PDTZRZF	PZTZRZF
	multiply by Z	PSORMRZ	PCUNMRZ	PDORMRZ	PZUNMRZ

or

$$R = n \begin{array}{c} \overset{n \quad\quad m-n}{\left(R_{11} \quad R_{12} \right)}, \end{array} \quad \text{if} \quad n < m,$$

where R_{11} is upper triangular. T has the form

$$T = n \begin{array}{c} \overset{p-n \quad n}{\left(0 \quad T_{12} \right)}, \end{array} \quad \text{if} \quad n \le p,$$

or

$$T = \begin{array}{c} n-p \\ p \end{array} \overset{p}{\left(\begin{array}{c} T_{11} \\ T_{21} \end{array} \right)}, \quad \text{if} \quad n > p,$$

where T_{12} or T_{21} is upper triangular.

Note that if B is square and nonsingular, the GQR factorization of A and B implicitly gives the QR factorization of the matrix $B^{-1}A$:

$$B^{-1}A = Z^T(T^{-1}R)$$

without explicitly computing the matrix inverse B^{-1} or the product $B^{-1}A$.

The routine PxGGQRF computes the GQR factorization by computing first the QR factorization of A and then the RQ factorization of $Q^T B$. The orthogonal (or unitary) matrices Q and Z can be formed explicitly or can be used just to multiply another given matrix in the same way as the orthogonal (or unitary) matrix in the QR factorization (see section 3.3.2).

The GQR factorization was introduced in [73, 100]. The implementation of the GQR factorization here follows that in [5]. Further generalizations of the GQR factorization can be found in [36].

3.3.3.2   Generalized RQ factorization

The **generalized RQ (GRQ) factorization** of an m-by-n matrix A and a p-by-n matrix B is given by the pair of factorizations

$$A = RQ \quad \text{and} \quad B = ZTQ,$$

where Q and Z are respectively n-by-n and p-by-p orthogonal matrices (or unitary matrices if A and B are complex). R has the form

$$R = m \begin{pmatrix} \overset{n-m}{0} & \overset{m}{R_{12}} \end{pmatrix}, \quad \text{if} \quad m \leq n,$$

or

$$R = \begin{matrix} m-n \\ n \end{matrix} \begin{pmatrix} \overset{n,}{R_{11}} \\ R_{21} \end{pmatrix}, \quad \text{if} \quad m > n,$$

where R_{12} or R_{21} is upper triangular. T has the form

$$T = \begin{matrix} n \\ p-n \end{matrix} \begin{pmatrix} \overset{n}{T_{11}} \\ 0 \end{pmatrix}, \quad \text{if} \quad p \geq n,$$

or

$$T = p \begin{pmatrix} \overset{p}{T_{11}} & \overset{n-p}{T_{12}} \end{pmatrix}, \quad \text{if} \quad p < nn$$

where T_{11} is upper triangular.

Note that if B is square and nonsingular, the GRQ factorization of A and B implicitly gives the RQ factorization of the matrix AB^{-1}:

$$AB^{-1} = (RT^{-1})Z^T$$

without explicitly computing the matrix inverse B^{-1} or the product AB^{-1}.

The routine PxGGRQF computes the GRQ factorization by computing first the RQ factorization of A and then the QR factorization of BQ^T. The orthogonal (or unitary) matrices Q and Z can be formed explicitly or can be used just to multiply another given matrix in the same way as the orthogonal (or unitary) matrix in the RQ factorization (see section 3.3.2).

3.3.4 Symmetric Eigenproblems

Let A be a real symmetric or complex Hermitian n-by-n matrix. A scalar λ is called an **eigenvalue** and a nonzero column vector z the corresponding **eigenvector** if $Az = \lambda z$. λ is always real when A is real symmetric or complex Hermitian.

The basic task of the symmetric eigenproblem routines is to compute values of λ and, optionally, corresponding vectors z for a given matrix A.

This computation proceeds in the following stages:

1. The real symmetric or complex Hermitian matrix A is reduced to **real tridiagonal form** T. If A is real symmetric, the decomposition is $A = QTQ^T$ with Q orthogonal and T symmetric tridiagonal. If A is complex Hermitian, the decomposition is $A = QTQ^H$ with Q unitary and T, as before, *real* symmetric tridiagonal.

2. Eigenvalues and eigenvectors of the real symmetric tridiagonal matrix T are computed. If all eigenvalues and eigenvectors are computed, this process is equivalent to factorizing T as $T = S\Lambda S^T$, where S is orthogonal and Λ is diagonal. The diagonal entries of Λ are the eigenvalues of T, which are also the eigenvalues of A, and the columns of S are the eigenvectors of T; the eigenvectors of A are the columns of $Z = QS$, so that $A = Z\Lambda Z^T$ ($Z\Lambda Z^H$ when A is complex Hermitian).

In the real case, the decomposition $A = QTQ^T$ is computed by the routine PxSYTRD (see table 3.8). The complex analogue of this routine is called PxHETRD. The routine PxSYTRD (or PxHETRD) represents the matrix Q as a product of elementary reflectors, as described in section 3.4. The routine PxORMTR (or in the complex case PxUNMTR) is provided to multiply another matrix by Q without forming Q explicitly; this can be used to transform eigenvectors of T, computed by PxSTEIN, back to eigenvectors of A.

The following routines compute eigenvalues and eigenvectors of T.

xSTEQR2 This routine [77] is a modified version of LAPACK routine xSTEQR. It has fewer iterations than the LAPACK routine due to a modification in the look-ahead technique described in [77]. Some additional modifications allow each process to perform partial updates to matrix Q. This routine computes all eigenvalues and, optionally, eigenvectors of a symmetric tridiagonal matrix using the implicit QL or QR method.

PxSTEBZ This routine uses bisection to compute some or all of the eigenvalues. Options provide for computing all the eigenvalues in a real interval or all the eigenvalues from the ith to the jth largest. It can be highly accurate but may be adjusted to run faster if lower accuracy is acceptable.

PxSTEIN Given accurate eigenvalues, this routine uses inverse iteration to compute some or all of the eigenvectors.

Without any reorthogonalization, inverse iteration may produce vectors that have large dot products. To cure this, most implementations of inverse iteration such as LAPACK's xSTEIN reorthogonalize when eigenvalues differ by less than $10^{-3}\|T\|$. As a result, the eigenvectors computed by xSTEIN are almost always orthogonal, but the increase in cost can result in $O(n^3)$ work. On some rare examples, xSTEIN may still fail to deliver accurate answers; see [43, 44]. The orthogonalization done by PxSTEIN is limited by the amount of workspace provided; whenever it performs less reorthogonalization than xSTEIN, there is a danger that the dot products may not be satisfactory.

Table 3.8: Computational routines for the symmetric eigenproblem

Type of Matrix	Operation	Single Precision		Double Precision	
		Real	Complex	Real	Complex
dense symmetric (or Hermitian)	tridiagonal reduction	PSSYTRD	PCHETRD	PDSYTRD	PZHETRD
orthogonal/ unitary	multiply matrix after reduction by PxSYTRD	PSORMTR	PCUNMTR	PDORMTR	PZUNMTR
symmetric tridiagonal	eigenvalues/ eigenvectors via look-ahead QR	SSTEQR2		DSTEQR2	
symmetric tridiagonal	eigenvalues only via bisection	PSSTEBZ		PDSTEBZ	
	eigenvectors by inverse iteration	PSSTEIN	PCSTEIN	PDSTEIN	PZSTEIN

3.3.5   Nonsymmetric Eigenproblems

3.3.5.1   Eigenvalues, Eigenvectors, and Schur Factorization

Let A be a square n-by-n matrix. A scalar λ is called an **eigenvalue** and a nonzero column vector v the corresponding **right eigenvector** if $Av = \lambda v$. A nonzero column vector u satisfying $u^H A = \lambda u^H$ is called the **left eigenvector**. The first basic task of the routines described in this section is to compute, for a given matrix A, all n values of λ and, if desired, their associated right eigenvectors v and/or left eigenvectors u.

A second basic task is to compute the **Schur factorization** of a matrix A. If A is complex, then its Schur factorization is $A = ZTZ^H$, where Z is unitary and T is upper triangular. If A is real, its Schur factorization is $A = ZTZ^T$, where Z is orthogonal, and T is upper quasi-triangular (1-by-1 and 2-by-2 blocks on its diagonal). The columns of Z are called the **Schur vectors** of A. The eigenvalues of A appear on the diagonal of T; complex conjugate eigenvalues of a real A correspond to 2-by-2 blocks on the diagonal of T.

These two basic tasks can be performed in the following stages:

1. A general matrix A is reduced to **upper Hessenberg form** H which is zero below the first subdiagonal. The reduction may be written $A = QHQ^T$ with Q orthogonal if A is real, or $A = QHQ^H$ with Q unitary if A is complex. The reduction is performed by subroutine PxGEHRD, which represents Q in a factored form, as described in section 3.4. The routine PxORMHR (or in the complex case PxUNMHR) is provided to multiply another matrix by Q without forming Q explicitly.

2. The upper Hessenberg matrix H is reduced to Schur form T, giving the Schur factorization $H = STS^T$ (for H real) or $H = STS^H$ (for H complex). The matrix S (the Schur vectors of H) may optionally be computed as well. Alternatively S may be postmultiplied into the matrix Q determined in stage 1, to give the matrix $Z = QS$, the Schur vectors of A. The eigenvalues are obtained from the diagonal of T. All this is done by subroutine PxLAHQR.

The algorithm used in PxLAHQR is similar to the LAPACK routine xLAHQR. Unlike xLAHQR, however, instead of sending one double shift through the largest unreduced submatrix, this algorithm sends multiple double shifts and spaces them apart so that there can be parallelism across several processor row/columns. Another critical difference is that this algorithm applies multiple double shifts in a block fashion, as opposed to xLAHQR which applies one double shift at a time, and xHSEQR from LAPACK which attempts to achieve a blocked code by combining the double shifts into one single large multi-shift. For complete details, please refer to [79].

See table 3.9 for a complete list of the routines.

Table 3.9: Computational routines for the nonsymmetric eigenproblem

Type of Matrix	Operation	Single Precision		Double Precision	
		Real	Complex	Real	Complex
general orthogonal/ unitary	Hessenberg reduction multiply matrix after Hessenberg reduction	PSGEHRD PSORMHR	PCGEHRD PCUNMHR	PDGEHRD PDORMHR	PZGEHRD PZUNMHR
Hessenberg	eigenvalues and Schur decomposition	PSLAHQR		PDLAHQR	

3.3.6 Singular Value Decomposition

Let A be a general real m-by-n matrix. The **singular value decomposition (SVD)** of A is the factorization $A = U\Sigma V^T$, where U and V are orthogonal and $\Sigma = \text{diag}(\sigma_1, \ldots \sigma_r)$, $r = \min(m, n)$, with $\sigma_1 \geq \cdots \geq \sigma_r \geq 0$. If A is complex, its SVD is $A = U\Sigma V^H$, where U and V are unitary and Σ is as before with real diagonal elements. The σ_i are called the **singular values**, the first r columns of V the **right singular vectors**, and the first r columns of U the **left singular vectors**.

The routines described in this section, and listed in table 3.10, are used to compute this decomposition. The computation proceeds in the following stages:

1. The matrix A is reduced to bidiagonal form: $A = U_1 B V_1^T$ if A is real ($A = U_1 B V_1^H$ if A is complex), where U_1 and V_1 are orthogonal (unitary if A is complex), and B is real and upper-bidiagonal when $m \geq n$ and lower bidiagonal when $m < n$, so that B is nonzero only on the main diagonal and either on the first superdiagonal (if $m \geq n$) or the first subdiagonal (if $m < n$).

2. The SVD of the bidiagonal matrix B is computed: $B = U_2 \Sigma V_2^T$, where U_2 and V_2 are orthogonal and Σ is diagonal as described above. The singular vectors of A are then $U = U_1 U_2$ and $V = V_1 V_2$.

The reduction to bidiagonal form is performed by the subroutine PxGEBRD and the SVD of B is computed using the LAPACK routine xBDSQR.

The routine PxGEBRD represents U_1 and V_1 in factored form as products of elementary reflectors, as described in section 3.4. If A is real, the matrices U_1 and V_1 may be multiplied by other matrices without forming U_1 and V_1 using routine PxORMBR. If A is complex, one instead uses PxUNMBR.

If $m \gg n$, it may be more efficient to first perform a QR factorization of A, using the routine PxGEQRF, and then to compute the SVD of the n-by-n matrix R, since if $A = QR$ and $R = U\Sigma V^T$, then the SVD of A is given by $A = (QU)\Sigma V^T$. Similarly, if $m \ll n$, it may be more efficient to first perform an LQ factorization of A, using xGELQF. These preliminary QR and LQ factorizations are performed by the driver PxGESVD.

The SVD may be used to find a minimum norm solution to a (possibly) rank-deficient linear least squares problem (3.1). The effective rank, k, of A can be determined as the number of singular values which exceed a suitable threshold. Let $\hat{\Sigma}$ be the leading k-by-k submatrix of Σ, and $\hat{V}$ be the matrix consisting of the first k columns of V. Then the solution is given by

$$x = \hat{V}\hat{\Sigma}^{-1}\hat{c}_1,$$

where $\hat{c}_1$ consists of the first k elements of $c = U^T b = U_2^T U_1^T b$. $U_1^T b$ can be computed by using PxORMBR.

Table 3.10: Computational routines for the singular value decomposition

Type of Matrix	Operation	Single Precision		Double Precision	
		Real	Complex	Real	Complex
general	bidiagonal reduction	PSGEBRD	PCGEBRD	PDGEBRD	PZGEBRD
orthogonal/ unitary	multiply matrix after bidiagonal reduction	PSORMBR	PCUNMBR	PDORMBR	PZUNMBR

3.3.7 Generalized Symmetric Definite Eigenproblems

This section is concerned with the solution of the generalized eigenvalue problems $Az = \lambda Bz$, $ABz = \lambda z$, and $BAz = \lambda z$, where A and B are real symmetric or complex Hermitian and B is positive definite. Each of these problems can be reduced to a standard symmetric eigenvalue problem, using a Cholesky factorization of B as either $B = LL^T$ or $B = U^T U$ (LL^H or $U^H U$ in the Hermitian case).

With $B = LL^T$, we have

$$Az = \lambda Bz \quad \Rightarrow \quad (L^{-1}AL^{-T})(L^T z) = \lambda(L^T z).$$

Hence the eigenvalues of $Az = \lambda Bz$ are those of $Cy = \lambda y$, where C is the symmetric matrix $C = L^{-1}AL^{-T}$ and $y = L^T z$. In the complex case C is Hermitian with $C = L^{-1}AL^{-H}$ and $y = L^H z$.

Table 3.11 summarizes how each of the three types of problem may be reduced to standard form $Cy = \lambda y$, and how the eigenvectors z of the original problem may be recovered from the eigenvectors y of the reduced problem. The table applies to real problems; for complex problems, transposed matrices must be replaced by conjugate transposes.

Table 3.11: Reduction of generalized symmetric definite eigenproblems to standard problems

	Type of Problem	Factorization of B	Reduction	Recovery of Eigenvectors
1.	$Az = \lambda Bz$	$B = LL^T$	$C = L^{-1}AL^{-T}$	$z = L^{-T}y$
		$B = U^T U$	$C = U^{-T}AU^{-1}$	$z = U^{-1}y$
2.	$ABz = \lambda z$	$B = LL^T$	$C = L^T AL$	$z = L^{-T}y$
		$B = U^T U$	$C = UAU^T$	$z = U^{-1}y$
3.	$BAz = \lambda z$	$B = LL^T$	$C = L^T AL$	$z = Ly$
		$B = U^T U$	$C = UAU^T$	$z = U^T y$

Given A and a Cholesky factorization of B, the routines PxyyGST overwrite A with the matrix C of the corresponding standard problem $Cy = \lambda y$ (see table 3.12). This may then be solved by using the routines described in subsection 3.3.4. No special routines are needed to recover the eigenvectors z of the generalized problem from the eigenvectors y of the standard problem, because these computations are simple applications of Level 2 or Level 3 BLAS.

Table 3.12: Computational routines for the generalized symmetric definite eigenproblem

Type of Matrix	Operation	Single Precision		Double Precision	
		Real	Complex	Real	Complex
symmetric/Hermitian	reduction	PSSYGST	PCHEGST	PDSYGST	PZHEGST

3.4 Orthogonal or Unitary Matrices

A real orthogonal or complex unitary matrix (usually denoted Q) is often represented in ScaLA-PACK as a product of **elementary reflectors** — also referred to as **elementary Householder matrices** (usually denoted H_i). For example,

$$Q = H_1 H_2 \ldots H_k.$$

Most users need not be aware of the details, because ScaLAPACK routines are provided to work with this representation:

- routines whose names begin PSORG- (real) or PCUNG- (complex) can generate all or part of Q explicitly;

- routines whose name begin PSORM- (real) or PCUNM- (complex) can multiply a given matrix by Q or Q^H without forming Q explicitly.

The following details may occasionally be useful.

An elementary reflector (or elementary Householder matrix) H of order n is a unitary matrix of the form

$$H = I - \tau v v^H, \tag{3.4}$$

where τ is a scalar and v is an n-vector, with $|\tau|^2 \|v\|_2^2 = 2\mathrm{Re}(\tau)$; v is often referred to as the **Householder vector**. Often v has several leading or trailing zero elements, but for the purpose of this discussion assume that H has no such special structure.

Some redundancy in the representation (3.4) exists, which can be removed in various ways. Like LAPACK, the representation used in ScaLAPACK (which differs from that used in LINPACK or EISPACK) sets $v_1 = 1$; hence v_1 need not be stored. In real arithmetic, $1 \le \tau \le 2$, except that $\tau = 0$ implies $H = I$.

In complex arithmetic, τ may be complex and satisfies $1 \le \mathrm{Re}(\tau) \le 2$ and $|\tau - 1| \le 1$. Thus a complex H is not Hermitian (as it is in other representations), but it is unitary, which is the important property. The advantage of allowing τ to be complex is that, given an arbitrary complex vector x, H can be computed so that

$$H^H x = \beta(1, 0, \ldots, 0)^T$$

with *real* β. This is useful, for example, when reducing a complex Hermitian matrix to real symmetric tridiagonal form or a complex rectangular matrix to real bidiagonal form.

For further details, see Lehoucq [94].

3.5 Algorithmic Differences between LAPACK and ScaLAPACK

The following ScaLAPACK routines use different algorithms from their LAPACK counterparts. Refer to the relevant LAPACK working notes or the leading comments of the source code for details.

- PxDBSV, PxDBTRF, PxDBTRS: No LAPACK equivalent; refer to [32].

- PxGBSV, PxGBTRF, PxGBTRS: Refer to [32].

- PxLAHQR: Refer to [79].

- PxPBSV, PxPBTRF, PxPBTRS: Refer to [32].

- PxPTSV, PxPTTRF, PxPTTRS: Refer to [32].

- PxSYEV

- PxSYEVX/PxHEEVX: Refer to [40].

- PxSYGVX/PxHEGVX.

Chapter 4

Data Distributions and Software Conventions

The ScaLAPACK software library provides routines that operate on three types of matrices: in-core dense matrices, in-core narrow band matrices and out-of-core dense matrices. On entry, these routines assume that the data has been distributed on the processors according to a specific data decomposition scheme. Conventional arrays are used to store locally the data when it resides in the processors' memory. The data layout information as well as the local storage scheme for these different matrix operands is conveyed to the routines via a simple array of integers called an array descriptor. The first entry of this array identifies the type of the descriptor, i.e., the data distribution scheme it describes. This chapter first presents the fundamental concepts of process grids, communication contexts and array descriptors. Then, for each of the three possible matrix operand mentioned above, the data distribution scheme and the corresponding descriptor array used by ScaLAPACK are discussed in detail. Finally, the software conventions common to all ScaLAPACK routines are presented.

4.1 Basics

ScaLAPACK requires that all global data (vectors or matrices) be distributed across the processes prior to invoking the ScaLAPACK routines. The storage schemes of global data structures in ScaLAPACK are conceptually the same as for LAPACK.

Global data is mapped to the local memories of processes assuming specific data distributions. The local data on each process is referred to as the **local array**.

The layout of an application's data within the hierarchical memory of a concurrent computer is critical in determining the performance and scalability of the parallel code. On shared-memory concurrent computers (or *multiprocessors*) LAPACK seeks to make efficient use of the hierarchical memory by maximizing data reuse (e.g., on a cache-based computer LAPACK avoids having to reload the cache too frequently). Specifically, LAPACK casts linear algebra computations in terms of block-oriented, matrix-matrix operations through the use of the Level 3 BLAS whenever possible.

This approach generally results in maximizing the ratio of floating-point operations to memory references and enables data reuse as much as possible while it is stored in the highest levels of the memory hierarchy (e.g., vector registers or high-speed cache).

An analogous approach has been followed in the design of ScaLAPACK for distributed-memory machines. By using block-partitioned algorithms we seek to reduce the frequency with which data must be transferred between processes, thereby reducing the fixed startup cost (or latency) incurred each time a message is communicated.

The ScaLAPACK routines for solving dense linear systems and eigenvalue problems assume that all global data has been distributed to the processes with a one-dimensional or two-dimensional block-cyclic data distribution. This distribution is a natural expression of the block-partitioned algorithms present in ScaLAPACK. The ScaLAPACK routines for solving band linear systems and tridiagonal systems assume that all global data has been distributed to the processes with a one-dimensional block data distribution. Each of these distributions is supported in the High Performance Fortran standard [91]. Explanations for each distribution will be presented and accompanied by the appropriate HPF directives.

Our implementation of ScaLAPACK emphasizes the mathematical view of a matrix over its storage. In fact, it is even possible to reuse our interface for a different block data distribution that would not fit in the block-cyclic scheme.

4.1.1 Process Grid

The P processes of an abstract parallel computer are often represented as a one-dimensional linear array of processes labeled $0, 1, \ldots, P - 1$. For reasons described below, it is often more convenient to map this one-dimensional array of processes into a two-dimensional rectangular grid, or **process grid**. (A process grid is also referred to as a process mesh.) This grid will have P_r process rows and P_c process columns, where $P_r * P_c = P$. A process can now be referenced by its row and column coordinates, (p_r, p_c), within the grid, where $0 \leq p_r < P_r$, and $0 \leq p_c < P_c$). An example of such a mapping is shown in figure 4.1, where $P_r = 2$ and $P_c = 4$.

	0	1	2	3
0	0	1	2	3
1	4	5	6	7

Figure 4.1: Eight processes mapped to a 2×4 process grid

In figure 4.1, the processes are mapped to the process grid by using **row-major order**; in other words, the numbering of the processes increases sequentially across each row. Similarly, the processes can be mapped in a **column-major order** whereby the numbering of the processes proceeds down each column of the process grid. The BLACS routine BLACS_GRIDINIT performs this task of mapping the processes to the process grid. By default, BLACS_GRIDINIT assumes a row-major ordering, although a column-major ordering can also be specified. The companion routine BLACS_GRIDMAP is a more general form of BLACS_GRIDINIT and allows the user to define the mapping of the processes. Refer to the Appendix D.3 for further details.

All ScaLAPACK routines, with the exception of the band linear system routines, allow the processes to be viewed as a one-dimensional or two-dimensional process grid. The band routines support only one-dimensional process grids.

4.1.2 Contexts

In ScaLAPACK, and thus the BLACS, each process grid is enclosed in a context. Similarly, a context is associated with every global matrix in ScaLAPACK. The use of a context provides the ability to have separate "universes" of message passing. This means that a process grid can safely communicate even if other (possibly overlapping) process grids are also communicating. Thus, a context is a powerful mechanism for avoiding unintentional nondeterminism in message passing and provides support for the design of safe, modular software libraries. In MPI, this concept is referred to as a *communicator*.

A context partitions the communication space. A message sent from one context cannot be received in another context. The use of separate communication contexts by distinct libraries (or distinct library routine invocations) insulates communication internal to a specific library routine from external communication that may be going on within the user's program.

In most respects, we can use the terms *process grid* and *context* interchangeably. For example, we may say we perform an operation "in context X" or "in process grid X". The slight difference here is that the user may define two identical process grids (say, two 1×3 process grids, both of which use processes 0, 1, and 2), but each will be enclosed in its own context, so that they are distinct in operation, even though they are indistinguishable from a process grid standpoint.

Another example of the use of context might be to define a normal two-dimensional process grid within which most computation takes place. However, in certain portions of the code it may be more convenient to access the processes as a one-dimensional process grid, whereas at other times we may wish, for instance, to share information among nearest neighbors. In such cases, we will want each process to have access to three contexts: the two-dimensional process grid, the one-dimensional process grid, and a small process grid that contains the process and its nearest neighbors.

Therefore, we see that context allows us to

- create arbitrary groups of processes,

- create an indeterminate number of overlapping and/or disjoint process grids, and

- isolate the process grid so that they do not interfere with each other.

The BLACS has two process grid creation routines, BLACS_GRIDINIT and BLACS_GRIDMAP, that create a process grid and its enclosing context. These routines return context handles, which are simple integers, assigned by the BLACS to identify the context. Subsequent BLACS routines will be passed these handles, which allow the BLACS to determine from which context/process grid a routine is being called. The user *should never alter or change these handles;* they are opaque data objects that are only meaningful for the BLACS routines.

A defined context consumes resources. It is therefore advisable to release contexts when they are no longer needed. This release is done via the routine BLACS_GRIDEXIT. When the entire BLACS system is shut down (via a call to BLACS_EXIT), all outstanding contexts are automatically freed. Further details about these routines can be found in Appendix D.3.

4.1.3 Scoped Operations

An operation that involves more than just a sender and a receiver is called a *scoped* operation. All processes that participate in a scoped operation are said to be within the operation's scope.

On a system using a linear array of processes, the only natural scope is all processes. Using a two-dimensional rectangular process grid, we have three natural scopes, as shown in table 4.1. Refer to [24] for further details.

Table 4.1: Scopes provided by a two-dimensional process grid

Scope	Meaning
Row	All processes in a process row participate.
Column	All processes in a process column participate.
All	All processes in the process grid participate.

These groupings of processes are of particular interest to the linear algebra programmer, since distributed data decompositions of a two-dimensional array (a linear algebra matrix) tend to follow this process mapping. For instance, all of a distributed matrix row can be found on a *process row*.

4.2 Array Descriptors

An **array descriptor** is associated with each global array. This array stores the information required to establish the mapping between each global array entry and its corresponding process and memory location. The notations x_ used in the entries of the array descriptor denote the attributes of a global array. For example, M_ denotes the number of rows and M_A specifically denotes the number of rows in global matrix A. These descriptors assume different storage for the global data. The length of the array descriptor is specified by DLEN_ and varies according to the descriptor type DTYPE_.

Array descriptors are provided for

- dense matrices,

- band and tridiagonal matrices, and

- out-of-core matrices

and are differentiated by the DTYPE_ entry in the descriptor. At the present time the following values of DESC_(DTYPE_) are valid.

Table 4.2: Valid values of DESC_(DTYPE_)

DESC_(DTYPE_)	Designation
1	dense matrices
501	narrow band and tridiagonal coefficient matrices
502	narrow band and tridiagonal right-hand-side matrices
601	out-of-core dense matrices

4.3   In-core Dense Matrices

The choice of an appropriate data distribution heavily depends on the characteristics or flow of the computation in the algorithm. For dense matrix computations, ScaLAPACK assumes the data to be distributed according to the *two-dimensional block-cyclic* data layout scheme. This section presents this distribution and demonstrates how the ScaLAPACK software encodes this essential information as well as the related software conventions.

Dense matrix computations feature a large amount of parallelism, so that a wide variety of distribution schemes have the potential for achieving high performance. The *block-cyclic* data layout has been selected for the dense algorithms implemented in ScaLAPACK principally because of its scalability [51], load balance, and communication [76] properties. The block-partitioned computation proceeds in consecutive order just like a conventional serial algorithm. This essential property of the block cyclic data layout explains why the ScaLAPACK design has been able to reuse the numerical and software expertise of the sequential LAPACK library.

4.3.1   The Two-dimensional Block-Cyclic Distribution

In this section, we consider the data layout of dense matrices on distributed-memory machines, with the goal of making dense matrix computations as efficient as possible. We shall discuss a sequence of data layouts, starting with the most simple, obvious, and inefficient one and working up to the complicated but efficient ScaLAPACK ultimately uses. Even though our justification is based on Gaussian elimination, analysis of many other algorithms has led to the same set of layouts. As a result, these layouts have been standardized as part of the High Performance Fortran standard [91], with corresponding data declarations as part of that language.

The two main issues in choosing a data layout for dense matrix computations are

- load balance, or splitting the work reasonably evenly among the processors throughout the algorithm, and

- use of the Level 3 BLAS during computations on a single processor, to account for the memory hierarchy on each processor.

It will help to remember the pictorial representation of Gaussian elimination below. As the algorithm proceeds, it works on successively smaller square southeast corners of the matrix. In addition, there is extra Level 2 BLAS work to factorize the submatrix $A_{k:N,k:k+b-1}$.

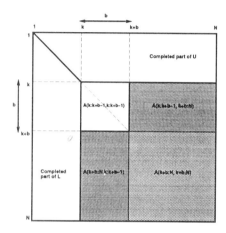

Figure 4.2: Gaussian elimination using Level 3 BLAS

For convenience we will number the processes from 0 to $P-1$, and matrix columns (or rows) from 1 to N. The following two figures shows a sequence of data layouts we will consider. In all cases, each submatrix is labeled with the number of the process (from 0 to 3) that contains it. Process 0 owns the shaded submatrices.

Consider the layout illustrated on the left of figure 4.3, the **one-dimensional block column distribution**. This distribution assigns a block of contiguous columns of a matrix to successive

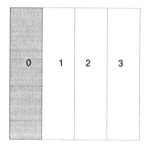
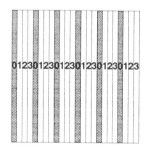

Figure 4.3: The one-dimensional block and cyclic column distributions

processes. Each process receives only one block of columns of the matrix. Column k is stored on process $\lfloor k/tc \rfloor$ where $tc = \lceil N/P \rceil$ is the maximum number of columns stored per process. In the figure $N = 16$ and $P = 4$. This layout does not permit good load balancing for the above Gaussian elimination algorithm because as soon as the first tc columns are complete, process 0 is idle for the rest of the computation. The transpose of this layout, the **one-dimensional block row distribution**, has a similar shortfall for dense computations.

The second layout illustrated on the right of figure 4.3, the **one-dimensional cyclic column distribution**, addressed this problem by assigning column k to process $(k - 1) \bmod P$. In the figure, $N = 16$ and $P = 4$. With this layout, each process owns approximately $1/P^{th}$ of the square southeast corner of the matrix, so the load balance is good. However, since single columns (rather than blocks) are stored, we cannot use the Level 2 BLAS to factorize $A_{k:N,k:k+b-1}$ and may not be able to use the Level 3 BLAS to update $A_{k+b:N,k+b:N}$. The transpose of this layout, the **one-dimensional cyclic row distribution**, has a similar shortfall.

The third layout shown on the left of figure 4.4, the **one-dimensional block-cyclic column distribution**, is a compromise between the distribution schemes shown in figure 4.3. We choose a block size NB, divide the columns into groups of size NB, and distribute these groups in a cyclic manner. This means column k is stored in process $\lfloor (k - 1)/NB \rfloor \bmod P$. In fact, this layout includes the first two as the special cases, $NB = tc = \lceil N/P \rceil$ and $NB = 1$, respectively. In the figure $N = 16$, $P = 4$ and $NB = 2$. For NB larger than 1, this has a slightly worse balance than the one-dimensional cyclic column distribution, but can use the Level 2 BLAS and Level 3 BLAS for the local computations. For NB less than tc, it has a better load balance than the one-dimensional block column distribution, but can call the BLAS only on smaller subproblems. Hence, it takes less advantage of the local memory hierarchy. Moreover, this layout has the disadvantage that the factorization of $A_{k:N,k:k+b-1}$ will take place on one process (in the natural situation where column blocks in the layout correspond to column blocks in Gaussian elimination), thereby representing a serial bottleneck.

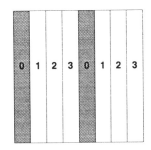

Figure 4.4: The one-dimensional block-cyclic column- and the two-dimensional block-cyclic distributions

This serial bottleneck is eased by the fourth layout shown on the right of figure 4.4, the **two-dimensional block cyclic distribution**. Here, we think of our P processes arranged in a $P_r \times P_c$ rectangular array of processes, indexed in a two-dimensional fashion by (p_r, p_c), with $0 <= p_r < P_r$ and $0 <= p_c < P_c$. All the processes (p_r, p_c) with a fixed p_c are referred to as process column p_c. All the processes (p_r, p_c) with a fixed p_r are referred to as process row p_r. Thus, this layout includes all the previous layouts, and their transposes, as special cases. In the figure, $N = 16$, $P = 4$, $P_r = P_c = 2$, and $MB = NB = 2$. This layout permits P_c-fold parallelism in any column, and calls to the Level 2 BLAS and Level 3 BLAS on local subarrays. Finally, this layout also features good scalability properties as shown in [61].

The two-dimensional block cyclic distribution scheme is the data layout that is used in the ScaLA-PACK library for dense matrix computations.

4.3.2 Local Storage Scheme and Block-Cyclic Mapping

The block-cyclic distribution scheme is a mapping of a set of blocks onto the processes. The previous section informally described this mapping as well as some of its properties. To be complete, we must now explain how the blocks that are mapped to the same process are arranged and stored in the local process memory. In other words, we shall describe the precise mapping that associates to a matrix entry identified by its global indexes the coordinates of the process that owns it and its local position within that process's memory.

Suppose we have an array of length N to be stored on P processes. By convention, the array entries are numbered 1 through N and the processes are numbered 0 through $P - 1$. First, the array is divided into contiguous blocks of size NB. When NB does not divide N evenly, the last block of array elements will only contain $\mod(N, NB)$ entries instead of NB. By convention, these blocks are numbered starting from zero and dealt out to the processes like a deck of cards. In other words, if we assume that the process 0 receives the first block, the kth block is assigned to the process of coordinate $\mod(k, P)$. The blocks assigned to the same process are stored contiguously in memory. The mapping of an array entry globally indexed by I is defined by the following analytical equation:

$$I = k\,NB + x = (l\,P + p) * NB + x,$$

where I is a global index in the array, l is the local block coordinate into which this entry resides, p is the coordinate of the process owning that block, and finally x is the coordinate within that block where the global array entry of index I is to be found. It is then fairly easy to establish the analytical relationship between these variables. One obtains:

$$p = \lfloor (I - 1)/NB \rfloor \bmod P, \ l = \lfloor (I - 1)/(P * NB) \rfloor, \text{ and } x = \mod(I - 1, NB) + 1. \tag{4.1}$$

These equations allow to determine the local information, i.e. the local index $l * NB + x$ as well as the process coordinate p corresponding to a global entry identified by its global index I and conversely. Table 4.3 illustrates this mapping for the block layout when $P = 2$ and $N = 16$, i.e., $NB = 8$. At most one block is assigned to each process.

Table 4.3: One-dimensional block mapping example for $P = 2$ and $N = 16$

I	1	2	3	4	5	6	7	8	9	10	11	12	13	14	15	16
p	0	0	0	0	0	0	0	0	1	1	1	1	1	1	1	1
l	0	0	0	0	0	0	0	0	0	0	0	0	0	0	0	0
x	1	2	3	4	5	6	7	8	1	2	3	4	5	6	7	8
$l * NB + x$	1	2	3	4	5	6	7	8	1	2	3	4	5	6	7	8

This example of the one-dimensional block distribution mapping can be expressed in HPF by using the following statements:

```
      REAL :: X( N )
!HPF$ PROCESSORS PROC( P )
!HPF$ DISTRIBUTE X( BLOCK( NB ) ) ONTO PROC
```

Table 4.4 illustrates Equation 4.1 for the cyclic layout, i.e., $NB = 1$ when $P = 2$ and $N = 16$.

Table 4.4: One-dimensional cyclic mapping example for $P = 2$ and $N = 16$

I	1	2	3	4	5	6	7	8	9	10	11	12	13	14	15	16
p	0	1	0	1	0	1	0	1	0	1	0	1	0	1	0	1
l	0	0	1	1	2	2	3	3	4	4	5	5	6	6	7	7
x	1	1	1	1	1	1	1	1	1	1	1	1	1	1	1	1
$l*NB+x$	1	1	2	2	3	3	4	4	5	5	6	6	7	7	8	8

This example of the one-dimensional cyclic distribution mapping can be expressed in HPF by using the following statements:

```
      REAL :: X( N )
!HPF$ PROCESSORS PROC( P )
!HPF$ DISTRIBUTE X( CYCLIC ) ONTO PROC
```

Table 4.5 illustrates Equation 4.1 for the block-cyclic layout when $P = 2$, $NB = 3$ and $N = 16$.

Table 4.5: One-dimensional block-cyclic mapping example for $P = 2$, $NB = 3$ and $N = 16$

I	1	2	3	4	5	6	7	8	9	10	11	12	13	14	15	16
p	0	0	0	1	1	1	0	0	0	1	1	1	0	0	0	1
l	0	0	0	0	0	0	1	1	1	1	1	1	2	2	2	2
x	1	2	3	1	2	3	1	2	3	1	2	3	1	2	3	1
$l*NB+x$	1	2	3	1	2	3	4	5	6	4	5	6	7	8	9	7

This example of the one-dimensional cyclic distribution mapping can be expressed in HPF by using the following statements:

```
      REAL :: X( N )
!HPF$ PROCESSORS PROC( P )
!HPF$ DISTRIBUTE X( CYCLIC( NB ) ) ONTO PROC
```

There is in fact no real reason to always deal out the blocks starting with the process 0. In fact, it is sometimes useful to start the data distribution with the process of arbitrary coordinate SRC, in which case Equation 4.1 becomes:

$$\begin{cases} p = (SRC + \lfloor (I-1)/NB \rfloor) \bmod P, \\ l = \lfloor (I-1)/(P*NB) \rfloor, \\ x = \bmod(I-1, NB) + 1. \end{cases} \qquad (4.2)$$

Table 4.6: One-dimensional block-cyclic mapping example for $P = 2$, $SRC = 1$, $NB = 3$ and $N = 16$

I	1	2	3	4	5	6	7	8	9	10	11	12	13	14	15	16
p	1	1	1	0	0	0	1	1	1	0	0	0	1	1	1	0
l	0	0	0	0	0	0	1	1	1	1	1	1	2	2	2	2
x	1	2	3	1	2	3	1	2	3	1	2	3	1	2	3	1
$l*NB+x$	1	2	3	1	2	3	4	5	6	4	5	6	7	8	9	7

Table 4.6 illustrates Equation 4.2 for the block-cyclic layout when $P = 2$, $SRC = 1$, $NB = 3$ and $N = 16$. This example of the one-dimensional block-cyclic distribution mapping can be expressed in HPF by using the following statements:

```
      REAL :: X( N )
!HPF$ PROCESSORS PROC( P )
!HPF$ TEMPLATE T( N + P*NB )
!HPF$ DISTRIBUTE T( CYCLIC( NB ) ) ONTO PROC
!HPF$ ALIGN X( I ) WITH T( SRC*NB + I )
```

In the two-dimensional case, assuming the matrix is partitioned in $MB \times NB$ blocks and that the first block is given to the process of coordinates $(RSRC, CSRC)$, the analytical formula given above for the one-dimensional case are simply reused independently in each dimension of the $P_r \times P_c$ process grid. For example, the matrix entry (I, J) is thus to be found in the process of coordinates (p_r, p_c) within the local (l, m) block at the position (x, y) given by:

$$\begin{cases} (l, m) & = & (\lfloor (I-1)/(P_r * MB) \rfloor, \lfloor (J-1)/(P_c * NB) \rfloor), \\ (p_r, p_c) & = & ((RSRC + \lfloor (I-1)/MB \rfloor) \bmod P_r, (CSRC + \lfloor (J-1)/NB \rfloor) \bmod P_c), \\ (x, y) & = & (\bmod(I-1, MB) + 1, \bmod(J-1, NB) + 1). \end{cases}$$

These formula specify how an M_A by N_A matrix A is mapped and stored on the process grid. It is first decomposed into MB_A by NB_A blocks starting at its upper left corner. These blocks are then uniformly distributed across the process grid in a cyclic manner.

Every process owns a collection of blocks, which are contiguously stored by column in a two-dimensional "column major" array.

This local storage convention allows the ScaLAPACK software to use efficiently the local memory hierarchy by calling the BLAS on subarrays that may be larger than a single MB_A by NB_A block. We present in figure 4.5 the mapping of a 5 × 5 matrix partitioned into 2 × 2 blocks mapped onto a 2 × 2 process grid (i.e., $M\_A = N\_A = 5$, $P_r = P_c = 2$, and $MB\_A = NB\_A = 2$). The local entries of every matrix column are contiguously stored in the processes' memories.

In figure 4.5, the process of coordinates $(0, 0)$ owns four blocks. The matrix entries of the global columns 1, 2 and 5 are contiguously stored in that process's memory. Finally, these columns are themselves continuously stored forming a conventional two-dimensional local array. In that local

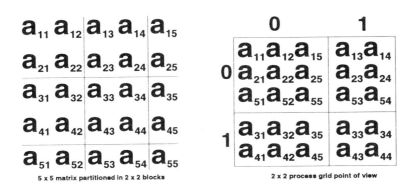

Figure 4.5: A 5×5 matrix decomposed into 2×2 blocks mapped onto a 2×2 process grid

array A, the entry $A(2,3)$ contains the value of the global matrix entry a_{25}. This example would be expressed in HPF as:

```
      REAL :: A( 5, 5 )
!HPF$ PROCESSORS PROC( 2, 2 )
!HPF$ DISTRIBUTE A( CYCLIC( 2 ), CYCLIC( 2 ) ) ONTO PROC
```

Determining the number of rows or columns of a global dense matrix that a specific process receives is an essential task for the user. ScaLAPACK provides a tool routine, NUMROC, to perform this function. The notation $LOC_r()$ and $LOC_c()$ is used to reflect these local quantities throughout the leading comments of the source code and is reflected in the sample argument description in section 4.3.5. The values of $LOC_r()$ and $LOC_c()$ computed by NUMROC are precise calculations.

However, if users want a general idea of the size of a local array, they can perform the following "back of the envelope" calculation to receive an upper bound on the quantity.

An upper bound on the value of $LOC_r()$ can be calculated as:

$$LOC_r() \approx \frac{\dfrac{M\_A + MB\_A - 1}{MB\_A} + P_r - 1}{P_r} * MB\_A$$

or equivalently as

$$LOC_r() \approx \lceil \lceil M\_A/MB\_A \rceil / P_r \rceil * MB\_A.$$

Similarly, an upper bound on the value of $LOC_c()$ can be calculated as

$$LOC_c() \approx \frac{\dfrac{N\_A + NB\_A - 1}{NB\_A} + P_c - 1}{P_c} * NB\_A$$

or equivalently as

$$LOC_c() \approx \lceil \lceil N\_A/NB\_A \rceil / P_c \rceil * NB\_A.$$

Note that this calculation can yield a gross overestimate of the amount of space actually required.

4.3.3 Array Descriptor for In-core Dense Matrices

The array descriptor **DESC_**, whose type is defined as **DESC_(DTYPE_)=1**, is an integer array of length 9. It is used for the ScaLAPACK routines solving dense linear systems and eigenvalue problems. All global vector and matrix operands are assumed to be distributed on the process grid according to the one- or two-dimensional block cyclic data distribution scheme. Refer to section 4.3.1 for further details on block cyclic data distribution.

A general M_ by N_ distributed matrix is defined by its dimensions, the size of the elementary MB_ by NB_ block used for its decomposition, the coordinates of the process having in its local memory the first matrix entry (RSRC_,CSRC_), and the BLACS context (CTXT_) in which this matrix is defined. Finally, a local leading dimension LLD_ is associated with the local memory address pointing to the data structure used for the local storage of this distributed matrix.

Let us assume, for example, that we have an array descriptor *DESCA* for a dense global matrix *A*. As previously mentioned, the notations x_ used in the entries of the array descriptor denote the attributes of a global array. For readability of the code, we have associated symbolic names for the descriptor entries. For example, M_ denotes the number of rows and M_A specifically denotes the number of rows in global matrix A.

Table 4.7: Content of the array descriptor for in-core dense matrices

DESC_()	Symbolic Name	Scope	Definition
1	DTYPE_A	(global)	Descriptor type DTYPE_A=1 for dense matrices.
2	CTXT_A	(global)	BLACS context handle, indicating the BLACS process grid over which the global matrix A is distributed. The context itself is global, but the handle (the integer value) may vary.
3	M_A	(global)	Number of rows in the global array A.
4	N_A	(global)	Number of columns in the global array A.
5	MB_A	(global)	Blocking factor used to distribute the rows of the array.
6	NB_A	(global)	Blocking factor used to distribute the columns of the array.
7	RSRC_A	(global)	Process row over which the first row of the array A is distributed.
8	CSRC_A	(global)	Process column over which the first column of the array A is distributed.
9	LLD_A	(local)	Leading dimension of the local array. LLD_A $\geq$ MAX(1,LOC$_r$(M_A)).

For a detailed description of LOC$_r$() notation, please refer to section 4.3.2.

4.3.4   Example

As mentioned in section 4.3.1, ScaLAPACK assumes a one-dimensional or two-dimensional block-cyclic distribution for the dense matrix computational routines. The **block-cyclic distribution** is a generalization of the block and cyclic distributions. In one dimension, blocks of rows of size MB or blocks of columns of size NB are cyclically distributed over the processes. In two dimensions, blocks of size $MB \times NB$ are distributed cyclically over the processes. Example programs can be found in section 2.3 and Appendix C.1.

According to the two-dimensional block cyclic data distribution, scheme an M_ by N_ dense matrix is first decomposed into MB_ by NB_ blocks starting at its upper left corner. These blocks are then uniformly distributed in each dimension of the process grid. Thus, every process owns a collection of blocks, which are locally and contiguously stored in a two-dimensional "column major" array. The partitioning of a 9×9 matrix into 2×2 blocks and the mapping of these blocks onto a 2×3 process grid are shown in figure 4.6. The local entries of every matrix column are contiguously stored in the processes' memories. The number of rows of a matrix and the number of columns

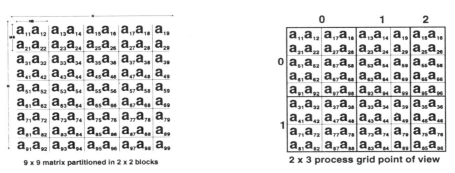

Figure 4.6: A 9×9 matrix decomposed into 2×2 blocks mapped onto a 2×3 process grid

of a matrix that a specific process owns, denoted LOC_r and LOC_c respectively, may differ from process to process in the process grid. Likewise, there is a local leading dimension LLD_ for each process in the process grid. This value may be different on each process in the process grid. For example, we can see on the right of figure 4.6 that the local array stored in process row 0 must have a local leading dimension LLD_ greater than or equal to 5, and greater than or equal to 4 in the process row 1. Table 4.8 gives the values of the local array sizes associated with figure 4.6.

Table 4.8: Sizes of the local arrays

Process Coordinates	LLD_	$\text{LOC}_r(\text{M}_-)$	$\text{LOC}_c(\text{N}_-)$
(0,0)	5	5	4
(0,1)	5	5	3
(0,2)	5	5	2
(1,0)	4	4	4
(1,1)	4	4	3
(1,2)	4	4	2

4.3.5 Submatrix Argument Descriptions

As previously mentioned, the ScaLAPACK routines that solve dense linear systems and eigenvalue problems assume that all global arrays are distributed in a one- or two-dimensional block cyclic fashion. After a global vector or matrix has been block-cyclicly distributed over a process grid, the user may choose to perform an operation on a portion of the global matrix. This subset of the global matrix is referred to as a "submatrix" and is referenced through the use of six arguments in the calling sequence: the number of rows of the submatrix M, the number of columns of the submatrix N, the local array A containing the global array, the row index IA, the column index JA and the array descriptor of the global array DESCA. This argument convention allows for a global view of the matrix operands and the global addressing of distributed matrices as illustrated in figure 4.7. This scheme allows the complete specification of the submatrix A(IA:IA+M-1,JA:JA+N-1) on which to be operated.

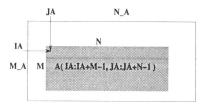

Figure 4.7: Global view of the matrix operands

The description of a global dense subarray consists of (M, N, A, IA, JA, DESCA)

- the number of rows and columns M and N of the global subarray,

- a pointer to the local array containing the entire global array (A, for example),

- the row and column indices, (IA, JA), in the global array, and

- the array descriptor, DESCA, for the global array.

The names of the row and column indices for the global array have the form I<array_name> and J<array_name>, respectively. The array descriptor has a name of the form DESC<array_name>. The length of the array descriptor is specified by DLEN_ and varies according to the descriptor type DTYPE_.

Included in the leading comments of each subroutine (immediately preceding the Argument section), is a brief note describing the **array descriptor** and some commonly used expressions in calculating workspace.

The style of the argument descriptions for dense matrices is illustrated by the following example. As previously mentioned, the notations x_ used in the entries of the array descriptor denote the attributes of a global array. For readability of the code, we have associated symbolic names for the descriptor entries. For example, M_ denotes the number of rows and M_A specifically denotes the number of rows in global matrix A. Complete details can be found in section 4.3.3.

M   (global input) INTEGER
The number of rows of the matrix $A(IA:IA+M-1,JA:JA+N-1)$ on which to be operated. $M \geq 0$ and $IA+M-1 \leq M\_A$.

N   (global input) INTEGER
The number of columns of the matrix $A(IA:IA+M-1,JA:JA+N-1)$ on which to be operated. $N \geq 0$ and $JA+N-1 \leq N\_A$.

NRHS   (global input) INTEGER
The number of right hand side vectors, i.e. the number of columns of the matrix $B(IB:IB+N-1,JB:JB+NRHS-1)$. $NRHS \geq 0$.

A   (local input/local output) REAL pointer into the local memory to an array of local dimension $(LLD\_A, LOC_c(JA+N-1))$

IA   (global input) INTEGER
The row index in the global array A indicating the first row of $A(IA:IA+M-1,JA:JA+N-1)$.

JA   (global input) INTEGER
The column index in the global array A indicating the first column of $A(IA:IA+M-1,JA:JA+N-1)$.

DESCA   (global and local input) INTEGER array of dimension DLEN_
The array descriptor for the global matrix A.

B   (local input/local output) REAL pointer into the local memory to an array of local dimension $(LLD\_B, LOC_c(JB+NRHS-1))$.

IB   (global input) INTEGER
The row index in the global array B indicating the first row of $B(IB:IB+N-1,JB:JB+NRHS-1)$.

JB   (global input) INTEGER
The column index in the global array B indicating the first column of $B(IB:IB+N-1,JB:JB+NRHS-1)$.

DESCB   (global and local input) INTEGER array of dimension DLEN_
The array descriptor for the global matrix B.

The description of each argument gives

- A classification of the argument as (local input), (global and local input), (local input/local output), (global input), (local output), (global output), (global input/global output), (local input or local output),[1] (local or global input),[2] (local workspace), or (local workspace/local output).

- The type of the argument;

- For an array, its dimension(s).

---

[1](Local input or local output) means that the argument may be either a local input argument or a local output argument, depending on the values of other arguments; for example, in the PxyySVX driver routines, some arguments are used either as local output arguments to return details of a factorization, or as local input arguments to supply details of a previously computed factorization.

[2](local or global input) is used to describe the length of the workspace arguments, e.g., LWORK, where the value can be local input specifying the size of the local WORK array, or global input LWORK=−1 specifying a global query for the amount of workspace required.

These dimensions are often expressed in terms of $LOC_r()$ and $LOC_c()$ calculations. For further details, please refer to section 4.3.2.

- A specification of the value(s) that must be supplied for the argument (if it is an input argument), or of the value(s) returned by the routine (if it is an output argument), or both (if it is an input/output argument). In the last case, the two parts of the description are introduced by the phrases "On entry" and "On exit".

- For a scalar input argument, any constraints that the supplied values must satisfy (such as $N \geq 0$ in the example above).

4.3.6 Matrix and Vector Storage Conventions

Whether a dense coefficient matrix operand is nonsymmetric, symmetric or Hermitian, the entire two-dimensional global array is distributed onto the process grid.

For symmetric and Hermitian matrix operands, only the upper (UPLO='U') triangle or the lower (UPLO='L') triangle of the global array is accessed. For triangular matrix operands, the argument UPLO defines whether the matrix is upper (UPLO='U') or lower (UPLO='L') triangular. Only the elements of the relevant triangle of the global array are accessed. Some ScaLAPACK routines have an option to handle unit triangular matrix operands (that is, triangular matrices with diagonal elements = 1). This option is specified by an argument DIAG. If DIAG = 'U' (Unit triangular), the local array elements corresponding to the diagonal elements of the matrix are not referenced by the ScaLAPACK routines.

If an input matrix operand is Hermitian, the imaginary parts of the diagonal elements are zero, and thus the imaginary parts of the corresponding local arrays need not be set, but are assumed to be zero. If an output matrix operand is Hermitian, the imaginary parts of the diagonal elements are set to zero (e.g., PCPOTRF and PCHETRD).

Similarly, if the matrix is upper Hessenberg, the local array elements corresponding to global array elements below the first subdiagonal are not referenced.

Vectors can be distributed across process rows or across process columns. A vector of length N distributed across process rows is distributed the same way that a N-by-1 matrix is. A vector of length N distributed across process columns is distributed the same way that a 1-by-N matrix is.

Within some ScaLAPACK routines, some vectors are replicated in one dimension and distributed in the other dimension. These vectors always aligned with one dimension of another distributed matrix. For example, in PDSYTRD, the vectors D, E, and TAU are replicated across process rows, distributed across process columns, and aligned with the distributed matrix operand A. The data distribution of these replicated vectors is inferred from the distribution of the matrix they are associated with. There is no specific array descriptors for these particular vectors at the present time.

4.4   In-Core Narrow Band and Tridiagonal Matrices

The ScaLAPACK routines solving narrow-band and tridiagonal linear systems assume their operands to be distributed according to the block-column and block-row data distribution schemes. Specifically, the narrow band or tridiagonal coefficient matrix is distributed in a block-column fashion, and the dense matrix of right hand side vectors is distributed in a block-row fashion. This section presents these distributions and demonstrates how the ScaLAPACK software encodes this essential information as well as the related software conventions.

The *block* data layout has been selected for narrow band matrices. Divide-and-conquer algorithms have been implemented in ScaLAPACK because these algorithms offer a much greater scope for exploiting parallelism than the corresponding adapted dense algorithms. The narrow band or tridiagonal coefficient matrix is partitioned into blocks. The inherent parallelism of these divide-and-conquer methods is limited by the number of these blocks because each block is processed independently; hence, it is necessary to choose the number of blocks at least equal to the desired parallelism. However, because the size of the reduced system is proportional to the number of blocks, and solving this reduced system is the major parallelism bottleneck, it follows that a *block* layout in which each process has exactly one block allows maximum exploitation of parallelism while minimizing the size of the reduced system.

4.4.1   The Block Column and Row Distributions

ScaLAPACK assumes a one-dimensional block distribution for the band and tridiagonal routines. The *block* distribution is used when the computational load is distributed homogeneously over the global data. This distribution leads to a highly efficient implementation of the divide-and-conquer algorithms used in ScaLAPACK.

For convenience we will number the processes from 0 to $P - 1$, and the matrix rows from 1 to M and the matrix columns from 1 to N. Figure 4.8 shows the two data layouts used in ScaLAPACK for solving narrow band linear systems. In all cases, each submatrix is labeled with the number of the process that contains it. Process 0 owns the shaded submatrices.

Consider the layout illustrated on the left of figure 4.8, the **one-dimensional block column distribution**. This distribution assigns a block of NB contiguous columns of a matrix to successive

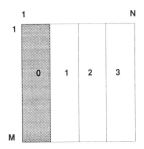
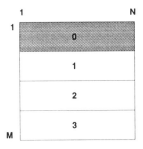

Figure 4.8: The one-dimensional block-column and block-row distributions

processes arranged in a $1 \times P$ one-dimensional process grid. Each process receives at most one block of columns of the matrix, i.e., $NB \geq \lceil N/P \rceil$. Column k is stored on process $\lfloor k/NB \rfloor$. The maximum number of columns stored per process is given by $\lceil N/P \rceil$. In the figure $M = N = 16$ and $P = 4$. This distribution assigns blocks of columns of size NB to successive processes. If the value of P evenly divides the value of N and $NB = N/P$, then each process owns a block of equal size. However, if this is not the case, then either the last process to receive a portion of the matrix will receive a smaller block than other processes, or some processes may receive an empty portion of the matrix. The transpose of this layout, the **one-dimensional block-row distribution**, is shown on the right of figure 4.8.

The block-column distribution scheme is the data layout that is used in the ScaLAPACK library for the coefficient matrix of the narrow band and tridiagonal solvers.

The block-row distribution scheme is the data layout that is used in the ScaLAPACK library for the right-hand-side matrix of the narrow band and tridiagonal solvers.

4.4.2 The Block Mapping

The one-dimensional distribution scheme is a mapping of a set of blocks onto the processes. The previous section informally described this mapping as well as some of its properties. To be complete, we shall describe the precise mapping that associates to a matrix entry identified by its global indexes the coordinates of the process that owns it and its local position within that process's memory.

Suppose we have a two dimensional array A of size $M \times N$ to be distributed on a $1 \times P$ process grid in a block-column fashion. By convention, the array columns are numbered 1 through N and the processes are numbered 0 through $P - 1$. First, the array is divided into contiguous blocks of NB columns with $NB \geq \lceil N/P \rceil$. When NB does not divide N evenly, the last block of columns will only contain $\mathrm{mod}(N, NB)$ columns instead of NB. By convention, these blocks are numbered starting from zero and dealt out to the processes. In other words, if we assume that the process 0 receives the first block, the pth block is assigned to the process of coordinate $(0, p)$. The mapping of a column of the array globally indexed by J is defined by the following analytical equation:

$$J = p\,NB + x,$$

where J is a global column index in the array, p is the column coordinate of the process owning that column, and finally x is the column coordinate within that block of columns where the global array column of index J is to be found. It is then fairly easy to establish the analytical relationship between these variables. One obtains:

$$p = \lfloor (J - 1)/NB \rfloor, \quad x = \mathrm{mod}(J - 1, NB) + 1. \tag{4.3}$$

These equations allow to determine the local information, i.e. the local index x as well as the process column coordinate p corresponding to a global column identified by its global index J and conversely. Table 4.9 illustrates this mapping layout when $P = 2$ and $N = 16$ and $NB = 8$. At most one block is assigned to each process.

Table 4.9: One-dimensional block-column mapping example for $P = 2$ and $N = 16$

J	1	2	3	4	5	6	7	8	9	10	11	12	13	14	15	16
p	0	0	0	0	0	0	0	0	1	1	1	1	1	1	1	1
x	1	2	3	4	5	6	7	8	1	2	3	4	5	6	7	8

This example of the one-dimensional block-column distribution mapping can be expressed in HPF by using the following statements:

```
      REAL :: A( M, N )
!HPF$ PROCESSORS PROC( 1, P )
!HPF$ DISTRIBUTE A( *, BLOCK( NB ) ) ONTO PROC
```

A similar example of block-row distribution can easily be constructed. For an $N \times NRHS$ array B, such an example can be expressed in HPF by using the following statements:

```
      REAL :: B( N, NRHS )
!HPF$ PROCESSORS PROC( P, 1 )
!HPF$ DISTRIBUTE B( BLOCK( NB ), * ) ONTO PROC
```

There is in fact no real reason to always deal out the blocks starting with the process 0. In fact, it is sometimes useful to start the data distribution with the process of arbitrary coordinate SRC, in which case Equation 4.3 becomes:

$$\begin{cases} p = (SRC + \lfloor (J-1)/NB \rfloor), \\ x = \mathrm{mod}(J-1, NB) + 1. \end{cases} \tag{4.4}$$

Table 4.10 illustrates Equation 4.4 for the block-cyclic layout when $P = 2$, $SRC = 1$, $NB = 3$ and $N = 16$. This example of the one-dimensional block-column distribution mapping can be

Table 4.10: One-dimensional block-column mapping for $P = 2$, $SRC = 1$, $N = 16$ and $NB = 8$

J	1	2	3	4	5	6	7	8	9	10	11	12	13	14	15	16
p	1	1	1	1	1	1	1	1	0	0	0	0	0	0	0	0
x	1	2	3	4	5	6	7	8	1	2	3	4	5	6	7	8

expressed in HPF by using the following statements:

```
      REAL :: A( M, N )
!HPF$ PROCESSORS PROC( 1, P )
!HPF$ TEMPLATE T( M, N + P*NB )
!HPF$ DISTRIBUTE T( *, BLOCK( NB ) ) ONTO PROC
!HPF$ ALIGN A( I, J ) WITH T( I, SRC*NB + J )
```

A similar example of block-row distribution can easily be constructed. For an $N \times NRHS$ array B, such an example can be expressed in HPF by using the following statements:

```
      REAL :: B( N, NRHS )
!HPF$ PROCESSORS PROC( P, 1 )
!HPF$ TEMPLATE T( N + P*NB, NRHS )
!HPF$ DISTRIBUTE T( BLOCK( NB ), * ) ONTO PROC
!HPF$ ALIGN A( I, J ) WITH T( SRC*NB + I, J )
```

In ScaLAPACK, the local storage convention of the one-dimensional block distributed matrix in every process's memory is assumed to be Fortran-like, that is, "column major".

Determining the number of rows or columns of a global band matrix that a specific process receives is an essential task for the user. The notation $LOC_r()$ is used for block-row distributions and $LOC_c()$ is used for block-column distributions. These local quantities occur throughout the leading comments of the source code, and are reflected in the sample argument description in section 4.4.7.

For block distribution, a matrix can be distributed unevenly. More specifically, one process in the process grid can receive an array that is smaller than other processes. It is also possible that some processes receive no data. For further information on one-dimensional block-column or block-row data distribution, please refer to section 4.4.1.

Block-Column Distribution: $LOC_c(N\_A)$ denotes the number of columns that a process would receive if N_A columns of a matrix is distributed over P_c columns of its process row.

For example, let us assume that the coefficient matrix A is band symmetric of order N and has been block-column distributed on a $1 \times P_c$ process grid.

In the ideal case where the matrix is evenly distributed to all processes in the process grid, $mod(N\_A, NB\_A) = 0$ and $\frac{N\_A}{NB\_A} = P_c$. Thus, each process receives a block of size NB_A of the matrix A. Therefore,

$LOC_c(N\_A) = NB\_A$.

However, if $mod(N\_A, NB\_A) \neq 0$, at least one of the processes in the process grid will receive a block of size smaller than NB_A. Thus,

if ($mod(N\_A, NB\_A) \neq 0$ and $int(\frac{N\_A}{NB\_A}) = K$) then
    processes $(0, 0), \ldots, (0, K - 1)$ receive
            $LOC_c(N\_A) = NB\_A$
    and process $(0, K)$ receives
            $LOC_c(N\_A) = N\_A - K * NB\_A$.
    if $K < P_c$ then processes $(0, K + 1), \ldots, (0, P_c - 1)$ do not receive any data.
end if

Block-Row Distribution: $LOC_r(M\_B)$ denotes the number of rows that a process would receive if M_B rows of a matrix is distributed over P_r rows of its process column.

Let us assume that the N-by-NRHS right-hand-side matrix B has been block-row distributed on a $P_r \times 1$ process grid.

In the ideal case where the matrix is evenly distributed to all processes in the process grid, $mod(M\_B, MB\_B) = 0$ and $\frac{M\_B}{MB\_B} = P_r$. Thus, each process receives a block of size MB_B of the matrix B. Therefore,

$LOC_r(M\_B) = MB\_B.$

However, if $mod(M\_B, MB\_B) \neq 0$, then at least one of the processes in the process grid will receive a block of size smaller than MB_B. Thus,

if ($mod(M\_B, MB\_B) \neq 0$ and $int(\frac{M\_B}{MB\_B}) = K$) then
   processes $(0,0), \ldots, (K-1, 0)$ receive
        $LOC_r(M\_B) = MB\_B$
   and process $(K, 0)$ receives
        $LOC_r(M\_B) = M\_B - K * MB\_B.$
   if $K < P_r$ then processes $(K+1, 0), \ldots, (P_r - 1, 0)$ do not receive any data.
end if

4.4.3 Local Storage Scheme for Narrow Band Matrices

Let us first discuss how to distribute a narrow band matrix A over a one-dimensional process grid using a block-column distribution. We assume that the coefficient band matrix A is of size 7×7 ($N\_A = 7$) with a bandwidth $BW = 2$ if the matrix A is symmetric positive definite, and $BWL = 2$ and $BWU = 2$ if the matrix A is nonsymmetric. The matrix A is represented by the following.

$$A = \begin{pmatrix} a_{11} & a_{12} & a_{13} & 0 & 0 & 0 & 0 \\ a_{21} & a_{22} & a_{23} & a_{24} & 0 & 0 & 0 \\ a_{31} & a_{32} & a_{33} & a_{34} & a_{35} & 0 & 0 \\ 0 & a_{42} & a_{43} & a_{44} & a_{45} & a_{46} & 0 \\ 0 & 0 & a_{53} & a_{54} & a_{55} & a_{56} & a_{57} \\ 0 & 0 & 0 & a_{64} & a_{65} & a_{66} & a_{67} \\ 0 & 0 & 0 & 0 & a_{75} & a_{76} & a_{77} \end{pmatrix}$$

If we assume that the matrix A is nonsymmetric band, the user may choose to perform partial pivoting or no pivoting during the factorization (PxGBTRF or PxDBTRF, respectively). Both strategies assume a block-column distribution of the coefficient matrix, but additional storage is required for fill-in if partial pivoting is selected. First, let us assume that we have selected no pivoting, and we distribute this matrix onto a 1×3 process grid with a block size of $NB\_A = 3$. The processes would contain the local arrays found in figure 4.9. Figure 4.9 also illustrates that the leading dimension of the local arrays containing the coefficient matrix must be at least $BWL + 1 + BWU$ for the non-pivoting narrow band linear solver.

If, however, we select partial pivoting and distribute this same matrix onto a 1×3 process grid with a block size of $NB\_A = 3$, the processes would contain the local arrays found in figure 4.10. The

Processes

	0		1			2
$*$	$*$	a_{13}	a_{24}	a_{35}	a_{46}	a_{57}
$*$	a_{12}	a_{23}	a_{34}	a_{45}	a_{56}	a_{67}
a_{11}	a_{22}	a_{33}	a_{44}	a_{55}	a_{66}	a_{77}
a_{21}	a_{32}	a_{43}	a_{54}	a_{65}	a_{76}	$*$
a_{31}	a_{42}	a_{53}	a_{64}	a_{75}	$*$	$*$

Figure 4.9: Mapping of local arrays for nonsymmetric band matrix A (no pivoting)

amount of additional storage required for fill-in is represented by F in the figure and is equal to the sum of the lower bandwidth (number of subdiagonals), BWL, and the upper bandwidth (number of superdiagonals), BWU. In this example, $BWL = 2$ and $BWU = 2$. Refer to the leading comments of the routine PxGBTRF for further details. Figure 4.10 also illustrates that the leading dimension of the local arrays containing the coefficient matrix must be at least $2 * (BWL + BWU) + 1$ for the partial pivoting narrow band linear solver.

Processes

	0		1			2
F	F	F	F	F	F	F
F	F	F	F	F	F	F
F	F	F	F	F	F	F
F	F	F	F	F	F	F
$*$	$*$	a_{13}	a_{24}	a_{35}	a_{46}	a_{57}
$*$	a_{12}	a_{23}	a_{34}	a_{45}	a_{56}	a_{67}
a_{11}	a_{22}	a_{33}	a_{44}	a_{55}	a_{66}	a_{77}
a_{21}	a_{32}	a_{43}	a_{54}	a_{65}	a_{76}	$*$
a_{31}	a_{42}	a_{53}	a_{64}	a_{75}	$*$	$*$

Figure 4.10: Mapping of local arrays for nonsymmetric band matrix A (partial pivoting)

Let us now assume that the matrix A is symmetric positive definite band with $BW = 2$, and we distribute this matrix assuming lower triangular storage (UPLO='L') onto a 1×3 process grid with a block size $NB\_A = 3$. The processes would contain the local arrays found in figure 4.11. We would then call the routine PxPBTRF with $BW = 2$ to perform the factorization, for example.

If we then distributed this same matrix assuming upper triangular storage (UPLO='U') onto a 1×3 process grid with a block size $NB\_A = 3$, the processes would contain the local arrays found in figure 4.12.

Figures 4.11 and 4.12 also illustrate that the leading dimension of the local arrays containing the coefficient matrix must be at least $BW + 1$ for the symmetric positive definite narrow band linear solver.

Processes

	0			1		2
a_{11}	a_{22}	a_{33}	a_{44}	a_{55}	a_{66}	a_{77}
a_{21}	a_{32}	a_{43}	a_{54}	a_{65}	a_{76}	*
a_{31}	a_{42}	a_{53}	a_{64}	a_{75}	*	*

Figure 4.11: Mapping of local arrays for symmetric positive definite band matrix A (UPLO='L')

Processes

	0			1		2
*	*	a_{31}	a_{42}	a_{53}	a_{64}	a_{75}
*	a_{21}	a_{32}	a_{43}	a_{54}	a_{65}	a_{76}
a_{11}	a_{22}	a_{33}	a_{44}	a_{55}	a_{66}	a_{77}

Figure 4.12: Mapping of local arrays for symmetric positive definite band matrix A (UPLO='U')

The * notation in figures 4.9, 4.10, 4.11, and 4.12 and the F notation in figure 4.10 signify an entry in which one need not store a value in that position of the local array. These storage positions, however, are required and overwritten during the computation.

The $N \times NRHS$ matrix of right-hand-side vectors B (for example, used in PxGBTRS, PxDBTRS, and PxPBTRS) is assumed to be a dense matrix distributed in a block-row manner across the process grid. Thus, consecutive blocks of rows of the matrix B are assigned to successive processes in the process grid, as described in section 4.4.1.

4.4.4   Local Storage Schemes for Tridiagonal Matrices

A global tridiagonal matrix A, represented as three vectors (DL, D, DU), should be distributed over a one-dimensional process grid assuming a block-column data distribution. We assume that the coefficient tridiagonal matrix A is of size 7×7 ($N\_A = 7$) and is represented by the following.

$$A = \begin{pmatrix} a_{11} & a_{12} & 0 & 0 & 0 & 0 & 0 \\ a_{21} & a_{22} & a_{23} & 0 & 0 & 0 & 0 \\ 0 & a_{32} & a_{33} & a_{34} & 0 & 0 & 0 \\ 0 & 0 & a_{43} & a_{44} & a_{45} & 0 & 0 \\ 0 & 0 & 0 & a_{54} & a_{55} & a_{56} & 0 \\ 0 & 0 & 0 & 0 & a_{65} & a_{66} & a_{67} \\ 0 & 0 & 0 & 0 & 0 & a_{76} & a_{77} \end{pmatrix}$$

If we first assume that the matrix A is nonsymmetric (diagonally dominant like), and it is known *a priori* that no pivoting is required for numerical stability, the user may choose to perform no pivoting during the factorization (PxDTTRF). If we distribute this matrix (assuming no pivoting) onto a 1×3 process grid with a block size of $NB\_A = 3$, the processes would contain the local arrays found in figure 4.13.

	Processes						
	0			1			2
DL	$*$	a_{21}	a_{32}	a_{43}	a_{54}	a_{65}	a_{76}
D	a_{11}	a_{22}	a_{33}	a_{44}	a_{55}	a_{66}	a_{77}
DU	a_{12}	a_{23}	a_{34}	a_{45}	a_{56}	a_{67}	

Figure 4.13: Mapping of local arrays for nonsymmetric tridiagonal matrix A

Finally, a global symmetric positive definite tridiagonal matrix A, represented as two vectors (D and E), should be distributed over a one-dimensional process grid assuming a block-column data distribution.

Let us now assume that this matrix A is symmetric positive definite and that we distribute this matrix assuming lower triangular storage (UPLO='L') onto a 1×3 process grid with a block size $NB\_A = 3$. The processes would contain the local arrays found in figure 4.14. We would then call the routine PxPTTRF to perform the factorization, for example.

	Processes						
	0			1			2
D	a_{11}	a_{22}	a_{33}	a_{44}	a_{55}	a_{66}	a_{77}
E	a_{21}	a_{32}	a_{43}	a_{54}	a_{65}	a_{76}	

Figure 4.14: Mapping of local arrays for symmetric positive definite tridiagonal matrix A (UPLO='L')

If we then distributed this same matrix assuming upper triangular storage (UPLO='U') onto a 1×3 process grid with a block size $NB\_A = 3$, the processes would contain the local arrays found in figure 4.15.

	Processes						
	0			1			2
D	a_{11}	a_{22}	a_{33}	a_{44}	a_{55}	a_{66}	a_{77}
E	a_{12}	a_{23}	a_{34}	a_{45}	a_{56}	a_{67}	

Figure 4.15: Mapping of local arrays for symmetric positive definite tridiagonal matrix A (UPLO='U')

Note that in the tridiagonal cases, it is not necessary to maintain the empty storage positions as designated by $*$ in the narrow band routines.

The matrix of right-hand-side vectors B (for example, used in PxDTTRS and PxPTTRS) is assumed to be a dense matrix distributed block-row across the process grid. Thus, consecutive blocks of rows of the matrix B are assigned to successive processes in the process grid, as described in section 4.4.1.

4.4.5　Array Descriptor for Narrow Band and Tridiagonal Matrices

The array descriptor **DESC_** whose type is defined as **DESC_(DTYPE_)=501**, is an integer array of length 7. This descriptor type is used in the ScaLAPACK narrow band routines and tridiagonal routines to specify a block-column distribution of a global array over a one-dimensional process grid. In the general and symmetric positive definite banded and tridiagonal routines, a one-dimensional block-column distribution is specified for the coefficient matrix. The matrix of right-hand-side vectors must be distributed over a one-dimensional process grid using a block-row data distribution. Refer to section 4.4.1 for further details on block data distribution.

Let us assume, for example, that we have an array descriptor $DESCA$ for a block-column distributed array A. For readability of the code, we have associated symbolic names with the descriptor entries. As previously mentioned, the notations x_ used in the entries of the array descriptor denote the attributes of a global array. For readability of the code, we have associated symbolic names for the descriptor entries. For example, N_ denotes the number of columns and N_A specifically denotes the number of columns in global array A.

When A is non-symmetric and factorized without pivoting, LLD_A must be at least $BWL + 1 + BWU$. When A is non-symmetric and factorized with partial pivoting, LLD_A must be at least $2(BWL + BWU) + 1$. When A is symmetric positive definite, LLD_A must be at least $BW + 1$. Finally, when A is tridiagonal, LLD_A is not referenced.

4.4.6　Array Descriptor for the Matrix of Right-Hand-Side Vectors

The array descriptor **DESC_** whose type is defined as **DESC_(DTYPE_)=502**, is an integer array of length 7. This descriptor type is used in the ScaLAPACK narrow band routines and tridiagonal routines to specify the block-row distribution of a global array containing the right-hand-side vectors over a one-dimensional process grid. In the narrow band and tridiagonal routines, a one-dimensional block-column distribution is specified for the coefficient matrix. The matrix of right-hand-side vectors however must be distributed over a one-dimensional process grid according to a block-row data distribution scheme. Refer to section 4.4.1 for further details on block data distribution.

Let us now assume that we have an array descriptor $DESCB$ for a block-row distributed matrix B. For readability of the code, we have associated symbolic names with the descriptor entries.

For a detailed description of $LOC_r()$ notation, please refer to section 4.4.2.

Table 4.11: Content of the array descriptor for in-core narrow band and tridiagonal coefficient matrices

DESC_()	Symbolic Name	Scope	Definition
1	DTYPE_A	(global)	The descriptor type (DTYPE_A=501) for $1 \times P_c$ process grid for band and tridiagonal matrices block-column distributed.
2	CTXT_A	(global)	The BLACS context handle, indicating the BLACS process grid over which the global matrix A is distributed. The context itself is global, but the handle (the integer value) may vary.
3	N_A	(global)	The number of columns in the global matrix A.
4	NB_A	(global)	The column block size.
5	CSRC_A	(global)	The process column over which the first column of the global matrix A is distributed.
6	LLD_A	(local)	The leading dimension of the local array. For the tridiagonal subroutines, this entry is ignored.
7			Unused, reserved

4.4.7 Argument Descriptions for Band and Tridiagonal Routines

All ScaLAPACK narrow band and tridiagonal routines assume that the global matrices are distributed in a one-dimensional block data distribution. Thus, each process has at most one block of data. With selective choices for the block size NB_ and the order N_ of the global matrix, it is possible that some processes in the process grid may not receive any data, or the last process receiving data will receive a smaller block of data than the other processes.

For further information on one-dimensional block-column or block-row data distribution, please refer to section 4.4.1.

The description of a block-column distributed band matrix consists of (N, A, JA, DESCA)

- the order N of the band matrix operand,

- a pointer to the local array containing the entire global array (A, for example),

- the column index, JA, of the global array, and

- the array descriptor, DESCA, of the global array.

The description of a block-row distributed right-hand-side matrix consists of (NRHS, B, IB, DESCB)

Table 4.12: Content of the array descriptor for right-hand-side dense matrices for narrow band and tridiagonal solvers

DESC_()	Symbolic Name	Scope	Definition
1	DTYPE_B	(global)	The descriptor type (DTYPE_B=502) for $P_r \times 1$ process grid for block-row distributed matrices.
2	CTXT_B	(global)	The BLACS context handle, indicating the BLACS process grid over which the global matrix B is distributed. The context itself is global, but the handle (the integer value) may vary.
3	M_B	(global)	The number of rows in the global matrix B.
4	MB_B	(global)	The row block size.
5	RSRC_B	(global)	The process row over which the first row of the global matrix B is distributed.
6	LLD_B	(local)	The leading dimension of the local array. LLD_B $\geq$ MAX(1,LOCr(M_B)). For the tridiagonal subroutines, this entry is ignored.
7			Reserved

- the number of right-hand-side vectors NRHS in the matrix,

- a pointer to the local array containing the entire global array (B, for example),

- the row index, IB, of the global array, and

- the array descriptor, DESCB, for the global array.

The description of a block-distributed diagonally dominant-like tridiagonal matrix consists of (N, DL, D, DU, JA, DESCA)

- the order N of the tridiagonal matrix operand,

- pointer to the local arrays, (DL, D, DU, for example),

- the column index, JA, of the global array, and

- the array descriptor, DESCA, for the global array.

The description of a block-distributed symmetric positive definite tridiagonal matrix consists of (N, D, E, JA, DESCA)

- the order N of the tridiagonal matrix operand,

- pointer to the local arrays, (D, E, for example),

- the column index, JA, of the global array,

- the array descriptor, DESCA, for the global array.

The name of the row or column index for the global array has the form I<array_name> or J<array_name>, respectively. The array descriptor has a name of the form DESC<array_name>.

The length of the array descriptor is specified by DLEN_ and varies according to the descriptor type DTYPE_.

Included in the leading comments of each subroutine (immediately preceding the Argument section), is a brief note describing the **array descriptor** and some commonly used expressions in calculating workspace.

The style of the argument descriptions for symmetric positive definite narrow band routines (PxP-Byyy) and diagonally dominant-like narrow band routines (PxDByyy) is illustrated by the following example:

N	(global input) INTEGER The number of rows and columns of the matrix A(JA:JA+N−1,JA:JA+N−1) on which to be operated. N ≥ 0.
NRHS	(global input) INTEGER The number of right hand sides, i.e., the number of columns of the matrix B(IB:IB+N-1,*). NRHS ≥ 0.
A	(local input/local output) REAL pointer into the local memory to an array of local dimension (LLD_A, LOC$_c$(JA+N−1)) On entry, the local part of the N-by-N global symmetric band matrix A(JA:JA+N−1,JA:JA+N−1).
JA	(global input) INTEGER The column index of the global matrix A.
DESCA	(global and local input) INTEGER array of dimension DLEN_ The array descriptor for the global matrix A.
B	(local input/local output) REAL array, dimension (LLD_B, NRHS) On entry, the local part of the N-by-NRHS right-hand-side matrix.
IB	(global input) INTEGER The row index of the global matrix B.
DESCB	(global and local input) INTEGER array of dimension DLEN_ The array descriptor for the global matrix B.

The style of the argument descriptions for diagonally dominant-like tridiagonal routines (PxDTyyy) is illustrated by the following example:

N   (global input) INTEGER
The number of rows and columns of the matrix $A(JA:JA+N-1,JA:JA+N-1)$ on which to be operated. $N \geq 0$.

NRHS   (global input) INTEGER
The number of right hand sides, i.e., the number of columns of the matrix $B(IB:IB+N-1,*)$. $NRHS \geq 0$.

DL   (local input/local output) REAL pointer into the local memory to an array of local dimension $(LOC_c(JA+N-1))$
On entry, the local part of the subdiagonal entries of the global tridiagonal matrix $A(JA:JA+N-1,JA:JA+N-1)$.

D   (local input/local output) REAL pointer into the local memory to an array of local dimension $(LOC_c(JA+N-1))$
On entry, the local part of the diagonal entries of the global tridiagonal matrix $A(JA:JA+N-1,JA:JA+N-1)$.

DU   (local input/local output) REAL pointer into the local memory to an array of local dimension $(LOC_c(JA+N-1))$
On entry, the local part of the superdiagonal entries of the global tridiagonal matrix $A(JA:JA+N-1,JA:JA+N-1)$.

JA   (global input) INTEGER
The column index of the global matrix A.

DESCA   (global and local input) INTEGER array of dimension DLEN_
The array descriptor for the global matrix A.

B   (local input/local output) REAL array, dimension (LLD_B, NRHS)
On entry, the local part of the N-by-NRHS right-hand-side matrix.

IB   (global input) INTEGER
The row index of the global matrix B.

DESCB   (global and local input) INTEGER array of dimension DLEN_
The array descriptor for the global matrix B.

The style of the argument descriptions for symmetric positive definite tridiagonal routines (PxP-Tyyy) is illustrated by the following example:

N   (global input) INTEGER
    The number of rows and columns of the matrix A(JA:JA+N−1,JA:JA+N−1) on which to be operated. N ≥ 0.

NRHS (global input) INTEGER
    The number of right hand sides, i.e., the number of columns of the matrix B(IB:IB+N-1,*). NRHS ≥ 0.

D   (local input/local output) REAL pointer into the local memory to an array of local dimension (LOC$_c$(JA+N−1))
    On entry, the local part of the diagonal entries of the global tridiagonal matrix A(JA:JA+N−1,JA:JA+N−1).

E   (local input/local output) REAL pointer into the local memory to an array of local dimension (LOC$_c$(JA+N−1))
    On entry, the local part of the off-diagonal entries of the global tridiagonal matrix A(JA:JA+N−1,JA:JA+N−1).

JA   (global input) INTEGER
    The column index of the global matrix A.

DESCA (global and local input) INTEGER array of dimension DLEN_
    The array descriptor for the global matrix A.

B   (local input/local output) REAL array, dimension (LLD_B, NRHS)
    On entry, the local part of the N-by-NRHS right-hand-side matrix.

IB   (global input) INTEGER
    The row index of the global matrix B.

DESCB (global and local input) INTEGER array of dimension DLEN_
    The array descriptor for the global matrix B.

The description of each argument contains the following information:

- A classification of the argument as (local input), (global and local input), (local input/local output), (global input), (local output), (global output), (global input/global output), (local or global input)[3] (local workspace), or (local workspace/local output).

- The type of the argument;

- For an array, its dimension(s).

  These dimensions are often expressed in terms of LOC$_r$() and LOC$_c$() calculations. For further details, please refer to section 4.4.2.

- A specification of the value(s) that must be supplied for the argument (if it is an input argument), or of the value(s) returned by the routine (if it is an output argument), or both

---

[3](local or global input) is used to describe the length of the workspace arguments, e.g., LWORK, where the value can be local input specifying the size of the local WORK array, or global input LWORK=−1 specifying a global query for the amount of workspace required.

(if it is an input/output argument). In the last case, the two parts of the description are introduced by the phrases "On entry" and "On exit".

- For a scalar input argument, any constraints that the supplied values must satisfy (such as "$N \geq 0$" in the example above).

4.4.8 Matrix Storage Conventions for Band and Tridiagonal Matrices

A general tridiagonal matrix of order n is stored globally in three one-dimensional arrays dl, d, du of length n containing the subdiagonal, diagonal, and superdiagonal elements, respectively. Note the mild change from LAPACK in which dl and du were actually of global length $n - 1$. To make the distribution of the vectors consistent, we have chosen to make them all of length n. Note that $dl(1) = du(n) = 0$.

Similarly, a symmetric tridiagonal matrix is stored globally in two one-dimensional arrays d, e of length n containing the diagonal and off-diagonal elements, respectively. Again, there is a slight departure from LAPACK in which e was of global length $n - 1$. Here, $e(n) = 0$.

The vectors (DL, D, DU) or (D, E) representing these matrices must be block distributed to a one-dimensional process grid. These vectors can be equivalently distributed block-row or block-column since vectors are one-dimensional data structures. Note that when inputting vectors to these special-purpose low-diagonal routines, LLD_ can be ignored, since it is assumed that the local portions of the vectors are of unit stride.

4.5 Out-of-Core Matrices

The ScaLAPACK software library provides routines for solving out-of-core linear systems, in which case the matrices are stored on disk. A particular array descriptor is required to specify such a data storage.

4.5.1 Array Descriptor for Out-core Dense Matrices

The array descriptor **DESC_** whose type is defined as **DESC_(DTYPE_)=601**, is an integer array of length 11. **DESC_(DTYPE_)=601** is used for the ScaLAPACK routines involved in the out-of-core solution of dense linear systems using LU, QR or Cholesky factorizations [55]. The matrix stored on disk is composed of records each record of which corresponds to an MMB × NNB **DESC_(DTYPE_)=1** ScaLAPACK matrix and these records are organized in a column major (Fortran array) manner. The array descriptor for out-of-core matrices has extra fields to store file parameters associated with the matrix, such as the I/O device unit number, MMB and NNB, and the amount of temporary buffer storage available.

Similar to **DESC_(DTYPE_)=1** symmetric ScaLAPACK matrices, only the upper (UPLO='U') triangle or the lower (UPLO='L') triangle is accessed. The entire coefficient matrix is stored on disk, regardless of whether the matrix is nonsymmetric or symmetric.

Table 4.13: Content of the array descriptor for out-of-core dense matrices

DESC_()	Symbolic Name	Scope	Definition
1	DTYPE_A	(global)	Descriptor type DTYPE_A=601 for an out-of-core matrix.
2	CTXT_A	(global)	BLACS context handle, indicating the $MP \times NQ$ BLACS process grid over which the global matrix A is distributed. Context itself is global, but the handle (the integer value) may vary.
3	M_A	(global)	Number of rows in the global array A.
4	N_A	(global)	Number of columns in the global array A.
5	MB_A	(global)	Blocking factor used to distribute the rows of the $MMB \times NNB$ submatrix.
6	NB_A	(global)	Blocking factor used to distribute the columns of the $MMB \times NNB$ submatrix.
7	RSRC_A	(global)	Process row over which the first row of the array A is distributed.
8	CSRC_A	(global)	Process column over which the first column of the array A is distributed.
9	LLD_A	(global)	The conceptual leading dimension of the global array.
10	IODEV_A	global	I/O unit device number associated with the out-of-core matrix A.
11	SIZE_A	local	Amount of local in-core memory available for the factorization of A.

4.6 Design and Documentation of Argument Lists

As in LAPACK, the argument lists of all ScaLAPACK routines conform to a single set of conventions for their design and documentation.

Specifications of all ScaLAPACK driver and computational routines are given in Part II of this users guide. These are derived from the specifications given in the leading comments in the code, but in Part II the specifications for real and complex versions of each routine have been merged in order to save space.

4.6.1 Structure of the Documentation

The documentation of each ScaLAPACK routine includes the following:

- The SUBROUTINE or FUNCTION statement, followed by statements declaring the type and dimensions of the arguments.

- A summary of the **Purpose** of the routine.

- Descriptions of each of the **Arguments** in the order of the argument list.

- (optionally) **Further Details** (only in the code, not in Part II of this users guide);

- (optionally) **Internal Parameters** (only in the code, not in Part II of this users guide).

4.6.2   Order of Arguments

Arguments of a ScaLAPACK routine appear in the following order:

- arguments specifying options,

- problem dimensions,

- array or scalar arguments defining the input data; some of them may be overwritten by results,

- other array or scalar arguments returning results,

- work arrays (and associated array dimensions), and

- diagnostic argument INFO.

Note that not every category is present in each of the routines.

When defining each of these categories of arguments, ScaLAPACK distinguishes between **local** and **global** data. On entry to a ScaLAPACK routine, **local input** arguments may have different values on each process in the process grid. Similarly, **local output** arguments may be assigned different values on different processes in the process grid on exit from the ScaLAPACK routine.

Global input arguments must have the same value on each process in the process grid on entry to a ScaLAPACK routine. If this is not the case, most routines will call PXERBLA and return. **Global output** arguments are assigned the same value on all processes in the process grid on exit from a ScaLAPACK routine.

4.6.3   Option Arguments

Arguments specifying options are usually of type CHARACTER∗1. The arguments that specify options are character arguments with the names SIDE, TRANS, UPLO, and DIAG. On entry to a ScaLAPACK routine, these arguments are **global input** and must have the same value on each process in the process grid.

SIDE is used by the routines as follows:

Value	Meaning
'L'	Multiply general distributed matrix by symmetric or triangular distributed matrix on the left.
'R'	Multiply general distributed matrix by symmetric or triangular distributed matrix on the right.

TRANS is used by the routines as follows:

Value	Meaning
'N'	Operate with the distributed matrix.
'T'	Operate with the transpose of the distributed matrix.
'C'	Operate with the conjugate transpose of the distributed matrix.

In the real case the values 'T' and 'C' have the same meaning, and in the complex case the value 'T' is not allowed.

UPLO is used by the Hermitian, symmetric, and triangular distributed matrix routines to specify whether the upper or lower triangle is being referenced as follows:

Value	Meaning
'U'	Upper triangle
'L'	Lower triangle

DIAG is used by the triangular distributed matrix routines to specify whether the distributed matrix is unit triangular, as follows:

Value	Meaning
'U'	Unit triangular
'R'	Nonunit triangular

When DIAG is supplied as 'U', the diagonal elements are not referenced. For example:

    UPLO   (global input) CHARACTER*1
              = 'U': Upper triangle of the matrix A(IA:IA+M−1,JA:JA+N−1);
              = 'L': Lower triangle of the matrix A(IA:IA+M−1,JA:JA+N−1).

The corresponding lower-case characters may be supplied (with the same meaning), but any other value is illegal (see section 4.6.6).

A longer character string can be passed as the actual argument, making the calling program more readable, but only the first character is significant; this is a standard feature of Fortran 77. For example:

```
CALL PSPOTRS('upper', . . . )
```

4.6.4   Problem Dimensions

The problem dimensions may be passed as zero, in which case the computation (or part of it) is skipped. Negative dimensions are regarded as erroneous.

4.6.5   Workspace Issues

4.6.5.1   WORK Arrays

Many ScaLAPACK routines require one or more work arrays to be passed as arguments. The name of a work array is usually WORK – sometimes IWORK or RWORK to distinguish work arrays of type integer or real. Immediately following the work array in the argument list is the specified length of the work array, LWORK, LIWORK, or LRWORK, respectively. LWORK is defined as the minimum amount of workspace necessary to perform the operation specified.

The first element of the work array is always used to return the correct value of LWORK for the computation. Whether or not an error is detected, the minimum value of LWORK is placed in $WORK(1)$ on exit from the routine.

If the user passes a value for $LWORK$ that is too small, an input error is detected and $INFO$ is set accordingly (see section 4.6.6), the correct value for $LWORK$ is placed in $WORK(1)$, and the routine PXERBLA is called. The user is thus strongly advised to always check the value of $INFO$ on exit from the called routine.

4.6.5.2   LWORK Query

If in doubt about the amount of workspace to supply to a ScaLAPACK routine, the user may choose to supply $LWORK = -1$ and use the returned value in $WORK(1)$ as the correct value for $LWORK$. Setting $LWORK = -1$ does not invoke an error message from PXERBLA and is defined as a global query.

4.6.5.3   LWORK $\geq$ WORK(1)

In some ScaLAPACK eigenvalue routines, such as the symmetric eigenproblems (PxSYEV and PxSYEVX/PxHEEVX) and the generalized symmetric eigenproblem (PxSYGVX/PxHEGVX), a larger value of $LWORK$ can guarantee the orthogonality of the returned eigenvectors at the risk of potentially degraded performance of the algorithm. The minimum amount of workspace required is returned in the first element of the work array, but a larger amount of workspace can allow for additional orthogonalization if desired by the user. Refer to section 5.3.6 and the leading comments of the source code for complete details.

4.6.6 Error Handling and the Diagnostic Argument INFO

All driver and computational routines have a diagnostic argument INFO that indicates the success or failure of the computation. It is recommended that the user always check the value of INFO on exit from calling a ScaLAPACK routine. The value of INFO is defined as follows:

- INFO = 0: successful exit

- INFO < 0: illegal value of one or more arguments — no computation performed

- INFO > 0: failure in the course of computation

The value of (INFO<0) is calculated as follows: if the error is detected in the jth entry of a descriptor array, which is the ith argument in the parameter list, the number passed to the error-handling routine PXERBLA() has been arbitrarily chosen to be $-(100 * i + j)$. This allows the user to distinguish an error on a descriptor entry from an error on a scalar argument.

The standard version of PXERBLA() only issues an error message and does not halt execution of the program. The main reason for this behavior is that some "errors" are deemed recoverable and we wanted to allow the user the flexibility to continue program execution if certain values were corrected. If user wish to change this behavior and additionally halt execution of the program, they may add a call to BLACS_ABORT() to their version of PXERBLA().

If an input error (INFO<0) is detected at a high-level routine (ScaLAPACK driver or computational routine), it is possible for the user to recover from such an error and proceed with the computation. An error message is printed by PXERBLA(), a RETURN is issued, and the program execution continues. However, if an error is detected in a low-level ScaLAPACK routine, this error is considered unrecoverable, a message is printed by PXERBLA(), and program execution is terminated by a call to BLACS_ABORT().

Likewise, if an error is detected at a low-level routine, such as a PBLAS or BLACS routine, this error is deemed fatal. An error message is printed, and the program execution is terminated by the specific error-handling routine.

All ScaLAPACK driver and computational routines perform global and local input error-checking. In general, no input error-checking is performed on the auxiliary routines. The exception to this rule is for the auxiliary routines which are Level 2 versions of computational routines (e.g., PxGETF2, PxGEQR2, PxORMR2, PxORM2R, etc.). For efficiency purposes, these specialized low-level routines perform only a local validity check of their argument list. If an error is detected in at least one process of the current context, the program execution is stopped.

4.6.7 Alignment Restrictions

Most routines in the present ScaLAPACK library have *alignment restrictions*. Alignment restrictions are constraints in the type of distributions and the indexing into the matrix that the user may utilize when calling a particular routine. For example, some routines will not accept submatrices whose starting index is not a multiple of the physical blocking factor.

More commonly, routines require that their various operand matrices have certain alignment commonalities. For instance, the solver routines generally require that row i of the matrix A be distributed across the same process row as row i of the right hand side matrix B.

Because of their idiosyncratic nature, it is almost impossible to give a full description of the alignment restrictions inherit in the present library without doing so on a routine-specific basis. All ScaLAPACK routines provide a description of the assumed alignment restrictions in the leading comments to the routine, and at this time the user must consult the actual code to find out what restrictions exist, if any.

We are working to remove the alignment restrictions, (with the exceptions noted below) so that the user will not have to worry about alignment, save as a performance issue.

Certain fundamental restrictions about data distributions are not currently being removed the library. Examples include the fact that the operands should be block-cyclicly distributed for the dense codes and one-dimensional block distributed for the banded codes. Also included here is the restriction that all operands be distributed across the same context (process grid).

Note that the ScaLAPACK library includes a redistribution/copy routine which allows the user to explicitly move matrices across contexts. Similar routines could be provided for distributions that do not match the ones presently employed in ScaLAPACK.

Finally, we note that the current *descriptor* structure does not accommodate the definition of replicated vectors. A *replicated* vector is a vector that is distributed across a row or column within the process grid and duplicated across subsequent process rows or columns, respectively. Such vectors occur, for example, as the IPIV vector in the LU factorization and the TAU vector in QR factorizations.

4.7   Extensions

Extensions to the library are under way and will remove the majority of the alignment restrictions in the PBLAS. The ScaLAPACK library and the PBLAS are also being modified to allow the possibility of a partial first block, as well as the incorporation of aggregate (algorithmic) blocking. The partial first block extension makes ScaLAPACK fully compatible with HPF and necessitates the establishment of a new matrix descriptor.

The incorporation of aggregate (algorithmic) blocking at the top-level ScaLAPACK routines, as well as in the PBLAS, removes the restriction of a user's performance being tied to his physical matrix distribution. Instead, the algorithms will perform at an optimal block size predetermined inside the PBLAS.

Chapter 5

Performance of ScaLAPACK

This chapter presents performance numbers for ScaLAPACK routines. The numbers are provided *for illustration only* and *should not* be regarded as a definitive up-to-date statement of performance. They have been selected from performance numbers obtained in 1996–1997 during the development of version 1.4 of ScaLAPACK. To obtain up-to-date performance figures, users should use the timing programs provided with ScaLAPACK.

5.1 Achieving High Performance with ScaLAPACK

ScaLAPACK achieves high performance on distributed memory computers, such as the SP2, T3D, T3E, and Paragon. ScaLAPACK can also achieve high performance on some networks of workstations.

Distributed memory computers are intended to be used primarily to run parallel programs. They typically include an efficient message-passing system, a one-to-one mapping of processes to processors, a gang scheduler and a well-connected communications network. Networks of workstations may be designed primarily for use as individual workstations and may not have all of these important features.

5.1.1 Achieving High Performance on a Distributed Memory Computer

Assuming that the ScaLAPACK installation was done correctly, the users need only make sure that they are using an appropriate number of processors and that their matrices are efficiently distributed. Here is a checklist to get started.

- Use the right number of processors.
    - Rule of thumb: $P = M \times N/1000000$ for an $M \times N$ matrix. This provides a local matrix of size approximately 1000 by 1000.
    - Do not try to solve a small problem on too many processors.

    – Do not exceed physical memory.

- Use an efficient data distribution.

    – Block size[1] (i.e., MB,NB) = 64.

    – Square processor grid, $P_r = P_c$.

- Use efficient machine-specific BLAS (not the Fortran 77 reference implementation BLAS) and BLACS (nondebug, $BLACSDBGLVL = 0$ in `Bmake.inc`)

If the performance is still below that expected, see section 5.3. For guidelines on tuning for higher performance, see section 5.4.

5.1.2   Achieving High Performance on a Network of Workstations

If a network meets the following guidelines, ScaLAPACK will perform well on it (see section 5.1.1). If a network of workstations does not meet one or more of these guidelines, read the rest of this chapter for more information.

- The bandwidth per node, if measured in Megabytes per second per node, should be no less than one tenth of the peak floating-point rate as measured in megaflops/second/node.

- The underlying network must allow simultaneous messages, that is, not standard ethernet and not FDDI.

- Message latency should be no more than 500 microseconds.

- All processors should be similar in architecture and performance. ScaLAPACK will be limited by the slowest processor. Data format conversion significantly reduces communication performance.

- No other jobs should be allowed to execute on the processors that are being used. If the processors are gang scheduled and there is enough physical memory for all jobs on all processors, this requirement may be relaxed, but we do not recommend doing so without careful study.

- No more than one process should be executed per processor.

Vendor specifications and actual performance often differ considerably, especially in communication latency and bandwidth. Users should make sure that they are using the most efficient BLAS and BLACS available on their system.

---

[1]The block size must be large enough that the local matrix multiply is efficient. A block size of 64 suffices for most computers that have only one processor per node. Computers that have multiple shared-memory processors on each node may require a larger block size.

5.2 Performance, Portability and Scalability

How can we provide **portable** software for dense linear algebra computations that is **efficient** on a wide range of modern distributed-memory computers? Answering this question — and providing the appropriate software — has been an objective of the ScaLAPACK project.

The ScaLAPACK software has been designed specifically to achieve high efficiency for a wide range of modern distributed-memory computers. Examples of such computers include the Cray T3D and T3E computers, the IBM Scalable POWERparallel SP series, the Intel iPSC and Paragon computers, the nCube-2/3 computer, networks and clusters of workstations (NoWs and CoWs), and "piles" of PCs (PoPCs).

5.2.1 The BLAS as the Key to (Trans)portable Efficiency

The total number of floating-point operations performed by most of the ScaLAPACK driver routines for dense matrices can be approximated by the quantity $C_f N^3$, where C_f is a constant and N is the order of the largest matrix operand. For solving linear equations or linear least squares, C_f is a constant depending solely on the selected algorithm. The algorithms used to find eigenvalues and singular values are iterative; hence, for these operations the constant C_f truly depends on the input data as well. It is, however, customary or "standard" to consider the values of the constants C_f for a fixed number of iterations. The "standard" constants C_f range from 1/3 to approximately 18, as shown in Table 5.8.

The performance of the ScaLAPACK drivers is thus bounded above by the performance of a computation that could be partitioned into P independent chunks of $C_f N^3/P$ floating-point operations each. This upper bound, referred to hereafter as the *peak performance*, can be computed as the product of $C_f N^3/P$ and the highest reachable local node flop rate. Hence, for a given problem size N and assuming a uniform distribution of the computational tasks, the most important factors determining the overall performance are the number P of nodes involved in the computation and the local node flop rate.

In a serial computational environment, *transportable efficiency* is the essential motivation for developing blocking strategies and block-partitioned algorithms [2, 3, 35, 90]. The linear algebra package (LAPACK) [3] is the archetype of such a strategy. The LAPACK software is constructed as much as possible out of calls to the BLAS. These kernels confine the impact of the computer architecture differences to a small number of routines. The efficiency and portability of the LAPACK software are then achieved by combining native and efficient BLAS implementations with portable high-level components.

The BLAS are subdivided into three levels, each of which offers increased scope for exploiting parallelism. This subdivision corresponds to three different types of basic linear algebra operations:

- Level 1 BLAS [93]: for vector operations, such as $y \leftarrow \alpha x + y$,

- Level 2 BLAS [59, 58]: for matrix-vector operations, such as $y \leftarrow \alpha A x + \beta y$,

- Level 3 BLAS [57, 56]: for matrix-matrix operations, such as $C \leftarrow \alpha A B + \beta C$.

Here, A, B, and C are matrices, x and y are vectors, and α and β are scalars.

The performance potential of the three levels of BLAS is strongly related to the ratio of floating-point operations to memory references, as well as to the reuse of data when it is stored in the higher levels of the memory hierarchy. Consequently, the Level 1 BLAS cannot achieve high efficiency on most modern supercomputers. The Level 2 BLAS can achieve near-peak performance on many vector processors; on RISC microprocessors, however, their performance is limited by the memory access bandwidth bottleneck. The greatest scope for exploiting the highest levels of the memory hierarchy as well as other forms of parallelism is offered by the Level 3 BLAS [3].

The previous reasoning applies to distributed-memory computational environments in two ways. First, in order to achieve overall high performance, it is necessary to express the bulk of the computation local to each node in terms of Level 3 BLAS operations. Second, designing and developing a set of parallel BLAS (PBLAS) for distributed-memory computers should lead to an efficient and straightforward port of the LAPACK software. This is the path followed by the ScaLAPACK initiative [25, 53] as well as others [1, 21, 30, 63]. As part of the ScaLAPACK project, a set of PBLAS has been early designed and developed [29, 26].

5.2.2 Two-Dimensional Block Cyclic Data Distribution as a Key to Load Balancing and Software Reuse

The way the data is distributed over the memory hierarchy of a computer is of fundamental importance to load balancing and software reuse. The block cyclic data distribution allows a reduction of the overhead due to load imbalance and data movement. Block-partitioned algorithms are used to maximize the local node performance.

Since the data decomposition largely determines the performance and scalability of a concurrent algorithm, a great deal of research [27, 65, 69, 78] has focused on different data decompositions [10, 20, 85]. In particular, the two-dimensional block cyclic distribution [92] has been suggested as a possible general-purpose basic decomposition for parallel dense linear algebra libraries [31, 76, 97, 17] such as ScaLAPACK.

Block cyclic distribution is beneficial because of its scalability [51], load balance, and communication [76] properties. The block-partitioned computation then proceeds in consecutive order just as a conventional serial algorithm does. This essential property of the block cyclic data distribution explains why the ScaLAPACK design has been able to reuse the numerical and software expertise of the sequential LAPACK library.

5.2.3 BLACS as an Efficient, Portable and Adequate Message-Passing Interface

The total volume of data communicated by most of the ScaLAPACK driver routines for dense matrices can be approximated by the quantity $C_v N^2$, where N is the order of the largest matrix operand. The number of messages, however, is proportional to N and can be approximated by the quantity $C_m N/NB$, where NB is the logical blocking factor used in the computation. Similar to the situation described above, the "standard" constants C_v for the communication volume depend upon the performed computation and are of the same order as the floating-point operation constants C_f shown in Table 5.8. The values of the "standard" constants C_v for a few selected ScaLAPACK drivers are presented in Table 5.8. As a result, a significant percentage of the ScaLAPACK software aims at exchanging messages between processes.

Developing an adequate message-passing interface specialized for linear algebra operations has been one of the first achievements of the ScaLAPACK project. The Basic Linear Algebra Communications Subprograms (BLACS) [50, 54] were thus specifically designed to facilitate the expression of the relevant communication operations. The simplicity of the BLACS interface, as well as the rigor of their specification, allows for an easy port of the entire ScaLAPACK software. Currently, the BLACS have been efficiently ported on machine-specific message-passing libraries such as the IBM (MPL) and Intel (NX) message-passing libraries, as well as more generic interfaces such as PVM and MPI. The BLACS overhead has been shown to be negligible [54].

The BLACS interface provides the user and library designer with an appropriate level of notation. Indeed, the BLACS operate on typed two-dimensional arrays. The computational model consists of a one- or two-dimensional grid of processes, where each process stores matrices and vectors. The BLACS include synchronous send/receive routines to send a matrix or submatrix from one process to another, to broadcast submatrices, or to perform global reductions (sums, maxima and minima). Other routines establish, change, or query the process grid. The BLACS provide an adequate interface level for linear algebra communication operations.

For ease of use and flexibility, the BLACS send operation is **locally blocking**; that is, the return from the send operation indicates that the resources may be reused. However, since this depends only on local information, it is unknown whether the receive operation has been called. Buffering is necessary on the sending or the receiving process. The BLACS receive operation is **globally blocking**. The return from the receive operation indicates that the message has been (sent and) received. On a system natively supporting globally blocking sends such as the IBM SP2 computer, nonblocking sends coupled with buffering are used to simulate locally blocking sends. This extra buffering operation may cause a slight performance degradation on those systems.

The BLACS broadcast and combine operations feature the ability of selecting different virtual network topologies. This easy-to-use built-in facility allows for the expression of various message scheduling approaches, such as a communication pipeline. This unique and distinctive BLACS characteristic is necessary for achieving the highest performance levels on distributed-memory platforms.

5.2.3.1 Parallel Efficiency

An important performance metric is *parallel efficiency*. Parallel efficiency, $E(N, P)$, for a problem of size N on P nodes is defined in the usual way [65, 92] by

$$E(N, P) = \frac{1}{P} \frac{T_{\text{seq}}(N)}{T(N, P)},$$

where $T(N, P)$ is the runtime of the parallel algorithm, and $T_{\text{seq}}(N)$ is the runtime of the best sequential algorithm. For dense matrix computations, an implementation is said to be *scalable* if the parallel efficiency is an increasing function of N^2/P, the problem size per node. The algorithms implemented in the ScaLAPACK library are scalable in this sense.

Figure 5.1 shows the scalability of the ScaLAPACK implementation of the *LU* factorization on the Intel XP/S Paragon computer. The nodes of the Intel XP/S Paragon computer are general-purpose (GP) or multiprocessor (MP) nodes, based on the Intel i860 XP RISC processors. Each Intel i860 processor is capable of a peak performance of 50 Mflop/s. On such a processor, however, the vendor-supplied BLAS matrix-matrix multiply routine DGEMM can achieve only approximately 45 Mflop/s. The computer used for obtaining the performance results presented in this chapter consisted of MP nodes configured as follows: each MP node had three Intel i860 XP processors — two to execute application code and a third used exclusively as a message coprocessor. On such a node, the vendor-supplied BLAS matrix-matrix multiply routine DGEMM can achieve approximately 90 Mflop/s.

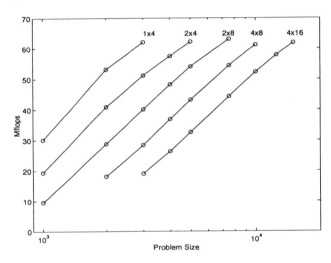

Figure 5.1: LU Performance per Intel XP/S MP Paragon node

Figure 5.1 shows the speed in Mflop/s per node of the ScaLAPACK *LU* factorization routine PDGETRF for different computer configurations. This figure illustrates that when the number of nodes is scaled by a constant factor, the same efficiency or speed per node is achieved for equidistant problem sizes on a logarithmic scale. In other words, maintaining a constant memory use per node allows efficiency to be maintained. (This scalability behavior is also referred to as *isoefficiency,* or *isogranularity.*) In practice, however, a slight degradation is acceptable. The ScaLAPACK driver routines, in general, feature the same scalability behavior up to a constant factor that depends on the exact number of floating-point operations and the total volume of data exchanged during the computation.

In large dense linear algebra computations, the computation cost dominates the communication cost. In the following, the time to execute one floating-point operation by one node is denoted by t_f. The time to communicate a message between two nodes is approximated by a linear function of the number of items communicated. The function is the sum of the time to prepare the message for transmission (t_m) and the time taken by the message to traverse the network to its destination, that is, the product of its length by the time to transfer one data item (t_v). Alternatively, t_m is also called the *latency,* since it is the time to communicate a message of zero length. On most modern interconnection networks, the order of magnitude of the latency varies between a microsecond and a millisecond.

The bandwidth of the network is also referred to as its *throughput.* It is proportional to the reciprocal of t_v. On modern networks, the order of magnitude of the bandwidth is the megabyte per second. For a scalable algorithm with N^2/P held constant, one expects the performance to be proportional to P. The algorithms implemented in ScaLAPACK are scalable in this sense. Table 5.1 summarizes the relevant constants used in our scalability analysis.

Table 5.1: Variable definitions

Variable	Description	Details
$C_f N^3$	Total number of floating-point operations	Table 5.8
$C_v N^2/\sqrt{P}$	Total number of data items communicated	Table 5.8
$C_m N/NB$	Total number of messages	Table 5.8
t_f	Time per floating-point operation (typically F_{mm})	Table 5.5
t_v	Time per data item communicated	Table 5.5
t_m	Time per message	Table 5.5
N	Matrix size	–
P	Number of processors	–
NB	Data distribution block size	–
$T()$	Parallel execution time (estimated)	Equation 5.1
$T_{\text{seq}}()$	Serial execution time (estimated)	Equation 5.1
$E()$	Efficiency (estimated)	Equation 5.2
F_{MM}	Time per floating-point operation in matrix-matrix multiply	Table 5.5

Using the notation presented in table 5.1, the execution time of the ScaLAPACK drivers can be approximated by

$$T(N,P) = \frac{C_f N^3}{P} t_f + \frac{C_v N^2}{\sqrt{P}} t_v + \frac{C_m N}{NB} t_m, \qquad T_{seq}(N,P) = C_f N^3 t_f. \tag{5.1}$$

The corresponding parallel efficiency can then be approximated by

$$E(N,P) = \left(1 + \frac{1}{NB} \frac{C_m t_m}{C_f t_f} \frac{P}{N^2} + \frac{C_v t_v}{C_f t_f} \frac{\sqrt{P}}{N}\right)^{-1}. \tag{5.2}$$

Equation 5.2 illustrates, in particular, that the communication versus computation performance ratio of a distributed-memory computer significantly affects parallel efficiency. The ratio of the latency to the time per flop (t_m/t_f) greatly affects the parallel efficiency of small problems. The ratio of the network throughput to the flop rate (t_f/t_v) significantly affects the parallel efficiency of medium-sized problems. For large problems, the node flop rate $(1/t_f)$ is the dominant factor contributing to the parallel efficiency of the parallel algorithms implemented in ScaLAPACK.

5.2.4 ScaLAPACK Performance

In this section, we present performance data for Version 1.4 of ScaLAPACK on four distributed memory computers and two networks of workstations. The four distributed memory computers are the Cray T3E computer, the IBM Scalable POWERparallel 2 computer, the Intel XP/S MP Paragon computer, and the Intel ASCI Option Red Supercomputer. One of the networks of workstations consists of Sun Ultra Enterprise 2 (Model 2170s) connected via switched ATM. The other network of workstations, the Berkeley NOW [34], consists of 100+ Sun UltraSPARC-1 workstations and 40+ Myricom crossbar switches and LANai 4.1 network interface cards. ScaLAPACK on the NOW uses MPI BLACS, where the MPI is a port of the freely-available MPICH reference code. MPI uses Active Messages as its underlying communications layer. Active Messages [98] provide ultra-lightweight remote-procedure calls for processes on the NOW. The system currently uses AM-II, a generalized active message layer that supports more than SPMD parallel programs, e.g., client-server programs and distributed filesystems. It retains the simple request/response paradigm common to all previous active message implementations as well as its high-performance. These six computers are a collection of processing nodes interconnected via a network. Each node has local memory and one or more processors. Tables 5.2, 5.3, and 5.4 describe the characteristics of these six computers.

Table 5.2: Characteristics of the Cray T3E and IBM SP2 computers timed

	Cray T3E	IBM SP2
Processor	Dec Alpha EV5	POWER2 590
Clock speed (MHz)	300	66
Processors per node	1	1
Memory per node (MB)	256	128
Operating system	UNICOS/mk 1.4.1	AIX 4.1.4
BLAS	LIBSCI 3.0	ESSL 2.2.2.2
BLACS	MPI BLACS 1.1α	MPI BLACS 1.1α
Communication Software	Cray MPI	POE (2.1.0.12)
C compiler	cc	mpcc (3.1.4.3)
C flags	-O3	-qarch=pwr2 -qtune=pwr2s -O2
Fortran compiler	f90	mpxlf (3.2.4.2)
Fortran flags	-O3	-qarch=pwr2 -qtune=pwr2s -O2
Precision	single (64-bit)	double (64-bit)

As noted in Tables 5.2, 5.3, and 5.4, a machine-specific optimized BLAS implementation was used for all the performance numbers reported in this chapter. For the IBM Scalable POWERparallel 2 (SP2) computer, the IBM Engineering and Scientific Subroutine Library (ESSL) was used [88]. On the Intel XP/S MP Paragon computer, the Intel Basic Math Library Software (Release 5.0) [89] was used. The Intel ASCI Option Red Supercomputer was tested using a pre-alpha version of the Cougar operating system and using an unoptimized functional version of the dual processor Basic Math Library from Kuck and Associates, Inc. The communication performance and library performance was still being enhanced. On the Sun Ultra Enterprise 2 workstation, the Dakota

Table 5.3: Characteristics of the Intel computers timed

	Intel XP/S MP Paragon	Intel ASCI Red Supercomputer
Processor	i860 XP	Intel Pentium Pro
Clock speed (MHz)	50	200
Processors per node	2	2
Memory per node (MB)	64	128
Operating system	R1.4-2	Alpha Release
BLAS	Basic Math Library Software (Release 5.0)	Alpha Release
BLACS	NX BLACS 1.0	NX BLACS 1.0
Communication Software	NX	NX on Portals
C compiler	icc (R5.0.2)	pgcc Rel 1.3-4a
C flags	-O4	-O
Fortran compiler	if77 (R5.0.2)	pgf77 Rel 1.3-4a
Fortran flags	-O4	-O
Precision	double (64-bit)	double (64-bit)

Scientific Software Library (DSSL)[2] was used. The DSSL BLAS implementation used only one processor per node. On the Berkeley NOW, the Sun Performance Library, version 1.2, was used. It should also be noted that for the IBM Scalable POWERparallel 2 (SP2) the communication layer used was the IBM Parallel Operating Environment (POE), which is a combination of MPI and MPL libraries.

Several data distributions were tried for $N = 2000$. The fastest data distribution for $N = 2000$ was used for all problem sizes, although this data distribution may not be optimal for all problem sizes. Whenever applicable, only the options UPLO='U' and TRANS='N' were timed. The test matrices were generated with randomly distributed entries. All runtimes are reported in seconds. Block size is denoted by NB.

This section first reports performance data for a relevant selection of BLAS and BLACS routines. Then, timing results obtained for some PBLAS routines are presented. Finally, performance numbers for selected ScaLAPACK driver routines are shown.

5.2.5 Performance of Selected BLACS and Level 3 BLAS Routines

The efficiency of the ScaLAPACK software depends on efficient implementations of the BLAS and the BLACS being provided by computer vendors (or others) for their computers. The BLAS and the BLACS form a low-level interface between ScaLAPACK software and different computer architectures. Table 5.5 presents performance numbers indicating how well the BLACS and Level 3 BLAS perform on different distributed-memory computers. For each computer this table shows the

---

[2]Dakota Scientific Software, Inc., 501 East Saint Joseph Street, Rapid City, SD 57701-3995 USA, (605) 394-2471

Table 5.4: Characteristics of the networks of workstations timed

	Cluster of Sun Ultra 2's	Berkeley NOW
Processor	UltraSPARC-1	UltraSPARC-1
Clock speed (MHz)	167	167
Processors per node	2	1
Memory per node (MB)	256	128
Operating system	SunOS 2.5.1	SunOS 2.5.1
BLAS		Sun Performance Library v1.2
BLACS	1.1α	1.1α
Communication Software	MPICH 1.0.13	MPI on AM-II
C compiler	gcc (v2.7.2.1)	cc (Sun Workshop 4.2)
C flags	-O3	-dalign -xarch=v8plusa -xchip=ultra -fast -xtarget=native -xO5
Fortran compiler	f77 (SC4.0)	f77 (SC4.0)
Fortran flags	-O4 -f	-dalign -xarch=v8plusa -xchip=ultra -fast -xtarget=native -xO5
Precision	double (64-bit)	double (64-bit)

flop rate achieved by the matrix-matrix multiply Level 3 BLAS routine SGEMM/DGEMM (F_{MM}) on a node versus the theoretical peak performance of that node, the underlying message-passing library called by the BLACS, and the approximated values of the latency (t_m) and the bandwidth ($1/t_v$) achieved by the BLACS versus the underlying message-passing software for the machine.

Table 5.5: BLACS and Level 3 BLAS performance indicators

	Mflop/s		t_m (μs)		$1/t_v$ (MB/s)	
	F_{MM}	Peak	BLACS	Native	BLACS	Native
Cray T3E (MPI)	360	600	17.8	15.8	117.5	117.5
IBM SP2 (MPL)	200	266	63.7	49.7	21.8	30.8
Intel XP/S MP Paragon (NX)	90	100	37.7	31.5	133.6	143.0
Intel ASCI Option Red Supercomputer (NX)	264	400	29.8	28.7	391.2	391.2
Cluster of Sun Ultra 2s (MPI)	125	334	369.0	313.0	9.8	10.9
Berkeley NOW (MPI on AM-II)	129	334	48.0	43.8	27.9	28.1

The values for latency in table 5.5 were obtained by timing the cost of a 0-byte message. The bandwidth numbers table 5.5 were obtained by increasing message length until message bandwidth was saturated. We used the same timing mechanism for both the BLACS and the underlying message-passing library.

These numbers are actual timing numbers, not values based on hardware peaks, for instance. Therefore, they should be considered as approximate values or indicators of the observed performance between two nodes, as opposed to precise evaluations of the interconnection network capabilities. On the CRAY, the numbers reported are for MPI and the MPIBLACS, instead of the more optimal shmem library with CRAY's native BLACS.

For all four computers, a machine-specific optimized BLAS implementation was used for all the performance numbers reported in this chapter. For the IBM Scalable POWERparallel 2 (SP2) computer, the IBM Engineering and Scientific Subroutine Library (ESSL) was used [88]. On the Intel XP/S MP Paragon computer, the Intel Basic Math Library Software (Release 5.0) [89] was used. On the Sun Ultra Enterprise 2 workstation, the Dakota Scientific Software Library (DSSL)[3] was used. The DSSL BLAS implementation used only one processor per node. The speed of the BLAS matrix-matrix multiply routine shown in Table 5.5 has been obtained for the following operation $C \leftarrow C + AB$, where A, B, and C are square matrices of order 500.

5.2.5.1 Performance of Selected PBLAS routines

The performance of Level 2 PBLAS routines is dependent on the performance of Level 2 BLAS routines which is dependent on the bulk transfer rate from main memory. Table 5.6 shows execution rates for the 64-bit matrix-vector multiply PBLAS routine PSGEMV/PDGEMV. The rates listed are for a matrix-vector product $y \leftarrow y + Ax$, where A is a square matrix of order N and x and y are vectors that are both distributed over a process column.

The Level 3 PBLAS are not necessarily limited by memory bandwidth because they perform many flops for each word involved. The flop rate is correspondingly higher. Table 5.7 shows the performance results obtained by the general matrix-matrix multiply PBLAS routine PSGEMM/PDGEMM. These results have been obtained for the matrix-matrix multiply operation $C \leftarrow C + AB$, where A, B, and C are square matrices of order N.

5.2.5.2 Solution of Common Numerical Linear Algebra Problems

This section contains performance numbers for selected driver routines. These routines provide complete solutions for common linear algebra problems.

- Solve a general N-by-N system of linear equations with one right-hand side using the routine PSGESV/PDGESV.

- Solve a symmetric positive definite N-by-N system of linear equations with one right-hand side, using PSPOSV/PDPOSV.

- Solve an N-by-N linear least squares problem with one right-hand side using the routine PSGELS/PDGELS.

---

[3]Dakota Scientific Software, Inc., 501 East Saint Joseph Street, Rapid City, SD 57701-3995 USA, (605) 394-2471

Table 5.6: Speed in Mflop/s for the PBLAS matrix-vector multiply routine PSGEMV/PDGEMV

	Process grid	Block size	Values of N				
			2000	4000	6000	8000	10000
Cray T3E[a]	2×2	32	516	475	477	459	
PSGEMV	4×4	32	1715	1422	2039	1892	1896
	8×8	32	2986	6129	7157	5253	7887
IBM SP2[b]	2×2	50	552	643	663		
PDGEMV	4×4	50	1497	2081	2244	2493	2380
	8×8	50	1379	2178	2355	2708	2575
Intel XP/S MP Paragon[c]	2×2	32	162				
PDGEMV	4×4	32	543	620	666		
	8×8	32	1597	2117	2356	2461	2563
Cluster of Sun Ultra 2s[d]	2×2	64	107	121	122	112	
PDGEMV	2×4	64	198	233	237	219	204
	3×4	64	275	332	345	261	307
Berkeley NOW[e]	2×2	32	118	123	121		
PDGEMV	2×4	32	232	221	240	220	
	4×4	32	226	406	401	439	474
	4×8	32	711	645	612	723	836
	8×8	32	328	816	1033	1274	1532

[a]A ScaLAPACK process is running on each node. Each node consists of one computational DEC Alpha EV5 processor, which is used by the BLAS.

[b]A ScaLAPACK process is running on each node. Each node consists of one computational IBM POWER2 590 processor, which is used by the BLAS.

[c]A ScaLAPACK process is running on each node. Each node consists of two computational Intel i860 XP RISC processors, which are used by the multi-threaded BLAS.

[d]A ScaLAPACK process is running on each node. Each node consists of two computational UltraSPARC-1 processors, but only one processor is used by the BLAS.

[e]A ScaLAPACK process is running on each node. Each node consists of one computational UltraSPARC-1 processor, which is used by the BLAS.

- Find the eigenvalues and optionally the corresponding eigenvectors of an N-by-N symmetric matrix, using the routine PSSYEVX/PDSYEVX.

- Find the eigenvalues and optionally the corresponding eigenvectors of an N-by-N symmetric matrix, using the routine PSSYEV/PDSYEV.

- Find the singular values and optionally the corresponding right and left singular vectors of an N-by-N matrix, using PSGESVD/PDGESVD.

- Find the eigenvalues and optionally the corresponding right eigenvectors of an N-by-N Hessenberg matrix, using the routine PSLAHQR/PDLAHQR.[4]

---

[4]Strictly speaking, PSLAHQR/PDLAHQR is an auxiliary routine for computing the eigenvalues and optionally the corresponding eigenvectors of the more general case of nonsymmetric matrices.

Table 5.7: Speed in Mflop/s for the PBLAS matrix-matrix multiply routine PSGEMM/PDGEMM

	Process grid	Block size	Values of N				
			2000	4000	6000	8000	10000
Cray T3E[a]	2×2	32	1055	1070	1075		
PSGEMM	4×4	32	3630	4005	4258	4171	4292
	8×8	32	13456	14287	15419	15858	16755
IBM SP2[b]	2×2	50	755				
PDGEMM	4×4	50	2514	2850	3040		
	8×8	50	6205	8709	9862	10468	10774
Intel XP/S MP Paragon[c]	2×2	32	330				
PDGEMM	4×4	32	1233	1281	1334		
	8×8	32	4496	4864	5030	5103	5257
Cluster of Sun Ultra 2s[d]	2×2	64	406	442			
PDGEMM	2×4	64	529	770	851		
	3×4	64	500	940	1119	1220	
Berkeley NOW[e]	2×2	32	463	470			
PDGEMM	2×4	32	496	524			
	4×4	32	926	1031	632	1754	
	4×8	32	2490	2822	3316	3306	3450
	8×8	32	4130	5457	6041	6360	6647

[a]A ScaLAPACK process is running on each node. Each node consists of one computational DEC Alpha EV5 processor, which is used by the BLAS.

[b]A ScaLAPACK process is running on each node. Each node consists of one computational IBM POWER2 590 processor, which is used by the BLAS.

[c]A ScaLAPACK process is running on each node. Each node consists of two computational Intel i860 XP RISC processors, which are used by the multi-threaded BLAS.

[d]A ScaLAPACK process is running on each node. Each node consists of two computational UltraSPARC-1 processors, but only one processor is used by the BLAS.

[e]A ScaLAPACK process is running on each node. Each node consists of one computational UltraSPARC-1 processor, which is used by the BLAS.

Table 5.8 presents "standard" floating-point operation costs ($C_f N^3$) for selected ScaLAPACK drivers for matrices of order N. Approximate values of the constants C_m and C_v defined in section 5.2.3 are also provided. The operation counts given for the eigenvalue and SVD drivers are incomplete. They do not include any of the $O(N^2)$ computation costs (i.e., the entire tridiagonal eigendecomposition is ignored in PxxxEVX). Furthermore, the reductions involved require matrix-vector multiplies, which are less efficient than the matrix-matrix multiplies required by the other drivers listed here. Hence this table greatly underestimates the execution time of the eigenvalue and SVD drivers, especially the expert symmetric eigensolver drivers. For PxLAHQR, when only eigenvalues are computed, C_m and C_v look the same as the full Schur form case, in terms of "order of magnitude". There is actually $\frac{1}{4}$ to $\frac{1}{2}$ the number of messages/volume depending on the circumstances.

Table 5.8: "Standard" floating-point operation (C_f) and communication costs (C_v, C_m) for selected ScaLAPACK drivers

Driver	Options	C_f	C_v	C_m
PxGESV	1 right hand side	2/3	$3 + 1/4 \log_2 P$	$NB\,(6 + \log_2 P)$
PxPOSV	1 right hand side	1/3	$2 + 1/2 \log_2 P$	$4 + \log_2 P$
PxGELS	1 right hand side	4/3	$3 + \log_2 P$	$2\,(NB \log_2 P + 1)$
PxSYEVX	eigenvalues only	4/3	$5/2 \log_2 P$	$17/2\,NB + 2$
PxSYEVX	eigenvalues and eigenvectors	10/3	$5 \log_2 P$	$17/2\,NB + 2$
PxSYEV	eigenvalues only	4/3	$5/2 \log_2 P$	$17/2\,NB + 2$
PxSYEV	eigenvalues and eigenvectors	22/3	$5 \log_2 P$	$17/2\,NB + 2$
PxGESVD	singular values only	26/3	$10 \log_2 P$	$17NB$
PxGESVD	singular values and left and right singular vectors	38/3	$14 \log_2 P$	$17NB$
PxLAHQR	eigenvalues only	5	$9/2\,(\sqrt{P}) * \log_2 P$ $+ 8\,N/NB$	$9\,(2 + \log_2 P)\,N$
PxLAHQR	full Schur form	18	$9/2\,(\sqrt{P}) * \log_2 P$ $+ 8\,N/NB$	$9\,(2 + \log_2 P)\,N$

5.2.6   Solving Linear Systems of Equations

Table 5.9 illustrates the speed of the ScaLAPACK driver routine PSGESV/PDGESV for solving a square linear system of order N by LU factorization with partial row pivoting of a real matrix. For all timings, 64-bit floating-point arithmetic was used. Thus, single-precision timings are reported for the Cray T3E, and double precision timings are reported on all other computers. The distribution block size is also used as the partitioning unit for the computation and communication phases.

Table 5.10 illustrates the speed of the ScaLAPACK routine PSPOSV/PDPOSV for solving a symmetric positive definite linear system of order N via the Cholesky factorization.

Right-looking variants of the LU and Cholesky factorizations were chosen for ScaLAPACK because they minimize total communication volume, that is, the aggregated amount of data transferred between processes during the operation.

5.2.6.1   Solving Linear Least Squares Problems

Table 5.11 summarizes performance results obtained for the ScaLAPACK routine PSGELS/PDGELS that solves full-rank linear least squares problems. Solving such problems of the form $\min_x \|A\,x - b\|$, where x and b are vectors and A is a rectangular matrix having full rank is traditionally achieved via the computation of the QR factorization of the matrix A. In ScaLAPACK, the QR factorization is based on the use of elementary Householder matrices of the general form

$$H = I - \tau v v^T,$$

Table 5.9: Speed in Mflop/s of PSGESV/PDGESV for square matrices of order N

	Process Grid	Block Size	Values of N				
			2000	5000	7500	10000	15000
Cray T3E[a]	1×4	32	702	884	932		
PSGESV	2×8	32	1608	2680	3218	3356	3602
	4×16	32	2419	6912	9028	10299	12547
IBM SP2[b]	1×4	50	421	603			
PDGESV	2×8	50	722	1543	1903	2149	
	4×16	50	924	3017	4295	5596	7057
Intel XP/S MP Paragon[c]	1×4	32	212	282			
PDGESV	2×8	32	460	865	1010		
	4×16	32	721	2084	2837	3344	3963
Intel ASCI Red Supercomputer[d]	1×4	32	566				
PDGESV	2×8	32	1379	2497			
	4×8	32	1912	4030			
	4×16	32	2389	6329		9731	
Cluster of Sun Ultra 2s[e]	2×2	64	203	340	380		
PDGESV	2×4	64	196	499	618	689	
	2×6	64	212	608	816	934	
Berkeley NOW[f]	1×4	32	350				
PDGESV	1×8	32	560	744			
	2×8	32	811	1310	1472	1547	
	4×8	32	948	1913	2149	2324	2436
	4×16	32	1171	3091	3937	4263	4560

[a]A ScaLAPACK process is running on each node. Each node consists of one computational DEC Alpha EV5 processor, which is used by the BLAS.

[b]A ScaLAPACK process is running on each node. Each node consists of one computational IBM POWER2 590 processor, which is used by the BLAS.

[c]A ScaLAPACK process is running on each node. Each node consists of two computational Intel i860 XP RISC processors, which are used by the multi-threaded BLAS.

[d]A ScaLAPACK process is running on each node. Each node consists of two computational Intel Pentium Pro processors, which are used by the multi-threaded BLAS.

[e]A ScaLAPACK process is running on each node. Each node consists of two computational UltraSPARC-1 processors, but only one processor is used by the BLAS.

[f]A ScaLAPACK process is running on each node. Each node consists of one computational UltraSPARC-1 processor, which is used by the BLAS.

where v is a column vector and τ is a scalar. This leads to an algorithm with excellent vector performance, especially if coded to use Level 2 PBLAS.

The key to developing a distributed block form of this algorithm is to represent a product of K elementary Householder matrices of order N as a block form of a Householder matrix. This can be done in various ways. ScaLAPACK uses the form [108]

$$H_1 H_2 \ldots H_K = I - VTV^T,$$

where V is an N-by-K matrix whose columns are the individual vectors $v_1, v_2, \ldots, v_K$ associated

Table 5.10: Speed in Mflop/s of PSPOSV/PDPOSV for matrices of order N with UPLO='U'

	Process Grid	Block Size	Values of N				
			2000	5000	7500	10000	15000
Cray T3E[a]	2×2	32	822	988	1018	1039	
PSPOSV	4×4	32	2203	3322	3586	3828	4050
	8×8	32	4664	9520	11534	12809	14048
IBM SP2[b]	2×2	50	462	615			
PDPOSV	4×4	50	1081	1811	2118	2312	
	8×8	50	1807	4431	5727	6826	8084
Intel XP/S MP Paragon[c]	2×2	32	193	255			
PDPOSV	4×4	32	499	822	942	964	
	8×8	32	1092	2258	2829	3220	3725
Cluster of Sun Ultra 2s[d]	2×2	64	177	347	397		
PDPOSV	2×4	64	141	415	569	669	
	3×4	64	100	410	610	755	
Berkeley NOW[e]	2×2	32	341				
PDPOSV	2×4	32	491	703	769		
	4×4	32	828	1301	1446	1542	
	4×8	32	890	1914	2332	2602	2889
	8×8	32	1487	3263	4047	4667	5369

[a]A ScaLAPACK process is running on each node. Each node consists of one computational DEC Alpha EV5 processor, which is used by the BLAS.

[b]A ScaLAPACK process is running on each node. Each node consists of one computational IBM POWER2 590 processor, which is used by the BLAS.

[c]A ScaLAPACK process is running on each node. Each node consists of two computational Intel i860 XP RISC processors, which are used by the multi-threaded BLAS.

[d]A ScaLAPACK process is running on each node. Each node consists of two computational UltraSPARC-1 processors, but only one processor is used by the BLAS.

[e]A ScaLAPACK process is running on each node. Each node consists of one computational UltraSPARC-1 processor, which is used by the BLAS.

with the Householder matrices $H_1, H_2, \ldots, H_K$ and T is an upper triangular matrix of order K. Extra work is required to compute the elements of T, but this is compensated for by the greater speed of applying the block form.

Table 5.11: Speed in Mflop/s of PSGELS/PDGELS for square matrices of order N

	Process	Block	Values of N				
	Grid	Size	2000	5000	7500	10000	15000
Cray T3E[a]	1×4	32	568	650	659	692	
PSGELS	2×8	32	1652	2290	2434	2594	2674
	4×16	32	3433	7020	8072	9012	9746
IBM SP2[b]	1×4	50	387	564			
PDGELS	2×8	50	725	1506	1829	2041	
	4×16	50	1157	3367	4715	5836	7182
Intel XP/S MP Paragon[c]	1×4	32	201	255			
PDGELS	2×8	32	528	825	898	930	
	4×16	32	1004	2354	2937	3263	3598
Cluster of Sun Ultra 2s[d]	2×2	64	206	281	318		
PDGELS	2×4	64	220	519	585	612	
	2×6	64	241	616	794	905	
Berkeley NOW[e]	1×4	32	238	363			
PDGELS	1×8	32	482	635	682		
	2×8	32	836	1246	1351	1415	
	4×8	32	1213	2026	2575	2433	
	8×8	32	1513	3446	4296	4754	5242

[a]A ScaLAPACK process is running on each node. Each node consists of one computational DEC Alpha EV5 processor, which is used by the BLAS.

[b]A ScaLAPACK process is running on each node. Each node consists of one computational IBM POWER2 590 processor, which is used by the BLAS.

[c]A ScaLAPACK process is running on each node. Each node consists of two computational Intel i860 XP RISC processors, which are used by the multi-threaded BLAS.

[d]A ScaLAPACK process is running on each node. Each node consists of two computational UltraSPARC-1 processors, but only one processor is used by the BLAS.

[e]A ScaLAPACK process is running on each node. Each node consists of one computational UltraSPARC-1 processor, which is used by the BLAS.

5.2.6.2 Eigenvalue Problems

ScaLAPACK includes block algorithms for solving symmetric and nonsymmetric eigenvalue problems as well as for computing the singular value decomposition.

The first step in solving many types of eigenvalue problems is to reduce the original matrix to a "condensed form" by orthogonal transformations. In the reduction to condensed forms, the unblocked algorithms all use elementary Householder matrices and have good vector performance. Block forms of these algorithms have been developed [28], but all require additional operations, and a significant proportion of the work must still be performed by the Level 2 PBLAS. Thus, there is less possibility of compensating for the extra operations.

The algorithms concerned are listed below:

- Reduction of a symmetric matrix to tridiagonal form to solve a symmetric eigenvalue problem: ScaLAPACK routine PSSYTRD/PDSYTRD applies a symmetric block update of the form

$$A \leftarrow A - UX^T - XU^T,$$

  using the Level 3 PBLAS routine PSSYR2K/PDSYR2K; Level 3 PBLAS account for at most half the work.

- Reduction of a rectangular matrix to bidiagonal form to compute a singular value decomposition: ScaLAPACK routine PSGEBRD /PDGEBRD applies a block update of the form

$$A \leftarrow A - UX^T - YV^T,$$

  using two calls to the Level 3 PBLAS routine PSGEMM/PDGEMM; Level 3 PBLAS account for at most half the work.

- Reduction of a nonsymmetric matrix to Hessenberg form to solve a nonsymmetric eigenvalue problem: ScaLAPACK routine PSGEHRD/PDGEHRD applies a block update of the form

$$A \leftarrow (I - VT^TV^T)(A - XV^T).$$

  Level 3 PBLAS account for at most three-quarters of the work.

Extra work must be performed to compute the N-by-K matrices X and Y that are required for the block updates (K is the block size), and extra workspace is needed to store them.

Following the reduction of a dense symmetric matrix to tridiagonal form T, one must compute the eigenvalues and (optionally) eigenvectors of T. The current version of ScaLAPACK includes two different routines PSSYEVX/PDSYEVX and PSSYEV/PDSYEV for solving symmetric eigenproblems. PSSYEVX/PDSYEVX uses bisection and inverse iteration. PSSYEV/PDSYEV uses the QR algorithm. Table 5.12 and Table 5.13 show the execution time in seconds of the routines PSSYEVX/PDSYEVX and PSSYEV/PDSYEV, respectively, for computing the eigenvalues and eigenvectors of symmetric matrices of order N. The performance of PSSYEVX/PDSYEVX deteriorates in the face of large clusters of eigenvalues. ScaLAPACK uses a nonscalable definition of clusters (because we chose to remain consistent with LAPACK). Hence, matrices larger than $N = 1000$ tend to have at least one very large cluster (see section 5.3.6). This needs further study. More detailed information concerning the performance of these routines may be found in [40]. Table 5.14 shows the execution time in seconds of the routines PSGESVD/PDGESVD for computing the singular values and the corresponding right and left singular vectors of a general matrix of order N.

Table 5.12: Execution time in seconds of PSSYEVX/PDSYEVX for square matrices of order N

	Process Grid	Block Size	Values of N				
			1000	2000	3000	4000	5000
Cray T3E[a]	1×4	32	15	76			
PSSYEVX	2×8	32	7	29	76	164	
	4×16	32	5	17	34	64	123
IBM SP2[b]	1×4	50	36				
PDSYEVX	2×8	50	18	76			
	4×16	50	16	52	112	213	397
Intel XP/S MP Paragon[c]	1×4	32	96				
PDSYEVX	2×8	32	35	142			
	4×16	32	19	60	132	260	
Cluster of Sun Ultra 2s[d]	2×2	64	45	263			
PDSYEVX	2×4	64	44	389			
	2×6	64	79	849	1486		
Berkeley NOW[e]	1×4	32	28				
PDSYEVX	1×8	32	23				
	2×8	32	15	64	172		
	4×8	32	14	47	128	608	663

[a]A ScaLAPACK process is running on each node. Each node consists of one computational DEC Alpha EV5 processor, which is used by the BLAS.

[b]A ScaLAPACK process is running on each node. Each node consists of one computational IBM POWER2 590 processor, which is used by the BLAS.

[c]A ScaLAPACK process is running on each node. Each node consists of two computational Intel i860 XP RISC processors, which are used by the multi-threaded BLAS.

[d]A ScaLAPACK process is running on each node. Each node consists of two computational UltraSPARC-1 processors, but only one processor is used by the BLAS.

[e]A ScaLAPACK process is running on each node. Each node consists of one computational UltraSPARC-1 processor, which is used by the BLAS.

For computing the eigenvalues and eigenvectors of a Hessenberg matrix—or rather, for computing its Schur factorization— two flavors of block algorithms have been developed. The first algorithm implemented in the routine PSLAHQR/PDLAHQR results from the parallelization of the QR algorithm. The key idea is to generate many shifts at once rather than two at a time, thereby allowing all bulges to carry out up-to-date shifts. The second algorithm that is currently implemented as a prototype code is based on the computation of the matrix sign function [14, 13, 12]. In this section, however, only performance results of the first approach are reported. Table 5.15 summarizes performance results obtained for the ScaLAPACK routine PDLAHQR doing a full Schur decomposition of an order N upper Hessenberg matrix. The supercomputers the table gives timings

Table 5.13: Execution time in seconds of PSSYEV/PDSYEV for square matrices of order N

	Process Grid	Block Size	Values of N				
			1000	2000	3000	4000	5000
Cray T3E[a]	1×4	32	38	263			
PSSYEV	2×8	32	15	81	239	552	
	4×16	32	9	34	102	183	366
IBM SP2[b]	1×4	50	49				
PDSYEV	2×8	50	27	121			
	4×16	50	23	75	162	344	
Intel XP/S MP Paragon[c]	1×4	32	189				
PDSYEV	2×8	32	84	645			
	4×16	32	60	344	980		
Cluster of Sun Ultra 2s[d]	2×2	64	81				
PDSYEV	2×4	64	66				
	2×6	64	110	933	2308		
Berkeley NOW[e]	1×4	32	42				
PDSYEV	1×8	32	30				
	2×8	32	19	99	285		
	4×8	32	15	67	271	829	650

[a]A ScaLAPACK process is running on each node. Each node consists of one computational DEC Alpha EV5 processor, which is used by the BLAS.

[b]A ScaLAPACK process is running on each node. Each node consists of one computational IBM POWER2 590 processor, which is used by the BLAS.

[c]A ScaLAPACK process is running on each node. Each node consists of two computational Intel i860 XP RISC processors, which are used by the multi-threaded BLAS.

[d]A ScaLAPACK process is running on each node. Each node consists of two computational UltraSPARC-1 processors, but only one processor is used by the BLAS.

[e]A ScaLAPACK process is running on each node. Each node consists of one computational UltraSPARC-1 processor, which is used by the BLAS.

for are the Intel XP/S MP Paragon supercomputer and technology from the Intel ASCI Option Red Supercomputer. For both machines, we assume only one CPU is being used for computation on this code. The Schur decomposition is based on iteratively applying orthogonal similarity transformations on a Hessenberg matrix H such as

$$T = Q^T H Q$$

until T becomes pseudo-upper triangular (i.e., in the real case, having one by one or two by two subdiagonal blocks.) The serial performance (assuming roughly $18N^3$ flops) of the LAPACK routine DLAHQR for computing a complex Schur decomposition is around 8.5 Mflops on the Intel MP Paragon supercomputer. The enhanced performance shown in Table 5.15 is slightly faster, a bit above 9 Mflops, and ends up peaking around 10 Mflops because of the block application of Householder transforms found in the ScaLAPACK serial auxiliary routine DLAREF. For the technology behind the Intel ASCI Option Red Supercomputer, it peaks at several times the speed of the Paragon, and has a slightly faster drop off in efficiency. For further details and timings, please see [79]. A more detailed performance analysis of the eigensolvers included in the ScaLAPACK software

Table 5.14: Execution time in seconds of PSGESVD/PDGESVD for square matrices of order N

	Process Grid	Block Size	Values of N				
			1000	2000	3000	4000	5000
Cray T3E[a]	1×4	32	71	508	1810		
PSGESVD	2×8	32	24	145	454	1020	1970
	4×16	32	13	55	141	314	542
IBM SP2[b]	1×4	50	86	1031			
PDGESVD	2×8	50	37	190	547		
	4×16	50	28	104	242		
Cluster of Sun Ultra 2s[c]	2×2	64	473	4417			
PDGESVD	2×4	64	283	2375			
	2×6	64	225	1819			
Berkeley NOW[d]	1×4	32	153	1250			
PDGESVD	1×8	32	88	671	1999		
	2×8	32	46	324	1004		
	4×8	32	25	189	584		

[a]A ScaLAPACK process is running on each node. Each node consists of one computational DEC Alpha EV5 processor, which is used by the BLAS.

[b]A ScaLAPACK process is running on each node. Each node consists of one computational IBM POWER2 590 processor, which is used by the BLAS.

[c]A ScaLAPACK process is running on each node. Each node consists of two computational UltraSPARC-1 processors, but only one processor is used by the BLAS.

[d]A ScaLAPACK process is running on each node. Each node consists of one computational UltraSPARC-1 processor, which is used by the BLAS.

library can be found in [48, 79]. Finally, we note that research into parallel algorithms for symmetric and nonsymmetric eigenproblems continues [11, 86, 45], and future versions of ScaLAPACK will be updated to contain the best algorithms available.

Table 5.15: Execution time in seconds of PDLAHQR for square matrices of order N

	Process Grid	Block Size	Values of N				
			100	300	500	1000	1500
Intel XP/S MP Paragon[a]	1×1	50	3.2	53.7			
	2×2	50	2.9	25.9	89.8		
	3×3	50	3.4	17.9	60.2	300.5	
	4×4	50	4.5	17.7	47.8	200.2	514.5
Intel ASCI Option Red Supercomputer[b]	1×1	50	0.5	11.2	54.0		
	2×2	50	0.4	4.7	19.5	131.2	442.5
	3×3	50	0.5	3.2	12.3	74.2	220.4
	4×4	50	0.7	3.0	8.7	43.8	137.1
	5×5	50	0.8	3.0	6.5	31.0	95.6

[a] A ScaLAPACK process is running on each node. Only one computational Intel i860 XP RISC processor cpu was used per node for these timings.

[b] A ScaLAPACK process is running on each node. Only one computational Intel Pentium Pro processor cpu was used per node for these timings.

5.3 Performance Evaluation

5.3.1 Obtaining High Performance with ScaLAPACK Codes

We suggest the following approach to obtain high performance with ScaLAPACK codes:

- Use the best BLAS and BLACS libraries available.

- Start with a standard data distribution.

  - A square processor grid $(P_r = P_c = \lfloor \sqrt{P} \rfloor)$ if $P \geq 9$
  - A one dimensional processor grid $(P_r=1, P_c=P)$ if $P < 9$
  - Block size = 64

- Determine whether reasonable performance is being achieved.

- Identify the performance bottleneck(s), if any,

- Tune the distribution or routine parameters to improve performance further.

The standard data distribution will typically achieve 25–50% of the peak performance possible (depending in part on how many processors are ignored, i.e., the difference between $\lfloor \sqrt{P} \rfloor$ and $\sqrt{P}$). We do not recommend experimenting with different data distributions until performance that is acceptable (or nearly so) has been achieved. If each individual node requires a block size larger than 64 to achieve near-peak performance on local matrix-matrix multiply, the block size may have to be increased. This step is unlikely, however, unless the computer has a shared-memory multiprocessor with more than four processors on each node.

5.3.2 Checking the BLAS and BLACS Libraries

The best way to determine whether one is using efficient BLAS and BLACS libraries is to time them. ScaLAPACK provides a rudimentary BLAS and BLACS timer in the examples/timers directory on *netlib* and on the CD-ROM. This directory also contains pointers and instructions for more complete timers, such as the LAPACK BLAS timer and the message-passing benchmark program [46]. We encourage users to use these timers to measure the performance of the BLAS.

This ScaLAPACK examples/timers directory also contains pointers to some benchmark results and some pointers on interpreting the results of the timers.

To determine which BLAS and BLACS libraries are being linked in, users should check the output of the linker. If the `Makefile` includes `SLmake.inc`, the BLAS library is given by the macro BLASLIB in `SLmake.inc` while the BLACS library is given by the macro BLACSLIB. If the BLAS library name is of the form `blas_LINUX.a`, this is probably the (slow) reference implementation BLAS. If the BLAS library name is `-lblas`, `-lessl`, `-ldxml` or the like, this may be an optimized BLAS library.

5.3.3   Estimate Execution Time

This section describes how one can estimate the execution time of a ScaLAPACK routine on a given platform, using Equation 5.1 and the values provided in table 5.5 and table 5.8. By comparing this estimate with experimental data, the user can determine whether reasonable performance has been achieved and can (possibly) identify the performance bottlenecks, if any.

For linear system solvers, the estimate typically is accurate to within 50% for moderate-sized problems (i.e., 160,000 or more matrix elements per node). For eigensolvers, the estimate may be low by a factor of 2 for moderate-sized problems and by more than that for smaller problems. The eigensolvers take longer because they involve matrix-vector flops, as well as matrix-matrix flops, and involve substantial numbers of o(N^3) flops that are not included in the approximation. The accuracy of performance estimates increases with the problem size. Unfortunately, because ScaLAPACK eigensolvers require more memory than the other ScaLAPACK drivers, large problems cannot be solved; hence, execution times for small and medium-sized problems (rather than medium-sized and large problems) are reported.

Table 5.16: Estimated (Est) versus obtained (Obt) Mflop/s rates of PDGESV and PDPOSV on P nodes of the IBM SP2 computer for matrices of order N and a block size (NB) equal to 50

IBM SP2[a]	P	Values of N									
		2000		5000		7500		10000		15000	
		Est	Obt	Est	Obt	Est	Obt	Est	Obt	Est	Obt
PDGESV	4	357	421	632	603						
	16	497	722	1581	1543	2116	1903	2424	2149		
	64	502	924	2432	3017	4235	4295	5793	5596	7992	7057
PDPOSV	4	530	462	669	615						
	16	1315	1081	2083	1811	2366	2118	2535	2312		
	64	2577	1807	5327	4431	6709	5727	7661	6826	8887	8084

[a]One process spawned per node and one computational IBM POWER2 590 processor per node.

Table 5.16 shows the estimated versus obtained Mflop/s rates for two ScaLAPACK driver routines solving linear systems of equations on the IBM Scalable POWERparallel 2 computer. The results show that for these drivers the estimated execution times are within approximately 35 % of the experimental data on the SP2. (The estimated times for the symmetric eigensolvers and SVD codes would not be as accurate.)

5.3.4   Determine Whether Reasonable Performance Is Achieved

This chapter contains performance results for some ScaLAPACK drivers on a range of different platforms. We recommend that users compare their performance results against the ones presented in the previous tables. For those users whose computer is not listed, the BLAS timing program [3] and the message-passing benchmark program [46] can be used to estimate the values of F_{MM}, t_m

and t_v for a specific computer. [5] The execution time can then be estimated by using the material presented in the previous section.

If this chapter does not contain performance data for the ScaLAPACK routine a user is using, we recommend verifying the performance with one of the drivers whose performance is presented in this chapter. The publicly available ScaLAPACK distribution contains timing programs for each driver. This performance sanity check should convince the user not only that the library has been correctly installed, but also that it is being used properly. Users may also send questions, suggestions, or comments regarding ScaLAPACK performance issues to `scalapack@cs.utk.edu`.

5.3.5 Identify Performance Bottlenecks

The formulas mentioned in section 5.3.3, in addition to providing an estimate of performance, can help one identify whether the performance is limited by computation, by the number of messages, or by the volume of communication. Even if the estimate is far from correct, the user may get some information about the performance bottleneck by studying the computation and communication estimates provided by those formulas.

Comparing the execution times of a problem of size N and one of size $N/2$ may also provide insight into the performance of the ScaLAPACK routine being used. Let t_N and $t_{N/2}$ be the time required for a problem of size N and size $N/2$, respectively, on P processors.

- If $t_N \gg 8\,t_{N/2}$, the physical memory of each node may be exceeded.

- If $t_N \approx 8\,t_{N/2}$, the performance may be limited by the rate at which flops are performed. If the flop rate is significantly less than expected, the user should check the data distribution (try the standard data distribution suggested in section 5.1.1) and the underlying BLAS.

- If $t_N \approx 4\,t_{N/2}$, the major performance factor may be bandwidth ($1/t_v$). This is what one should obtain for medium values of N.

- If $t_N \approx 2\,t_{N/2}$, the major performance factor may be latency (t_m). This is what one should obtain for small values of N.

This performance analysis suggests which computer characteristic is most likely limiting the performance. It cannot say whether one is getting good performance.

5.3.6 Performance Bottlenecks in the Expert Symmetric Eigenproblem Drivers

Large clusters of eigenvalues in the input matrix may cause poor performance in the expert symmetric eigenproblem drivers, PSSYEVX/PDSYEVX. If the execution time observed for the ScaLAPACK drivers PxSYGVX and PxSYEVX is more than double the estimate in section 5.3.3 and

---

[5] The ScaLAPACK sample timer page (contained within the ScaLAPACK examples directory on *netlib* and on the CD-ROM) has a BLACS port of the message-passing program and instructions for building the BLAS timing program.

more than the minimum LWORK is provided, we recommend that this part of the code be retimed after relaxing the orthogonalization requirements. This can be achieved either by setting the value of the formal parameter LWORK to the minimum allowed by the driver as specified in the leading comments of the source code, or by calling the driver with the value of the parameter ORTOL set to the machine epsilon multiplied by the norm of the matrix. These last two values may be obtained by calling respectively the ScaLAPACK routines PxLAMCH and PxLANSY. If the execution time obtained for the driver after relaxing the orthogonalization requirements is substantially reduced, it is likely that the spectrum of the matrix or matrix pencil has a large cluster of eigenvalues that the driver attempts to reorthogonalize. Otherwise, it is likely that the performance bottleneck is caused by other factors as mentioned in section 5.3.5. If the matrix or matrix pencil has a large cluster of eigenvalues, we recommend using the corresponding simple driver PxSYEV, instead. If the application can tolerate loss of orthogonality, the drivers PxSYGVX and PxSYEVX may achieve good performance by relaxing the orthogonalization requirements using the method suggested above. Please check the value of the INFO parameter returned by these and all ScaLAPACK drivers.

5.4 Performance Improvement

Before experimenting with different data layouts, users should make sure that they are using the fastest BLACS and BLAS libraries.

Three major factors influence the performance of a ScaLAPACK routine: the flop rate achieved by the BLAS on each node, the computational load balance, and the communication costs.

5.4.1 Choosing a Faster BLACS Library

Users should choose vendor-supplied BLACS optimized for their computer; these BLAS will be the fastest BLACS implementation. If no vendor-supplied BLACS exists, users will have to choose among the publicly available BLACS libraries.

Many distributed-memory computers offer several communication libraries. The SP2, for example, offers MPI, PVM and MPL communication libraries. Since implementations of the BLACS exist on each of several communication libraries, one may have a choice of several different BLACS implementations. On the SP2, for example, the user can run the BLACS MPI, BLACS MPL, or BLACS PVM version.

Unfortunately, no hard rule exists as to which BLACS implementation will be fastest. However, since the BLACS cannot be faster than the communication library upon which it is built, and since the BLACS typically add little overhead, it is usually best to choose the BLACS implementation that is based on the fastest communication library.

Identifying the fastest communication library may not be trivial. The speed of communication libraries may be reported in different ways. Moreover, although the speed of blocking sends is reported because they are faster than nonblocking sends, the BLACS must use the nonblocking sends or provide its own buffering. Those who are using one of the computers listed in this chapter should refer to Tables 5.2 and 5.3 to see which library we used for timing. Our experience is that the fastest communication library was the library that is native to that particular computer.

5.4.2 Choosing a Faster BLAS Library

Highly efficient machine-specific implementations of the BLAS are available for many modern high-performance computers. Users who cannot obtain an efficient BLAS for your architecture may be able to create one from by using a set of BLAS that requires only an efficient implementation of the matrix-matrix multiply BLAS routine xGEMM [35, 90], combined with an automatically generated machine-specific and efficient implementation of xGEMM [16].

Users who are using one of the computers listed in this chapter should refer to Tables 5.2 and 5.3 to see which library we used for timing. Otherwise, the computer vendor may be able to provide information about optimized BLAS for a specific computer.

A reference Fortran 77 implementation of the BLAS is available from the *blas* directory on *netlib*.

```
http://www.netlib.org/blas/blas.shar
```

5.4.3 Tuning the Distribution Parameters for Better Performance

By adjusting the data distribution of the matrices, users may be able to achieve 10–50 % greater performance than by using the standard data distribution suggested in section 5.1.1.

The performance attained using the standard data distribution is usually fairly close to optimal; hence, if one is getting poor performance, it is unlikely that modifying the data distribution will solve the performance problem.

An optimal data distribution depends upon several factors including the performance characteristics of the hardware, the ScaLAPACK routine invoked, and (to a certain extent) the problem size. The algorithms currently implemented in ScaLAPACK fall into two main classes.

The first class of algorithms is distinguished by the fact that at each step a block of rows or columns is replicated in all process rows or columns. Furthermore, the process row or column source of this broadcast operation is the one immediately following — or preceding depending on the algorithm — the process row or column source of the broadcast operation performed at the previous step of the algorithm. The QR factorization and the right looking variant of the LU factorization are typical examples of such algorithms, where it is thus possible to establish and maintain a communication pipeline in order to overlap computation and communication. The direction of the pipeline determines the best possible shapes of the process grid. For instance, the LU, QR, and QL factorizations perform better for "flat" process grids ($P_r < P_c$). These factorizations perform a reduction operation for each matrix column for pivoting in the LU factorization and for computing the Householder transformation in the QR and QL decompositions. Moreover, after this reduction has been performed, it is important to update the next block of columns as fast as possible. This update is done by broadcasting the current block of columns using a ring topology, that is, feeding the ongoing communication pipe. Similarly, the performance of the LQ and RQ factorizations take advantage of "tall" grids ($P_r > P_c$) for the same, but transposed, reasons.

The second group of algorithms is characterized by the physical transposition of a block of rows and/or columns at each step. Square or near square grids are more adequate from a performance point of view for these transposition operations. Examples of such algorithms implemented in ScaLAPACK include the right-looking variant of the Cholesky factorization, the matrix inversion algorithm, and the reductions to bidiagonal form (PxGEBRD), to Hessenberg form (PxGEHRD), and to tridiagonal form (PxSYTRD). It is interesting to note that if square grids are more efficient for these matrix reduction operations, the corresponding eigensolver usually prefers flatter grids.

Table 5.17 summarizes this paragraph and provides suggestions for selecting the most appropriate shape of the logical $P_r \times P_c$ process grid from a performance point of view. The results presented in this table may need to be refined depending on the physical characteristics of the physical interconnection network.

Table 5.17: Process grid suggestions for some ScaLAPACK drivers

Driver	Process Grid	Comments
PxGESV	$P_r \ll P_c$	1D distribution for small P (≤ 8)
PxPOSV	$P_r = P_c$	
PxGELS	$P_r \ll P_c$ if $M \geq N$	1D distribution for small P (≤ 8)
	$P_r \gg P_c$ otherwise	
PxSYEVX	$P_r \leq P_c$	$lcm(P_r, P_c) < 10 \times max(P_r, P_c)$
PxSYEV	$P_r \leq P_c$	$lcm(P_r, P_c) < 10 \times max(P_r, P_c)$
PxGESVD	$P_r = P_c$	$lcm(P_r, P_c) < 10 \times max(P_r, P_c)$
PxLAHQR	$P_r = P_c$	

Assume that at most P nodes are available. A natural question is: Could we decide which $P_r \times P_c \leq P$ process grid should be used? Similarly, depending on the value of P, it is not always possible to factor $P = P_r \times P_c$ to create an appropriate grid shape. For example, if the number of nodes available is a prime number and a square grid is suitable with respect to performance, it may be beneficial to let some nodes remain idle so that the remaining nodes can be arranged in a "squarer" grid.

If the BLACS implementation or the interconnection network features high latency, a one-dimensional data distribution will improve the performance for small and medium problem sizes. The number of messages significantly impacts the performance achieved for small problem sizes, whereas the total message volume becomes a dominant factor for medium-sized problems. The performance cost due to floating-point operations dominates for large problem sizes. One-dimensional data distributions reduce the total number of messages exchanged on the interconnection network but increase the total volume of message traffic. Therefore, one-dimensional data distributions are better for small problem sizes but are worse for large problem sizes, especially when one is using eight or more processors.

Determining optimal, or near-optimal, distribution block sizes with respect to performance for a given platform is a difficult task. However, it is empirically true that as soon as a good block size or even a set of good distribution parameters is found, the performance is not highly sensitive to small changes of the values of these parameters.

5.5 Performance of Banded and Out-of-Core Drivers

ScaLAPACK provides *LU* and Cholesky factorizations for band matrices. For small bandwidth, divide and conquer algorithms have been chosen even though they require more floating-point operations. A more detailed performance analysis can be found in [18].

ScaLAPACK also provides prototype out-of-core linear system solvers. Information on these particular routines as well as the algorithms that have been selected can be found in [47, 55, 18]. In particular, it is shown in [55] that these out-of-core solvers incur approximately a 20% overhead over the corresponding in-core ScaLAPACK solvers.

Chapter 6

Accuracy and Stability

In this chapter we explain our overall approach to obtaining error bounds and provide enough information to use the software. The comments at the beginning of the individual routines should be consulted for more details. It is beyond the scope of this chapter to justify all the bounds we present. Instead, we give references to the literature. For example, standard material on error analysis can be found in [71, 114, 84, 38].

To make this chapter easy to read, we have labeled parts not essential for a first reading as **Further Details**. The sections not labeled as **Further Details** should provide all the information needed to understand and use the main error bounds computed by ScaLAPACK. The **Further Details** sections provide mathematical background, references, and tighter but more expensive error bounds, and may be read later.

Since ScaLAPACK uses the same overall algorithmic approach as LAPACK, its error bounds are essentially the same as those for LAPACK. Therefore, this chapter is largely analogous to Chapter 4 of the LAPACK Users' Guide [3]. Significant differences between LAPACK and ScaLAPACK include the following:

- Section 6.1 discusses how machine constants in a heterogeneous network of machines with differing floating-point arithmetics must be redefined. ScaLAPACK can also exploit arithmetic with $\pm\infty$, which is available in IEEE standard floating-point arithmetic.

- Section 6.2 discusses reliability problems that can arise on heterogeneous networks of machines and how to guarantee reliability on a homogeneous network.

- Section 6.5 discusses some routines that do Gaussian elimination on band matrices with the pivot order chosen for parallelism rather than numerical stability. These routines are numerically stable only when applied to matrices that are do not require partial pivoting for stability (such as diagonally dominant and symmetric positive definite matrices).

- Section 6.7 discusses PxSYEVX.[1] In contrast to its LAPACK analogue, xSYEVX, PxSYEVX allows the user to trade off orthogonality of computed eigenvectors and runtime.

---

[1] See section 3.1.3 for explanation of the naming convention used for ScaLAPACK routines.

In section 6.1 we discuss the sources of numerical error, in particular roundoff error. We also briefly discuss IEEE arithmetic. Section 6.2 discusses the new sources of numerical error specific to parallel libraries, and the restrictions they impose on the reliable use of ScaLAPACK. Section 6.3 discusses how to measure errors, as well as some standard notation. Section 6.4 discusses further details of how error bounds are derived. Sections 6.5 through 6.9 present error bounds for linear equations, linear least squares problems, the symmetric eigenproblem, the singular value decomposition, and the generalized symmetric definite eigenproblem, respectively.

6.1   Sources of Error in Numerical Calculations

The effects of two sources of error can be measured by the bounds in this chapter: *roundoff error* and *input error*. Roundoff error arises from rounding results of floating-point operations during the algorithm. Input error is error in the input to the algorithm from prior calculations or measurements. We describe roundoff error first, and then input error.

Almost all the error bounds ScaLAPACK provides are multiples of *relative machine precision*, which we abbreviate by ϵ. Relative machine precision (epsilon) bounds the roundoff in individual floating-point operations. It may be loosely defined as the largest relative error in any floating-point operation that neither overflows nor underflows. (Overflow means the result is too large to represent accurately, and underflow means the result is too small to represent accurately.) Relative machine precision (epsilon) is available either by the function call PSLAMCH(ICTXT, 'Epsilon') (or simply PSLAMCH(ICTXT, 'E'))[2] in single precision, or by the function call PDLAMCH(ICTXT, 'Epsilon') (or PDLAMCH(ICTXT, 'E')) in double precision. See section 6.1 and Table 6.1 for a discussion of common values of machine epsilon.

PDLAMCH(ICTXT,'E') returns a single value for the selected machine parameter 'E' on all processes within the context ICTXT. If these processes are running on a network of heterogeneous processors, with different floating-point arithmetics, then a "safe" common value is returned, the maximum value of machine epsilon for all the processors.

In case of overflow, there are two common system responses: stopping with an error message, or returning $\pm\infty$ and continuing to compute. The latter is the default response of IEEE standard floating-point arithmetic [7, 8] , the most commonly used arithmetic. It is possible to change this default to abort with an error message, which is often useful for debugging.

In contrast to LAPACK, ScaLAPACK can take advantage of arithmetic with $\pm\infty$ to accelerate the routines that compute eigenvalues of symmetric matrices using PxLAIECT (the drivers PxSYEVX and PxSYGVX, and their complex counterparts).   PxLAIECT comes in two different versions, one in which arithmetic with $\pm\infty$ is available (the default) and one in which it is not. When $\pm\infty$ is available, the inner loop of PxLAIECT is accelerated by removing a branch to test for and avoid division by zero. This speed advantage is realized only when arithmetic with $\pm\infty$ is as fast as arithmetic with normalized floating-point numbers; this is usually but not always the case [42]. The compile time flag NO_IEEE can be used during installation to run without using $\pm\infty$; see the ScaLAPACK Installation Guide for details [24].

---

[2]ICTXT refers to the BLACS CONTEXT parameter. Refer to section 4.1.2 for further details.

Since underflow is almost always less significant than roundoff, we will not consider it further in this section (but see section 6.1).

Bounds on *input errors* may be easily incorporated into most ScaLAPACK error bounds. Suppose the input data is accurate to, say, five decimal digits (we discuss exactly what this means in section 6.3). Then one simply replaces ϵ by $\max(\epsilon, 10^{-5})$ in the error bounds.

Further Details: Floating-Point Arithmetic

Roundoff error is bounded in terms of the *relative machine precision* ϵ, which is the smallest value satisfying

$$|fl(a \oplus b) - (a \oplus b)| \leq \epsilon \cdot |a \oplus b| \ ,$$

where a and b are floating-point numbers, $\oplus$ is any one of the four operations $+$, $-$, $\times$, and $\div$, and $fl(a \oplus b)$ is the floating-point result of $a \oplus b$. Relative machine precision, ϵ, is the smallest value for which this inequality is true for all $\oplus$, and for all a and b such that $a \oplus b$ is neither too large (magnitude exceeds the overflow threshold) nor too small (is nonzero with magnitude less than the underflow threshold) to be represented accurately in the machine. We also assume ϵ bounds the relative error in unary operations such as square root:

$$|fl(\sqrt{a}) - (\sqrt{a})| \leq \epsilon \cdot |\sqrt{a}| \ .$$

A precise characterization of ϵ depends on the details of the machine arithmetic and sometimes even of the compiler. For example, if addition and subtraction are implemented without a guard digit,[3] we must redefine ϵ to be the smallest number such that

$$|fl(a \pm b) - (a \pm b)| \leq \epsilon \cdot (|a| + |b|).$$

In order to assure portability, machine parameters such as relative machine precision (epsilon), the overflow threshold and underflow threshold are computed at runtime by the auxiliary routine PxLAMCH. The alternative, keeping a fixed table of machine parameter values, would degrade portability because the table would have to be changed when moving from one machine or combination of machines, or even one compiler, to another.

Most machines (but not yet all) do have the same machine parameters because they implement IEEE Standard Floating Point Arithmetic [7, 8], which exactly specifies floating-point number representations and operations. For these machines, including all modern workstations and PCs,[4] the values of these parameters are given in Table 6.1.

Unfortunately, machines claiming to implement IEEE arithmetic may still compute different results from the same program and input. Here are some examples. Intel processors have 80-bit floating-point registers, and the fastest way to use them is to evaluate all results to 80-bit accuracy until they are stored back to memory in 32-bit or 64-bit format. The IBM RS/6000 has a fused multiply-add instruction that evaluates $a + b*c$ with one rounding error instead of two. The DEC Alpha's default (fast) mode is to flush underflowed values to zero instead of returning subnormal numbers, which is

---

[3]This is the case on the Cray C90 and its predecessors and emulators.

[4]Important machines that do not implement the IEEE standard include the CRAY X-MP, CRAY Y-MP, CRAY 2, CRAY C90, IBM 370, DEC Vax, and their emulators.

the default demanded by the IEEE standard; in this mode the DEC Alpha aborts if it encounters a subnormal number. In all these cases machines may be made to operate absolutely identically, for example, by rounding all intermediate results back to single or double on an Intel machine, or by doing subnormal arithmetic carefully and slowly on a DEC Alpha. These heterogeneities lead to errors encountered only in parallel computing; see section 6.2 for further discussion.

Table 6.1: Values of machine parameters in IEEE floating-point arithmetic

Machine Parameter	Single Precision (32 bits)	Double Precision (64 bits)
Relative machine precision ϵ = PxLAMCH(ICTXT, 'E')	$2^{-24} \approx 5.96 \cdot 10^{-8}$	$2^{-53} \approx 1.11 \cdot 10^{-16}$
Underflow threshold = PxLAMCH(ICTXT, 'U')	$2^{-126} \approx 1.18 \cdot 10^{-38}$	$2^{-1022} \approx 2.23 \cdot 10^{-308}$
Overflow threshold = PxLAMCH(ICTXT, 'O')	$2^{128}(1 - \epsilon) \approx 3.40 \cdot 10^{38}$	$2^{1024}(1 - \epsilon) \approx 1.79 \cdot 10^{308}$

As stated above, we will ignore underflow in discussing error bounds. Reference [37] discusses extending error bounds to include underflow and shows that for many common computations, when underflow occurs, it is less significant than roundoff.

Overflow historically resulted in an error message and stopped execution, in which case our error bounds would not apply. But with IEEE floating-point arithmetic, the default is that overflow returns $\pm\infty$ and execution continues. Indeed, with IEEE arithmetic machines can continue to compute past overflows, even division by zero, square roots of negative numbers, etc., by producing $\pm\infty$ and NaN ("Not a Number") symbols according to special rules of arithmetic. The default on many systems is to continue computing with these symbols. Routine PxLAIECT exploits this arithmetic to accelerate the computations of eigenvalues, as discussed above. It is also possible to stop with an error message when overflow occurs, a feature that is often useful for debugging. The user should consult the system manual to see how to turn error messages on or off.

Most of our error bounds will simply be proportional to relative machine precision (epsilon). This means, for example, that if the same problem in solved in double precision and single precision, the error bound in double precision will be smaller than the error bound in single precision by a factor of $\epsilon_{\text{double}}/\epsilon_{\text{single}}$. In IEEE arithmetic, this ratio is $2^{-53}/2^{-24} \approx 10^{-9}$, meaning that one expects the double-precision answer to have approximately nine more decimal digits correct than the single-precision answer.

Like their counterparts in LAPACK. ScaLAPACK routines are generally insensitive to the details of rounding, provided all processes perform arithmetic identically. The one exception is PxLAIECT, as mentioned above. The next section discusses what can happen when processes do not perform arithmetic identically, that is, are *heterogeneous*.

6.2 New Sources of Error in Parallel Numerical Computations

An important difference between ScaLAPACK and LAPACK is that a parallel computing environment, possibly consisting of a heterogeneous collection of processors, introduces new sources of possible errors not found in the serial environment in which LAPACK runs. These errors could indeed afflict any parallel algorithm that uses floating-point arithmetic. For example, consider the following pseudocode, executed in parallel by several processors:

$s=$ global_sum(x) ... each processor receives the sum s of global array x
if $s < thresh$ then
    return my part of answer 1
else
    do more computations
    return my part of answer 2
end if

It is possible for the value of s to differ from processor to processor; we call this *incoherence*. This can happen if the floating-point arithmetic varies from processor to processor (we call this *heterogeneity*), since processors may not even share the same set of floating-point numbers. The value of s can also vary if global_sum accumulates the sum in different orders on different processors, since floating-point addition is not associative. In either case, the test $s < thresh$ may be true on one processor but not another, so that the program may inconsistently return answer 1 on some processors and answer 2 on others. If the "more computations" include communication with synchronization, even deadlock could result.

Deadlock can also result if the floating-point numbers communicated from one processor to another cause fatal floating-point errors on the receiving processor. For example, if an IBM RS/6000, running in its default mode, sends a message containing a denormalized number [7, 8] to a DEC Alpha running in its default mode, then the DEC Alpha aborts [19].[5]

It is also possible for global_sum to compute the same s on all processors but compute a different s from run to run of the program, for example, if global_sum computes the sum in a nondeterministic order on one processor and broadcasts the result to all processors. We call this *nonrepeatability*. If this happens, debugging the overall code can be more difficult.

Coherence and repeatability are independent properties of an algorithm. It is possible in principle for an algorithm running on a particular platform to be incoherent and repeatable, coherent and nonrepeatable, or any other combination. On a different platform, the same algorithm may have different properties.

Reference [19] contains a more extensive discussion of these possible errors.

One run of a ScaLAPACK routine is designed to be as reliable as LAPACK, so that errors due to incoherence cannot occur as long as ScaLAPACK is executed on a *homogeneous network* of

---

[5]Running either machine in non-default mode to avoid this problem, either so that the IBM RS/6000 flushes denormalized numbers to zero or so that the DEC Alpha handles denormalized numbers correctly by doing gradual underflow, slows down the machine significantly [42].

processors. The following conditions apply:

- The processors are completely identical. This also means that relevant flags, like those controlling the way overflow and underflow are handled in IEEE floating-point arithmetic, must be identical.

- The communication library used by the BLACS may only "copy bits" and not modify any floating-point numbers (by translation to a different internal floating-point format, as XDR [111] may do).

- The identical ScaLAPACK object code must be executed by each processor.

The above conditions guarantee that a single ScaLAPACK call is as reliable as its LAPACK counterpart. If, in addition, identical answers from one run to another are desired (i.e., *repeatability*), this can be guaranteed at runtime by calling BLACS_SET to enforce repeatability of the BLACS, and the ScaLAPACK routines that use them, by using an appropriate topology (see the BLACS users guide [54] for details).

Maintaining coherence on a heterogeneous network is harder, and not always possible. If floating-point formats differ (say, on a Cray C90 and IBM RS/6000, which uses IEEE arithmetic), there is no cost-effective way to guarantee coherence. If floating-point formats are the same, however, operations such as global sums can accumulate the result on one processor and broadcast it to guarantee coherence (except for the problem of DEC Alphas and denormalized numbers mentioned above). The BLACS do this, except when using the "bidirectional exchange" topology. One can avoid using "bidirectional exchange" and so guarantee coherence whenever possible, by calling BLACS_SET to enforce coherence (see the BLACS users guide [54] for details).

Still other ScaLAPACK routines are guaranteed to work only on homogeneous networks (PxGESVD and PxSYEV). These routines do large numbers of redundant calculations on all processors and depend on the results of these calculations being the same. There are too many of these calculations to cost-effectively compute them all on one processor and broadcast the results.

The user may wonder why ScaLAPACK and the BLACS are not designed to guarantee coherence and repeatability in the most general possible situations, so that calling BLACS_SET would not be necessary. The reason is that the possible bugs described above are quite rare, and so ScaLAPACK and the BLACS were designed to maximize performance instead. Provided the mere sending of floating-point numbers does not cause a fatal error, these bugs cannot occur at all in most ScaLAPACK routines, because branches depending on a supposedly identical floating-point value like s do not occur. For most other ScaLAPACK routines where such branches do occur, we have not seen these bugs despite extensive testing, including attempts to cause them to occur. Complete understanding and cost-effective elimination of such possible bugs are future work.

In the meantime, to get repeatability when running on a homogeneous network, we recommend calling BLACS_SET as described above when using the following ScaLAPACK drivers: PxGESVX, PxPOSVX, PxSYEV, PxSYEVX, PxGESVD, and PxSYGVX.

6.3   How to Measure Errors

ScaLAPACK routines return four types of floating-point output arguments:

- *Scalar*, such as an eigenvalue of a matrix,

- *Vector*, such as the solution x of a linear system $Ax = b$,

- *Matrix*, such as a matrix inverse A^{-1}, and

- *Subspace*, such as the space spanned by one or more eigenvectors of a matrix.

This section provides measures for errors in these quantities, which we need in order to express error bounds.

First, consider *scalars*. Let the scalar $\hat{\alpha}$ be an approximation of the true answer α. We can measure the difference between α and $\hat{\alpha}$ either by the **absolute error** $|\hat{\alpha} - \alpha|$, or, if α is nonzero, by the **relative error** $|\hat{\alpha} - \alpha|/|\alpha|$. Alternatively, it is sometimes more convenient to use $|\hat{\alpha} - \alpha|/|\hat{\alpha}|$ instead of the standard expression for relative error. If the relative error of $\hat{\alpha}$ is, say, 10^{-5}, we say that $\hat{\alpha}$ is *accurate to 5 decimal digits*.

To measure the error in *vectors*, we need to measure the *size* or *norm* of a vector x. A popular norm is the magnitude of the largest component, $\max_{1 \leq i \leq n} |x_i|$, which we denote by $\|x\|_\infty$. This is read *the infinity norm of* x. See Table 6.2 for a summary of norms.

Table 6.2: Vector and matrix norms

Norm	Vector	Matrix
One-norm	$\|x\|_1 = \sum_i |x_i|$	$\|A\|_1 = \max_j \sum_i |a_{ij}|$
Two-norm	$\|x\|_2 = (\sum_i |x_i|^2)^{1/2}$	$\|A\|_2 = \max_{x \neq 0} \|Ax\|_2 / \|x\|_2$
Frobenius norm	$\|x\|_F = \|x\|_2$	$\|A\|_F = (\sum_{ij} |a_{ij}|^2)^{1/2}$
Infinity-norm	$\|x\|_\infty = \max_i |x_i|$	$\|A\|_\infty = \max_i \sum_j |a_{ij}|$

If $\hat{x}$ is an approximation to the exact vector x, we will refer to $\|\hat{x} - x\|_p$ as the absolute error in $\hat{x}$ (where p is one of the values in Table 6.2)   and refer to $\|\hat{x} - x\|_p / \|x\|_p$ as the relative error in $\hat{x}$ (assuming $\|x\|_p \neq 0$). As with scalars, we will sometimes use $\|\hat{x} - x\|_p / \|\hat{x}\|_p$ for the relative error. As above, if the relative error of $\hat{x}$ is, say 10^{-5}, we say that $\hat{x}$ is accurate to 5 decimal digits. The following example illustrates these ideas.

$$x = \begin{pmatrix} 1 \\ 100 \\ 9 \end{pmatrix} \ , \quad \hat{x} = \begin{pmatrix} 1.1 \\ 99 \\ 11 \end{pmatrix}$$

$$\|\hat{x} - x\|_\infty = 2 \ , \quad \frac{\|\hat{x} - x\|_\infty}{\|x\|_\infty} = .02 \ , \quad \frac{\|\hat{x} - x\|_\infty}{\|\hat{x}\|_\infty} = .0202$$

$$\|\hat{x} - x\|_2 = 2.238 \;, \quad \frac{\|\hat{x} - x\|_2}{\|x\|_2} = .0223 \;, \quad \frac{\|\hat{x} - x\|_2}{\|\hat{x}\|_2} = .0225$$

$$\|\hat{x} - x\|_1 = 3.1 \;, \quad \frac{\|\hat{x} - x\|_1}{\|x\|_1} = .0282 \;, \quad \frac{\|\hat{x} - x\|_1}{\|\hat{x}\|_1} = .0279$$

Thus, we would say that $\hat{x}$ approximates x to 2 decimal digits.

Errors in *matrices* may also be measured with norms. The most obvious generalization of $\|x\|_\infty$ to matrices would appear to be $\|A\| = \max_{i,j} |a_{ij}|$, but this does not have certain important mathematical properties that make deriving error bounds convenient. Instead, we will use $\|A\|_\infty = \max_{1 \le i \le m} \sum_{j=1}^{n} |a_{ij}|$, where A is an m-by-n matrix, or $\|A\|_1 = \max_{1 \le j \le n} \sum_{i=1}^{m} |a_{ij}|$; see Table 6.2 for other matrix norms. As before, $\|\hat{A} - A\|_p$ is the absolute error in $\hat{A}$, $\|\hat{A} - A\|_p / \|A\|_p$ is the relative error in $\hat{A}$, and a relative error in $\hat{A}$ of 10^{-5} means $\hat{A}$ is accurate to 5 decimal digits. The following example illustrates these ideas.

$$A = \begin{pmatrix} 1 & 2 & 3 \\ 4 & 5 & 6 \\ 7 & 8 & 10 \end{pmatrix} \;, \quad \hat{A} = \begin{pmatrix} .44 & 2.36 & 3.04 \\ 3.09 & 5.87 & 6.66 \\ 7.36 & 7.77 & 9.07 \end{pmatrix}$$

$$\|\hat{A} - A\|_\infty = 2.44 \;, \quad \frac{\|\hat{A} - A\|_\infty}{\|A\|_\infty} = .0976 \;, \quad \frac{\|\hat{A} - A\|_\infty}{\|\hat{A}\|_\infty} = .1008$$

$$\|\hat{A} - A\|_2 = 1.75 \;, \quad \frac{\|\hat{A} - A\|_2}{\|A\|_2} = .1007 \;, \quad \frac{\|\hat{A} - A\|_2}{\|\hat{A}\|_2} = .1020$$

$$\|\hat{A} - A\|_1 = 1.83 \;, \quad \frac{\|\hat{A} - A\|_1}{\|A\|_1} = .0963 \;, \quad \frac{\|\hat{A} - A\|_1}{\|\hat{A}\|_1} = .0975$$

$$\|\hat{A} - A\|_F = 1.87 \;, \quad \frac{\|\hat{A} - A\|_F}{\|A\|_F} = .1075 \;, \quad \frac{\|\hat{A} - A\|_F}{\|\hat{A}\|_F} = .1082$$

so $\hat{A}$ is accurate to 1 decimal digit.

We now introduce some related notation we will use in our error bounds. The **condition number of a matrix** A is defined as $\kappa_p(A) \equiv \|A\|_p \cdot \|A^{-1}\|_p$, where A is square and invertible, and p is ∞ or one of the other possibilities in Table 6.2. The condition number measures how sensitive A^{-1} is to changes in A; the larger the condition number, the more sensitive is A^{-1}. For example, for the same A as in the last example,

$$A^{-1} \approx \begin{pmatrix} -.667 & -1.333 & 1 \\ -.667 & 3.667 & -2 \\ 1 & -2 & 1 \end{pmatrix} \quad \text{and} \quad \kappa_\infty(A) = 158.33 \;.$$

ScaLAPACK error estimation routines typically compute a variable called RCOND, which is the reciprocal of the condition number (or an approximation of the reciprocal). The reciprocal of the condition number is used instead of the condition number itself in order to avoid the possibility of overflow when the condition number is very large. Also, some of our error bounds will use the vector of absolute values of x, $|x|$ ($|x|_i = |x_i|$), or similarly $|A|$ ($|A|_{ij} = |a_{ij}|$).

Now we consider errors in *subspaces*. Subspaces are the outputs of routines that compute eigenvectors and invariant subspaces of matrices. We need a careful definition of error in these cases for the following reason. The nonzero vector x is called a *(right) eigenvector* of the matrix A with *eigenvalue* λ if $Ax = \lambda x$. From this definition, we see that $-x$, $2x$, or any other nonzero multiple βx of x is also an eigenvector. In other words, eigenvectors are not unique. This means we cannot measure the difference between two supposed eigenvectors $\hat{x}$ and x by computing $\|\hat{x} - x\|_2$, because this may be large while $\|\hat{x} - \beta x\|_2$ is small or even zero for some $\beta \neq 1$. This is true even if we normalize x so that $\|x\|_2 = 1$, since both x and $-x$ can be normalized simultaneously. Hence, to define error in a useful way, we need instead to consider the set $\mathcal{S}$ of all scalar multiples $\{\beta x \,, \beta \text{ a scalar}\}$ of x. The set $\mathcal{S}$ is called the *subspace spanned by* x and is uniquely determined by any nonzero member of $\mathcal{S}$. We will measure the difference between two such sets by the *acute angle* between them. Suppose $\hat{\mathcal{S}}$ is spanned by $\{\hat{x}\}$ and $\mathcal{S}$ is spanned by $\{x\}$. Then the acute angle between $\hat{\mathcal{S}}$ and $\mathcal{S}$ is defined as

$$\theta(\hat{\mathcal{S}}, \mathcal{S}) = \theta(\hat{x}, x) \equiv \arccos \frac{|\hat{x}^T x|}{\|\hat{x}\|_2 \cdot \|x\|_2} \ .$$

One can show that $\theta(\hat{x}, x)$ does not change when either $\hat{x}$ or x is multiplied by any nonzero scalar. For example, if

$$x = \begin{pmatrix} 1 \\ 100 \\ 9 \end{pmatrix} \quad \text{and} \quad \hat{x} = \begin{pmatrix} 1.1 \\ 99 \\ 11 \end{pmatrix}$$

as above, then $\theta(\gamma \hat{x}, \beta x) = .0209$ for any nonzero scalars β and γ.

Let us consider another way to interpret the angle θ between $\hat{\mathcal{S}}$ and $\mathcal{S}$. Suppose $\hat{x}$ is a unit vector ($\|\hat{x}\|_2 = 1$). Then there is a scalar β such that

$$\|\hat{x} - \beta x\|_2 = \frac{\sqrt{2} \sin \theta}{\sqrt{1 + \cos \theta}} \approx \theta \ .$$

The approximation $\approx \theta$ holds when θ is much less than 1 (less than .1 will do nicely). If $\hat{x}$ is an approximate eigenvector with error bound $\theta(\hat{x}, x) \leq \bar{\theta} \ll 1$, where x is a true eigenvector, there is another true eigenvector βx satisfying $\|\hat{x} - \beta x\|_2 \lesssim \bar{\theta}$. For example, if

$$\hat{x} = \begin{pmatrix} 1.1 \\ 99 \\ 11 \end{pmatrix} \cdot (1.1^2 + 99^2 + 11^2)^{-1/2} \text{ so } \|\hat{x}\|_2 = 1 \text{ and } x = \begin{pmatrix} 1 \\ 100 \\ 9 \end{pmatrix}$$

then $\|\hat{x} - \beta x\|_2 \approx .0209$ for $\beta \approx .01$.

Finally, many of our error bounds will contain a factor $p(n)$ (or $p(m, n)$), which grows as a function of matrix dimension n (or dimensions m and n). It represents a potentially different function for each problem. In practice, the true errors usually grow at most linearly; using $p(n) = 1$ in the error bound formulas will often give a reasonable estimate; $p(n) = n$ is more conservative. Therefore, we will refer to $p(n)$ as a "modestly growing" function of n; however. it can occasionally be much larger. For simplicity, the error bounds computed by the code fragments in the following sections will use $p(n) = 1$. This means these computed error bounds may occasionally slightly underestimate the true error. For this reason we refer to these computed error bounds as "approximate error bounds."

Further Details: How to Measure Errors

The relative error $|\hat{\alpha} - \alpha|/|\alpha|$ in the approximation $\hat{\alpha}$ of the true solution α has a drawback: it often cannot be computed directly, because it depends on the unknown quantity $|\alpha|$. However, we can often instead estimate $|\hat{\alpha} - \alpha|/|\hat{\alpha}|$, since $\hat{\alpha}$ is known (it is the output of our algorithm). Fortunately, these two quantities are necessarily close together, provided either one is small, which is the only time they provide a useful bound anyway. For example, $|\hat{\alpha} - \alpha|/|\hat{\alpha}| \le .1$ implies

$$.9\frac{|\hat{\alpha} - \alpha|}{|\hat{\alpha}|} \le \frac{|\hat{\alpha} - \alpha|}{|\alpha|} \le 1.1\frac{|\hat{\alpha} - \alpha|}{|\hat{\alpha}|} \ ,$$

so they can be used interchangeably.

Table 6.2 contains a variety of norms we will use to measure errors. These norms have the properties that $\|Ax\|_p \le \|A\|_p \cdot \|x\|_p$, and $\|AB\|_p \le \|A\|_p \cdot \|B\|_p$, where p is one of 1, 2, ∞, and F. These properties are useful for deriving error bounds.

An error bound that uses a given norm may be changed into an error bound that uses another norm. This is accomplished by multiplying the first error bound by an appropriate function of the problem dimension. Table 6.3 gives the factors $f_{pq}(n)$ such that $\|x\|_p \le f_{pq}(n)\|x\|_q$, where n is the dimension of x.

Table 6.3: Bounding one vector norm in terms of another

Values of $f_{pq}(n)$ such that $\|x\|_p \le f_{pq}(n)\|x\|_q$, where x is an n-vector

		q		
		1	2	∞
	1	1	$\sqrt{n}$	n
p	2	1	1	$\sqrt{n}$
	∞	1	1	1

Table 6.4 gives the factors $f_{pq}(m,n)$ such that $\|A\|_p \le f_{pq}(m,n)\|A\|_q$, where A is m-by-n.

The two-norm of A, $\|A\|_2$, is also called the **spectral norm** of A and is equal to the **largest singular value** $\sigma_{\max}(A)$ of A. We shall also need to refer to the **smallest singular value** $\sigma_{\min}(A)$ of A; its value can be defined in a similar way to the definition of the two-norm in Table 6.2, namely, as $\min_{x \ne 0} \|Ax\|_2/\|x\|_2$ when A has at least as many rows as columns, and defined as $\min_{x \ne 0} \|A^T x\|_2/\|x\|_2$ when A has more columns than rows. The two-norm, Frobenius norm, and singular values of a matrix do not change if the matrix is multiplied by a real orthogonal (or complex unitary) matrix.

Now we define *subspaces* spanned by more than one vector, and *angles between subspaces*. Given a set of k n-dimensional vectors $\{x_1, ..., x_k\}$, they determine a subspace $\mathcal{S}$ consisting of all their possible linear combinations $\{\sum_{i=1}^{k} \beta_i x_i, \ \beta_i$ scalars $\}$. We also say that $\{x_1, ..., x_k\}$ *spans* $\mathcal{S}$. The difficulty in measuring the difference between subspaces is that the sets of vectors spanning them are not unique. For example, $\{x\}$, $\{-x\}$ and $\{2x\}$ all determine the same subspace. Therefore, we cannot simply compare the subspaces spanned by $\{\hat{x}_1, ..., \hat{x}_k\}$ and $\{x_1, ..., x_k\}$ by comparing each

Table 6.4: Bounding one matrix norm in terms of another

Values of $f_{pq}(m,n)$ such that $\|A\|_p \leq f_{pq}(m,n)\|A\|_q$, where A is m by n

		q			
		1	2	F	∞
p	1	1	$\sqrt{m}$	$\sqrt{m}$	m
	2	$\sqrt{n}$	1	1	$\sqrt{m}$
	F	$\sqrt{n}$	$\sqrt{\min(m,n)}$	1	$\sqrt{m}$
	∞	n	$\sqrt{n}$	$\sqrt{n}$	1

$\hat{x}_i$ to x_i. Instead, we will measure the *angle* between the subspaces, which is independent of the spanning set of vectors. Suppose subspace $\hat{S}$ is spanned by $\{\hat{x}_1, ..., \hat{x}_k\}$ and that subspace S is spanned by $\{x_1, ..., x_k\}$. If $k = 1$, we instead write more simply $\{\hat{x}\}$ and $\{x\}$. When $k = 1$, we define the angle $\theta(\hat{S}, S)$ between $\hat{S}$ and S as the acute angle between $\hat{x}$ and x. When $k > 1$, we define the acute angle between $\hat{S}$ and S as the largest acute angle between any vector $\hat{x}$ in $\hat{S}$ and the closest vector x in S to $\hat{x}$:

$$\theta(\hat{S}, S) \equiv \max_{\substack{\hat{x} \in \hat{S} \\ \hat{x} \neq 0}} \min_{\substack{x \in S \\ x \neq 0}} \theta(\hat{x}, x) .$$

ScaLAPACK routines that compute subspaces return vectors $\{\hat{x}_1, ..., \hat{x}_k\}$ spanning a subspace $\hat{S}$ that are *orthonormal*. This means the n-by-k matrix $\hat{X} = [\hat{x}_1, ..., \hat{x}_k]$ satisfies $\hat{X}^H \hat{X} = I$. Suppose also that the vectors $\{x_1, ..., x_k\}$ spanning S are orthonormal, so $X = [x_1, ..., x_k]$ also satisfies $X^H X = I$. Then there is a simple expression for the angle between $\hat{S}$ and S:

$$\theta(\hat{S}, S) = \arccos \sigma_{\min}(\hat{X}^H X) .$$

For example, if

$$\hat{X} = \begin{pmatrix} -.79996 & .60005 \\ -.59997 & -.79990 \\ -.01 & -.01 \end{pmatrix} \text{ and } X = \begin{pmatrix} 1 & 0 \\ 0 & 1 \\ 0 & 0 \end{pmatrix}$$

then $\theta(\hat{S}, S) = .01414$.

As stated above, all our bounds will contain a factor $p(n)$ (or $p(m, n)$), which measures how roundoff errors can grow as a function of matrix dimension n (or m and m). In practice, the true error usually grows just linearly with n, or even slower, but we can generally prove only much weaker bounds of the form $p(n) = O(n^3)$. This is because we cannot rule out the extremely unlikely possibility of rounding errors all adding together instead of canceling on average. Using $p(n) = O(n^3)$ would give pessimistic and unrealistic bounds, especially for large n, so we content ourselves with describing $p(n)$ as a "modestly growing" polynomial function of n. Using $p(n) = 1$ in the error bound formulas will often give a reasonable error estimate. For detailed derivations of various $p(n)$, see [71, 114, 84, 38].

There is also one situation where $p(n)$ can grow as large as 2^{n-1}: Gaussian elimination. This typically occurs only on specially constructed matrices presented in numerical analysis courses [114, p. 212]. However, the expert driver for solving linear systems, PxGESVX, provides error bounds incorporating $p(n)$, and so this rare possibility can be detected.

6.4 Further Details: How Error Bounds Are Derived

6.4.1 Standard Error Analysis

We illustrate standard error analysis with the simple example of evaluating the scalar function $y = f(z)$. Let the output of the subroutine that implements $f(z)$ be denoted alg(z); this includes the effects of roundoff. If alg$(z) = f(z + \delta)$, where δ is small, we say that alg is a **backward stable** algorithm for f, or that the **backward error** δ is small. In other words, alg(z) is the exact value of f at a slightly perturbed input $z + \delta$.[6]

Suppose now that f is a smooth function, so that we may approximate it near z by a straight line: $f(z + \delta) \approx f(z) + f'(z) \cdot \delta$. Then we have the simple error estimate

$$\text{alg}(z) - f(z) = f(z + \delta) - f(z) \approx f'(z) \cdot \delta.$$

Thus, if δ is small and the derivative $f'(z)$ is moderate, the error alg$(z) - f(z)$ will be small.[7] This is often written in the similar form

$$\left| \frac{\text{alg}(z) - f(z)}{f(z)} \right| \approx \left| \frac{f'(z) \cdot z}{f(z)} \right| \cdot \left| \frac{\delta}{z} \right| \equiv \kappa(f, z) \cdot \left| \frac{\delta}{z} \right|.$$

This approximately bounds the **relative error** $\left| \frac{\text{alg}(z) - f(z)}{f(z)} \right|$ by the product of the **condition number of f at z**, $\kappa(f, z)$, and the **relative backward error** $\left| \frac{\delta}{z} \right|$. Thus we get an error bound by multiplying a condition number and a backward error (or bounds for these quantities). We call a problem **ill-conditioned** if its condition number is large, and **ill-posed** if its condition number is infinite (or does not exist).[8]

If f and z are vector quantities, then $f'(z)$ is a matrix (the Jacobian). Hence, instead of using absolute values as before, we now measure δ by a vector norm $\|\delta\|$ and $f'(z)$ by a matrix norm $\|f'(z)\|$. The conventional (and coarsest) error analysis uses a norm such as the infinity norm. We therefore call this **normwise backward stability**. For example, a normwise stable method for solving a system of linear equations $Ax = b$ will produce a solution $\hat{x}$ satisfying $(A + E)\hat{x} = b + f$, where $\|E\|_\infty/\|A\|_\infty$ and $\|f\|_\infty/\|b\|_\infty$ are both small (close to machine epsilon). In this case the condition number is $\kappa_\infty(A) = \|A\|_\infty \cdot \|A^{-1}\|_\infty$ (see section 6.5).

---

[6] Sometimes our algorithms satisfy only alg$(z) = f(z + \delta) + \eta$ where both δ and η are small. This does not significantly change the following analysis.

[7] More generally, we need only Lipschitz continuity of f and may use the Lipschitz constant in place of f' in deriving error bounds.

[8] This is a different use of the term ill-posed from that used in other contexts. For example, to be well-posed (not ill-posed) in the sense of Hadamard, it is sufficient for f to be continuous, whereas we require Lipschitz continuity.

Almost all of the algorithms in ScaLAPACK (as well as LAPACK) are stable in the sense just described[9]: when applied to a matrix A they produce the exact result for a slightly different matrix $A + E$, where $\|E\|_\infty / \|A\|_\infty$ is of order ϵ.

Condition numbers may be expensive to compute exactly. For example, it costs about $\frac{2}{3}n^3$ operations to solve $Ax = b$ for a general matrix A, and computing $\kappa_\infty(A)$ *exactly* costs an additional $\frac{4}{3}n^3$ operations, or twice as much. But $\kappa_\infty(A)$ can be *estimated* in only $O(n^2)$ operations beyond those $\frac{2}{3}n^3$ necessary for solution, a tiny extra cost. Therefore, most of ScaLAPACK's condition numbers and error bounds are based on estimated condition numbers, using the method of [72, 80, 81].

The price one pays for using an estimated rather than an exact condition number is occasional (but very rare) underestimates of the true error; years of experience attest to the reliability of our estimators, although examples where they badly underestimate the error can be constructed [82]. Note that once a condition estimate is large enough (usually $O(1/\epsilon)$), it confirms that the computed answer may be completely inaccurate, and so the exact magnitude of the condition estimate conveys little information.

6.4.2   Improved Error Bounds

The standard error analysis just outlined has a drawback: by using the infinity norm $\|\delta\|_\infty$ to measure the backward error, entries of equal magnitude in δ contribute equally to the final error bound $\kappa(f, z)(\|\delta\|/\|z\|)$. This means that if z is sparse or has some tiny entries, a normwise backward stable algorithm may make large changes in these entries compared wit their original values. If these tiny values are known accurately by the user, these errors may be unacceptable, or the error bounds may be unacceptably large.

For example, consider solving a diagonal system of linear equations $Ax = b$. Each component of the solution is computed accurately by Gaussian elimination: $x_i = b_i / a_{ii}$. The usual error bound is approximately $\epsilon \cdot \kappa_\infty(A) = \epsilon \cdot \max_i |a_{ii}| / \min_i |a_{ii}|$, which can arbitrarily overestimate the true error, ϵ, if at least one a_{ii} is tiny and another one is large.

LAPACK addresses this inadequacy by providing some algorithms whose backward error δ is a tiny relative change in each component of z: $|\delta_i| = O(\epsilon)|z_i|$. This backward error retains both the sparsity structure of z as well as the information in tiny entries. These algorithms are therefore called **componentwise relatively backward stable**. Furthermore, computed error bounds reflect this stronger form of backward error.[10]

If the input data has independent uncertainty in each component, each component must have at

---

[9]There are some caveats to this statement. When computing the inverse of a matrix, the backward error E is small, taking the columns of the computed inverse one at a time, with a different E for each column [62]. The same is true when computing the eigenvectors of a nonsymmetric matrix. When computing the eigenvalues and eigenvectors of $A - \lambda B$, $AB - \lambda I$ or $BA - \lambda I$, with A symmetric and B symmetric and positive definite (using PxSYGVX or PxHEGVX), the method may not be backward normwise stable if B has a large condition number $\kappa_\infty(B)$, although it has useful error bounds in this case too (see section 6.9). Solving the Sylvester equation $AX + XB = C$ for the matrix X may not be backward stable, although there are again useful error bounds for X [83].

[10]For other algorithms, the answers (and computed error bounds) are as accurate as though the algorithms were componentwise relatively backward stable, even though they are not. These algorithms are called *componentwise relatively forward stable.*

least a small *relative* uncertainty, since each is a floating-point number. In this case, the extra uncertainty contributed by the algorithm is not much worse than the uncertainty in the input data, so one could say the answer provided by a componentwise relatively backward stable algorithm is as accurate as the data warrants [4].

When solving $Ax = b$ using expert driver PxyySVX or computational routine PxyyRFS, for example, we almost always compute $\hat{x}$ satisfying $(A + E)\hat{x} = b + f$, where e_{ij} is a small relative change in a_{ij} and f_k is a small relative change in b_k. In particular, if A is diagonal, the corresponding error bound is always tiny, as one would expect (see the next section).

ScaLAPACK can achieve this accuracy for linear equation solving, the bidiagonal singular value decomposition, and the symmetric tridiagonal eigenproblem and provides facilities for achieving this accuracy for least squares problems.

6.5 Error Bounds for Linear Equation Solving

Let $Ax = b$ be the system to be solved, and $\hat{x}$ the computed solution. Let n be the dimension of A. An approximate error bound for $\hat{x}$ may be obtained in one of the following two ways, depending on whether the solution is computed by a simple driver or an expert driver:

1. Suppose that $Ax = b$ is solved using the simple driver PSGESV (section 3.2.1). Then the approximate error bound[11]
$$\frac{\|\hat{x} - x\|_\infty}{\|x\|_\infty} \leq \text{ERRBD}$$
can be computed by the following code fragment.

```
      EPSMCH = PSLAMCH( ICTXT, 'E' )
*     Get infinity-norm of A
      ANORM = PSLANGE( 'I', N, N, A, IA, JA, DESCA, WORK )
*     Solve system; The solution X overwrites B
      CALL PSGESV( N, 1, A, IA, JA, DESCA, IPIV, B, IB, JB, DESCB, INFO )
      IF( INFO.GT.0 ) THEN
         PRINT *,'Singular Matrix'
      ELSE IF( N.GT.0 ) THEN
*        Get reciprocal condition number RCOND of A
         CALL PSGECON( 'I', N, A, IA, JA, DESCA, ANORM, RCOND, WORK,
     $                 LWORK, IWORK, LIWORK, INFO )
         RCOND = MAX( RCOND, EPSMCH )
         ERRBD = EPSMCH / RCOND
      END IF
```

---

[11] As discussed in section 6.3, this approximate error bound may underestimate the true error by a factor $p(n)$, which is a modestly growing function of the problem dimension n. Often $p(n) \leq 10n$.

For example, suppose $\texttt{PSLAMCH(ICTXT,'E')} = 2^{-24} = 5.961 \cdot 10^{-8}$,

$$A = \begin{pmatrix} 4 & 16000 & 17000 \\ 2 & 5 & 8 \\ 3 & 6 & 10 \end{pmatrix} \text{ and } b = \begin{pmatrix} 100.1 \\ .1 \\ .01 \end{pmatrix}.$$

Then (to 4 decimal places)

$$x = \begin{pmatrix} -.3974 \\ -.3349 \\ .3211 \end{pmatrix}, \quad \hat{x} = \begin{pmatrix} -.3968 \\ -.3344 \\ .3207 \end{pmatrix},$$

$\texttt{ANORM} = 3.300 \cdot 10^{4}$, $\texttt{RCOND} = 3.907 \cdot 10^{-6}$, the true reciprocal condition number $= 3.902 \cdot 10^{-6}$, $\texttt{ERRBD} = 1.5 \cdot 10^{-2}$, and the true error $= 1.5 \cdot 10^{-3}$.

2. Suppose that $Ax = b$ is solved using the expert driver PSGESVX (section 3.2.1). This routine provides an explicit error bound FERR, measured with the infinity-norm:

$$\frac{\|\hat{x} - x\|_\infty}{\|x\|_\infty} \leq \texttt{FERR}$$

For example, the following code fragment solves $Ax = b$ and computes an approximate error bound FERR:

```
      CALL PSGESVX( 'E', 'N', N, 1, A, IA, JA, DESCA, AF, IAF, JAF,
     $              DESCAF, IPIV, EQUED, R, C, B, IB, JB, DESCB, X, IX,
     $              JX, DESCX, RCOND, FERR, BERR, WORK, LWORK, IWORK,
     $              LIWORK, INFO )
      IF( INFO.GT.0 ) PRINT *,'(Nearly) Singular Matrix'
```

For the same A and b as above, $\hat{x} = \begin{pmatrix} -.3974 \\ -.3349 \\ .3211 \end{pmatrix}$, $\texttt{FERR} = 3.0 \cdot 10^{-5}$, and the actual error is $4.3 \cdot 10^{-7}$.

This example illustrates that the expert driver provides an error bound with less programming effort than the simple driver, and also that it may produce a significantly more accurate answer.

Similar code fragments, with obvious adaptations, may be used with all the driver routines PxPOSV and PxPOSVX in Table 3.2. For example, if a symmetric positive definite or Hermitian positive definite system is solved by using the simple driver PxPOSV, then PxLANSY or PxLANHE, respectively, must be used to compute ANORM, and PxPOCON must be used to compute RCOND.

The drivers PxGBSV (for solving general band matrices with partial pivoting), PxPBSV (for solving positive definite band matrices) and PxPTSV (for solving positive definite tridiagonal matrices), do not yet have the corresponding routines needed to compute error bounds, namely, PxLAnHE to compute ANORM and PxyyCON to compute RCOND.

The drivers PxDBSV (for solving general band matrices) and PxDTSV (for solving general tridiagonal matrices) do not pivot for numerical stability, and so may be faster but less accurate than their pivoting counterparts above. These routines may be used safely when any diagonal pivot sequence leads to a stable factorization; diagonally dominant matrices and symmetric positive definite matrices [71] have this property, for example.

Further Details: Error Bounds for Linear Equation Solving

The conventional error analysis of linear equation solving goes as follows. Let $Ax = b$ be the system to be solved. Let $\hat{x}$ be the solution computed by ScaLAPACK (or LAPACK) using any of their linear equation solvers. Let r be the residual $r = b - A\hat{x}$. In the absence of rounding error, r would be zero and $\hat{x}$ would equal x; with rounding error, one can only say the following:

The normwise backward error of the computed solution $\hat{x}$, with respect to the infinity norm, is the pair E, f, which minimizes

$$\max\left(\frac{\|E\|_\infty}{\|A\|_\infty}, \frac{\|f\|_\infty}{\|b\|_\infty}\right)$$

subject to the constraint $(A + E)\hat{x} = b + f$. The minimal value of $\max\left(\frac{\|E\|_\infty}{\|A\|_\infty}, \frac{\|f\|_\infty}{\|b\|_\infty}\right)$ is given by

$$\omega_\infty = \frac{\|r\|_\infty}{\|A\|_\infty \cdot \|\hat{x}\|_\infty + \|b\|_\infty} .$$

One can show that the computed solution $\hat{x}$ satisfies $\omega_\infty \leq p(n) \cdot \epsilon$, where $p(n)$ is a modestly growing function of n. The corresponding condition number is $\kappa_\infty(A) \equiv \|A\|_\infty \cdot \|A^{-1}\|_\infty$. The error $x - \hat{x}$ is bounded by

$$\frac{\|x - \hat{x}\|_\infty}{\|x\|_\infty} \lesssim 2 \cdot \omega_\infty \cdot \kappa_\infty(A) = \texttt{ERRBD} .$$

In the first code fragment in the preceding section, $2 \cdot \omega_\infty$, which is $4.504 \cdot 10^{-8}$ in the numerical example, is approximated by $\epsilon = 2^{-24} = 5.960 \cdot 10^{-8}$. Approximations of $\kappa_\infty(A)$ — or, strictly speaking, its reciprocal RCOND — are returned by computational routines PxyyCON (section 3.3.1) or driver routines PxyySVX (section 3.2.1). The code fragment makes sure RCOND is at least $\epsilon = $ EPSMCH to avoid overflow in computing ERRBD. This limits ERRBD to a maximum of 1, which is no loss of generality because a relative error of 1 or more indicates the same thing: a complete loss of accuracy. Note that the value of RCOND returned by PxyySVX may apply to a linear system obtained from $Ax = b$ by *equilibration*, namely, scaling the rows and columns of A in order to make the condition number smaller. This is the case in the second code fragment in the preceding section, where the program chose to scale the rows by the factors returned in R $= (5.882 \cdot 10^{-5}, .125, .1)$, resulting in RCOND $= 3.454 \cdot 10^{-3}$.

As stated in section 6.4.2, this approach does not respect the presence of zero or tiny entries in A. In contrast, the ScaLAPACK computational routines PxyyRFS (section 3.3.1) or driver routines PxyySVX (section 3.2.1) will (except in rare cases) compute a solution $\hat{x}$ with the following properties:

The componentwise backward error of the computed solution $\hat{x}$ is the pair E, f which minimizes

$$\max_{i,j,k} \left(\frac{|e_{ij}|}{|a_{ij}|}, \frac{|f_k|}{|b_k|} \right)$$

(where we interpret 0/0 as 0) subject to the constraint $(A + E)\hat{x} = b + f$. The minimal value of $\max_{i,j,k} \left(\frac{|e_{ij}|}{|a_{ij}|}, \frac{|f_k|}{|b_k|} \right)$ is given by

$$\omega_c = \max_i \frac{|r_i|}{(|A| \cdot |\hat{x}| + |b|)_i} \ .$$

One can show that for most problems the $\hat{x}$ computed by PxyySVX satisfies $\omega_c \leq p(n) \cdot \epsilon$, where $p(n)$ is a modestly growing function of n. In other words, $\hat{x}$ is the exact solution of the perturbed problem $(A + E)\hat{x} = b + f$, where E and f are small relative perturbations in each entry of A and b, respectively. The corresponding condition number is $\kappa_c(A, b, \hat{x}) \equiv \| |A^{-1}|(|A| \cdot |\hat{x}| + |b|) \|_\infty / \|\hat{x}\|_\infty$. The error $x - \hat{x}$ is bounded by

$$\frac{\|x - \hat{x}\|_\infty}{\|\hat{x}\|_\infty} \leq \omega_c \cdot \kappa_c(A, b, \hat{x}).$$

The routines PxyyRFS and PxyySVX return ω_c,   which is called BERR (for Backward ERRor), and a bound on the the actual error $\|x - \hat{x}\|_\infty / \|\hat{x}\|_\infty$, called FERR (for Forward ERRor), as in the second code fragment in the last section. FERR is actually calculated by the following formula, which can be smaller than the bound $\omega_c \cdot \kappa_c(A, b, \hat{x})$ given above:

$$\frac{\|x - \hat{x}\|_\infty}{\|\hat{x}\|_\infty} \leq \text{FERR} = \frac{\| |A^{-1}|(|\hat{r}| + n\epsilon(|A| \cdot |\hat{x}| + |b|)) \|_\infty}{\|\hat{x}\|_\infty} \ .$$

Here, $\hat{r}$ is the computed value of the residual $b - A\hat{x}$, and the norm in the numerator is estimated by using the same estimation subroutine used for RCOND.

The value of BERR for the example in the preceding section is $4.6 \cdot 10^{-8}$.

Even in the rare cases where PxyyRFS fails to make BERR close to its minimum ϵ, the error bound FERR may remain small. See [9] for details.

6.6   Error Bounds for Linear Least Squares Problems

The linear least squares problem is to find x that minimizes $\|Ax - b\|_2$. We discuss error bounds for the most common case where A is m-by-n with $m > n$, and A has full rank; this is called an *overdetermined least squares problem* (the following code fragments deal with $m = n$ as well).

Let $\hat{x}$ be the solution computed by the driver routine PxGELS (see section 3.2.2). An approximate error bound[11]

$$\frac{\|\hat{x} - x\|_2}{\|x\|_2} \lesssim \text{ERRBD}$$

may be computed in the following way:

```
      EPSMCH = PSLAMCH( ICTXT, 'E' )
*     Get the 2-norm of the right hand side B
      BNORM = PSLANGE( 'F', M, 1, B, IB, JB, DESCB, WORK )
*     Solve the least squares problem; the solution X overwrites B
      CALL PSGELS( 'N', M, N, 1, A, IA, JA, DESCA, B, IB, JB, DESCB,
     $             WORK, LWORK, INFO )
      IF( MIN( M, N ).GT.0 ) THEN
*        Get the 2-norm of the residual A*X-B
         RNORM = PSLANGE( 'F', M-N, 1, B, IB+N, JB, DESCB, WORK )
*        Get the reciprocal condition number RCOND of A
         CALL PSTRCON( 'I', 'U', 'N', N, A, IA, JA, DESCA, RCOND, WORK,
     $                 LWORK, IWORK, LIWORK, INFO )
         RCOND = MAX( RCOND, EPSMCH )
         IF( BNORM.GT.0.0 ) THEN
            SINT = RNORM / BNORM
         ELSE
            SINT = 0.0
         END IF
         COST = MAX( SQRT( ( 1.0E0 - SINT )*( 1.0E0 + SINT ) ), EPSMCH )
         TANT = SINT / COST
         ERRBD = EPSMCH*( 2.0E0 / ( RCOND*COST ) + TANT / RCOND**2 )
      END IF
```

For example, if $\text{PSLAMCH}(\text{ICTXT},'E') = 2^{-24} = 5.961 \cdot 10^{-8}$,

$$A = \begin{pmatrix} 4 & 3 & 5 \\ 2 & 5 & 8 \\ 3 & 6 & 10 \\ 4 & 5 & 11 \end{pmatrix} \text{ and } b = \begin{pmatrix} 100.1 \\ .1 \\ .01 \\ .01 \end{pmatrix},$$

then, to four decimal places,

$$x = \hat{x} = \begin{pmatrix} 38.49 \\ 21.59 \\ -23.88 \end{pmatrix},$$

BNORM = 100.1, RNORM = 8.843, RCOND = $4.712 \cdot 10^{-2}$, ERRBD = $4.9 \cdot 10^{-6}$, and the true error is $4.6 \cdot 10^{-7}$.

Note that in the preceding code fragment, the routine PSLANGE was used to compute the two-norm of the right hand side matrix B and the residual $A * X - B$. This routine was chosen because the result of the computation (BNORM or RNORM, respectively) is automatically known on all process columns within the process grid. The routine PSNRM2 could have also been used to perform this calculation; however, the use of PSNRM2 in this example would have required an additional communication broadcast, because the resulting value of BNORM or RNORM, respectively, is known only within the process column owning B(:,JB).

Further Details: Error Bounds for Linear Least Squares Problems

The conventional error analysis of linear least squares problems goes as follows. As above, let $\hat{x}$ be the solution to minimizing $\|Ax - b\|_2$ computed by ScaLAPACK using the least squares driver PxGELS (see section 3.2.2). We discuss the most common case, where A is overdetermined (i.e., has more rows than columns) and has full rank [71]:

The computed solution $\hat{x}$ has a small normwise backward error. In other words, $\hat{x}$ minimizes $\|(A + E)\hat{x} - (b + f)\|_2$, where E and f satisfy

$$\max\left(\frac{\|E\|_2}{\|A\|_2}, \frac{\|f\|_2}{\|b\|_2}\right) \leq p(n)\epsilon$$

and $p(n)$ is a modestly growing function of n. We take $p(n) = 1$ in the code fragments above. Let $\kappa_2(A) = \sigma_{\max}(A)/\sigma_{\min}(A)$ (approximated by 1/RCOND in the above code fragments), $\rho = \|A\hat{x} - b\|_2$ (= RNORM above), and $\sin(\theta) = \rho/\|b\|_2$ (SINT = RNORM / BNORM above). Here, θ is the acute angle between the vectors $A\hat{x}$ and b. Then when $p(n)\epsilon$ is small, the error $\hat{x} - x$ is bounded by

$$\frac{\|x - \hat{x}\|_2}{\|x\|_2} \lesssim p(n)\epsilon \left\{\frac{2\kappa_2(A)}{\cos(\theta)} + \tan(\theta)\kappa_2^2(A)\right\},$$

where $\cos(\theta) = $ COST and $\tan(\theta) = $ TANT in the code fragments above.

We avoid overflow by making sure RCOND and COST are both at least $\epsilon = $ EPSMCH, and by handling the case of a zero B matrix separately (BNORM = 0).

$\kappa_2(A) = \sigma_{\max}(A)/\sigma_{\min}(A)$ may be computed directly from the singular values of A returned by PxGESVD. It may also be approximated by using PxTRCON following calls to PxGELS. PxTRCON estimates κ_∞ or κ_1 instead of κ_2, but these can differ from κ_2 by at most a factor of n.

6.7 Error Bounds for the Symmetric Eigenproblem

The eigendecomposition of an n-by-n real symmetric matrix is the factorization $A = Z\Lambda Z^T$ ($A = Z\Lambda Z^H$ in the complex Hermitian case), where Z is orthogonal (unitary) and $\Lambda = \text{diag}(\lambda_1, \ldots, \lambda_n)$ is real and diagonal, with $\lambda_1 \leq \lambda_2 \leq \cdots \leq \lambda_n$. The λ_i are the **eigenvalues** of A, and the columns z_i of Z are the **eigenvectors**. This is also often written $Az_i = \lambda_i z_i$. The eigendecomposition of a symmetric matrix is computed by the driver routines PxSYEV and PxSYEVX. The complex counterparts of these routines, which compute the eigendecomposition of complex Hermitian matrices, are the driver routines PxHEEV and PxHEEVX (see section 3.2.3).

The approximate error bounds[11] for the computed eigenvalues $\hat{\lambda}_1 \leq \cdots \leq \hat{\lambda}_n$ are

$$|\hat{\lambda}_i - \lambda_i| \leq \text{EERRBD} .$$

The approximate error bounds for the computed eigenvectors $\hat{z}_i$, which bound the acute angles between the computed eigenvectors and true eigenvectors z_i, are

$$\theta(\hat{z}_i, z_i) \leq \text{ZERRBD}(i) .$$

These bounds can be computed by the following code fragment:

```
        EPSMCH = PSLAMCH( ICTXT, 'E' )
*       Compute eigenvalues and eigenvectors of A
*       The eigenvalues are returned in W
*       The eigenvector matrix Z overwrites A
        CALL PSSYEV( 'V', UPLO, N, A, IA, JA, DESCA, W, Z, IZ, JZ,
     $               DESCZ, WORK, LWORK, INFO )
        IF( INFO.GT.0 ) THEN
           PRINT *,'PSSYEV did not converge'
        ELSE IF( N.GT.0 ) THEN
*          Compute the norm of A
           ANORM = MAX( ABS( W( 1 ) ), ABS( W( N ) ) )
           EERRBD = EPSMCH * ANORM
*          Compute reciprocal condition numbers for eigenvectors
           CALL SDISNA( 'Eigenvectors', N, N, W, RCONDZ, INFO )
           DO 10 I = 1, N
              ZERRBD( I ) = EPSMCH * ( ANORM / RCONDZ( I ) )
10         CONTINUE
        END IF
```

For example, if $\texttt{PSLAMCH}(\texttt{ICTXT},'\texttt{E}') = 2^{-24} = 5.961 \cdot 10^{-8}$ and

$$A = \begin{pmatrix} 1 & 2 & 3 \\ 2 & 4 & 5 \\ 3 & 5 & 6 \end{pmatrix},$$

then the eigenvalues, approximate error bounds, and true errors are as follows.

i	$\hat{\lambda}_i$	EERRBD	true $\lvert\hat{\lambda}_i - \lambda_i\rvert$	ZERRBD(i)	true $\theta(\hat{z}_i, z_i)$
1	$-.5157$	$6.7 \cdot 10^{-7}$	$1.6 \cdot 10^{-7}$	$9.8 \cdot 10^{-7}$	$1.2 \cdot 10^{-7}$
2	$.1709$	$6.7 \cdot 10^{-7}$	$3.2 \cdot 10^{-7}$	$9.8 \cdot 10^{-7}$	$7.0 \cdot 10^{-8}$
3	11.34	$6.7 \cdot 10^{-7}$	$2.8 \cdot 10^{-6}$	$6.1 \cdot 10^{-8}$	$9.7 \cdot 10^{-8}$

Further Details: Error Bounds for the Symmetric Eigenproblem

The usual error analysis of the symmetric eigenproblem using ScaLAPACK driver PSSYEV (see subsection 3.2.3) is as follows [101]:

The computed eigendecomposition $\hat{Z}\hat{\Lambda}\hat{Z}^T$ is nearly the exact eigendecomposition of $A + E$, namely, $A + E = (\hat{Z} + \delta\hat{Z})\hat{\Lambda}(\hat{Z} + \delta\hat{Z})^T$ is a true eigendecomposition so that $\hat{Z} + \delta\hat{Z}$ is orthogonal, where $\|E\|_2/\|A\|_2 \leq p(n)\epsilon$ and $\|\delta\hat{Z}\|_2 \leq p(n)\epsilon$. Here $p(n)$ is a modestly growing function of n. We take $p(n) = 1$ in the above code fragment. Each computed eigenvalue $\hat{\lambda}_i$ differs from a true λ_i by at most

$$|\hat{\lambda}_i - \lambda_i| \leq p(n) \cdot \epsilon \cdot \|A\|_2 = \text{EERRBD} .$$

Thus, large eigenvalues (those near $\max_i |\lambda_i| = \|A\|_2$) are computed to high relative accuracy, while small ones may not be.

The angular difference between the computed unit eigenvector $\hat{z}_i$ and a true unit eigenvector z_i satisfies the approximate bound

$$\theta(\hat{z}_i, z_i) \lesssim \frac{p(n)\epsilon \|A\|_2}{\text{gap}_i} = \texttt{ZERRBD}(i)$$

if $p(n)\epsilon$ is small enough. Here, $\text{gap}_i = \min_{j \neq i} |\lambda_i - \lambda_j|$ is the **absolute gap** between λ_i and the nearest other eigenvalue. Thus, if λ_i is close to other eigenvalues, its corresponding eigenvector z_i may be inaccurate. The gaps may be easily computed from the array of computed eigenvalues by using subroutine `SDISNA`. The gaps computed by `SDISNA` are ensured not to be so small as to cause overflow when used as divisors.

Let $\hat{S}$ be the invariant subspace spanned by a collection of eigenvectors $\{\hat{z}_i,\, i \in \mathcal{I}\}$, where $\mathcal{I}$ is a subset of the integers from 1 to n. Let S be the corresponding true subspace. Then

$$\theta(\hat{S}, S) \lesssim \frac{p(n)\epsilon \|A\|_2}{\text{gap}_{\mathcal{I}}},$$

where

$$\text{gap}_{\mathcal{I}} = \min_{\substack{i \in \mathcal{I} \\ j \notin \mathcal{I}}} |\lambda_i - \lambda_j|$$

is the absolute gap between the eigenvalues in $\mathcal{I}$ and the nearest other eigenvalue. Thus, a cluster of close eigenvalues that is far away from any other eigenvalue may have a well-determined invariant subspace $\hat{S}$ even if its individual eigenvectors are ill-conditioned.

A small possibility exists that `PSSYEV` will fail to achieve the above error bounds on a heterogeneous network of processors for reasons discussed in section 6.2. On a homogeneous network, `PSSYEV` is as robust as the corresponding LAPACK routine `SSYEV`. A future release will attempt to detect heterogeneity and warn the user to use an alternative algorithm.

In contrast to LAPACK, where the same error analysis applied to the simple and expert drivers, the expert driver `PSSYEVX` satisfies slightly weaker error bounds than `PSSYEV`. The bounds $|\hat{\lambda}_i - \lambda_i| \leq$ `EERRBD` and $\theta(\hat{z}_i, z_i) \leq$ `ZERRBD` continue to hold, but the computed eigenvectors $\hat{z}_i$ are no longer guaranteed to be orthogonal to one another. The corresponding LAPACK routine `SSYEVX` tries to guarantee orthogonality by *reorthogonalizing* computed eigenvectors against one another provided their corresponding computed eigenvalues are close enough together: $|\hat{\lambda}_i - \hat{\lambda}_j| \leq$ `ORTOL` $\cdot \max_k |\hat{\lambda}_k|$, where the threshold `ORTOL` $= 10^{-3}$. If m eigenvalues lie in a cluster satisfying this closeness criterion, the `SSYEVX` requires $O(nm^2)$ serial time to execute. When m is a large fraction of n, this serial bottleneck is expensive and does not always improve orthogonality.

ScaLAPACK addresses this problem in two ways. First, it lets the user use more or less time and space to perform reorthogonalization, rather than have a fixed criterion. In particular, the user can set the threshold `ORTOL` used above, decreasing it to make reorthogonalization less frequent or increasing it to reorthogonalize more. Furthermore, since each processor computes a subset of the eigenvectors, ScaLAPACK permits reorthogonalization only with the local eigenvectors; that is, no communication is allowed. Hence, if a cluster of eigenvalues is small enough for the corresponding eigenvectors to fit on one processor, the same reorthogonalization will be done as in LAPACK. The user can supply more or less workspace to limit the size of a cluster on one processor. Hence, at one extreme, with a large cluster and lots of workspace, the algorithm will be essentially equivalent to SSYEVX. At the other extreme, with all small clusters or little workspace supplied, the algorithm will be perfectly load balanced and perform minimal communication to compute the eigenvectors.

The second way ScaLAPACK will deal with reorthogonalization is to introduce a new algorithm [103, 102, 44] that requires nearly *no* reorthogonalization to compute orthogonal eigenvectors in a fully parallel way. This algorithm will be introduced in future ScaLAPACK and LAPACK releases.

In the special case of a real symmetric tridiagonal matrix T, the eigenvalues can sometimes be computed much more accurately. PxSYEV (and the other symmetric eigenproblem drivers) computes the eigenvalues and eigenvectors of a dense symmetric matrix by first reducing it to tridiagonal form T and then finding the eigenvalues and eigenvectors of T. Reduction of a dense matrix to tridiagonal form T can introduce additional errors, so the following bounds for the tridiagonal case do not apply to the dense case.

> The eigenvalues of the symmetric tridiagonal matrix T may be computed with small componentwise relative backward error $(O(\epsilon))$ by using subroutine PxSTEBZ (section 3.3.4). To compute tighter error bounds for the computed eigenvalues $\hat{\lambda}_i$ we must make some assumptions about T. The bounds discussed here are from [15]. Suppose T is positive definite, and write $T = DHD$ where $D = \mathrm{diag}(t_{11}^{1/2}, \ldots, t_{nn}^{1/2})$ and $h_{ii} = 1$. Then the computed eigenvalues $\hat{\lambda}_i$ can differ from true eigenvalues λ_i by
>
> $$|\hat{\lambda}_i - \lambda_i| \leq p(n) \cdot \epsilon \cdot \kappa_2(H) \cdot \lambda_i,$$
>
> where $p(n)$ is a modestly growing function of n. Thus, if $\kappa_2(H)$ is moderate, each eigenvalue will be computed to high relative accuracy, no matter how tiny it is.

6.8 Error Bounds for the Singular Value Decomposition

The singular value decomposition (SVD) of a real m-by-n matrix A is defined as follows. Let $r = \min(m, n)$. The SVD of A is $A = U\Sigma V^T$ ($A = U\Sigma V^H$ in the complex case), where U and V are orthogonal (unitary) matrices and $\Sigma = \mathrm{diag}(\sigma_1, \ldots, \sigma_r)$ is diagonal, with $\sigma_1 \geq \sigma_2 \geq \cdots \geq \sigma_r \geq 0$. The σ_i are the **singular values** of A and the leading r columns u_i of U and v_i of V the **left and right singular vectors**, respectively. The SVD of a general matrix is computed by PxGESVD (see subsection 3.2.3).

The approximate error bounds[11] for the computed singular values $\hat{\sigma}_1 \geq \cdots \geq \hat{\sigma}_r$ are

$$|\hat{\sigma}_i - \sigma_i| \leq \texttt{SERRBD} \ .$$

The approximate error bounds for the computed singular vectors $\hat{v}_i$ and $\hat{u}_i$, which bound the acute angles between the computed singular vectors and true singular vectors v_i and u_i, are

$$\theta(\hat{v}_i, v_i) \ \leq \ \texttt{VERRBD}(i)$$
$$\theta(\hat{u}_i, u_i) \ \leq \ \texttt{UERRBD}(i) \ .$$

These bounds can be computing by the following code fragment:

```
      EPSMCH = PSLAMCH( ICTXT, 'E' )
*     Compute singular value decomposition of A
*     The singular values are returned in S
*     The left singular vectors are returned in U
*     The transposed right singular vectors are returned in VT
      CALL PSGESVD( 'V', 'V', M, N, A, IA, JA, DESCA, S, U, IU, JU,
     $                  DESCU, VT, IVT, JVT, DESCVT, WORK, LWORK, INFO )
      IF( INFO.GT.0 ) THEN
          PRINT *,'PSGESVD did not converge'
      ELSE IF( MIN( M, N ).GT.0 ) THEN
          SERRBD  = EPSMCH * S( 1 )
*         Compute reciprocal condition numbers for singular vectors
          CALL SDISNA( 'Left', M, N, S, RCONDU, INFO )
          CALL SDISNA( 'Right', M, N, S, RCONDV, INFO )
          DO 10 I = 1, MIN( M, N )
             VERRBD( I ) = EPSMCH*( S( 1 ) / RCONDV( I ) )
             UERRBD( I ) = EPSMCH*( S( 1 ) / RCONDU( I ) )
   10     CONTINUE
      END IF
```

For example, if $\text{PSLAMCH}(\text{ICTXT},'E') = 2^{-24} = 5.961 \cdot 10^{-8}$ and

$$A = \begin{pmatrix} 4 & 3 & 5 \\ 2 & 5 & 8 \\ 3 & 6 & 10 \\ 4 & 5 & 11 \end{pmatrix},$$

then the singular values, approximate error bounds, and true errors are given below.

i	$\hat{\sigma}_i$	SERRBD	true $\lvert\hat{\sigma}_i - \sigma_i\rvert$	VERRBD(i)	true $\theta(\hat{v}_i, v_i)$	UERRBD(i)	true $\theta(\hat{u}_i, u_i)$
1	21.05	$1.3 \cdot 10^{-6}$	$1.7 \cdot 10^{-6}$	$6.7 \cdot 10^{-8}$	$8.1 \cdot 10^{-8}$	$6.7 \cdot 10^{-8}$	$1.5 \cdot 10^{-7}$
2	2.370	$1.3 \cdot 10^{-6}$	$5.8 \cdot 10^{-7}$	$1.0 \cdot 10^{-6}$	$2.9 \cdot 10^{-7}$	$1.0 \cdot 10^{-6}$	$2.4 \cdot 10^{-7}$
3	1.143	$1.3 \cdot 10^{-6}$	$3.2 \cdot 10^{-7}$	$1.0 \cdot 10^{-6}$	$3.0 \cdot 10^{-7}$	$1.1 \cdot 10^{-6}$	$2.4 \cdot 10^{-7}$

6.8.1 Further Details: Error Bounds for the Singular Value Decomposition

The usual error analysis of the SVD algorithm PxGESVD in ScaLAPACK (see subsection 3.2.3)is as follows [71]:

The SVD algorithm is backward stable. This means that the computed SVD, $\hat{U}\hat{\Sigma}\hat{V}^T$, is nearly the exact SVD of $A + E$ where $\lVert E \rVert_2 / \lVert A \rVert_2 \le p(m,n)\epsilon$, and $p(m,n)$ is a modestly growing function of m and n. This means $A + E = (\hat{U} + \delta\hat{U})\hat{\Sigma}(\hat{V} + \delta\hat{V})$ is the true

SVD, so that $\hat{U} + \delta\hat{U}$ and $\hat{V} + \delta\hat{V}$ are both orthogonal, where $\|\delta\hat{U}\| \leq p(m,n)\epsilon$, and $\|\delta\hat{V}\| \leq p(m,n)\epsilon$. Each computed singular value $\hat{\sigma}_i$ differs from true σ_i by at most

$$|\hat{\sigma}_i - \sigma_i| \leq p(m,n) \cdot \epsilon \cdot \sigma_1 = \texttt{SERRBD},$$

(we take $p(m,n) = 1$ in the above code fragment). Thus, large singular values (those near σ_1) are computed to high relative accuracy and small ones may not be.

The angular difference between the computed left singular vector $\hat{u}_i$ and a true u_i satisfies the approximate bound

$$\theta(\hat{u}_i, u_i) \lesssim \frac{p(m,n)\epsilon\|A\|_2}{\text{gap}_i} = \texttt{UERRBD}(i)$$

where $\text{gap}_i = \min_{j \neq i} |\sigma_i - \sigma_j|$ is the **absolute gap** between σ_i and the nearest other singular value (we take $p(m,n) = 1$ in the above code fragment). Thus, if σ_i is close to other singular values, its corresponding singular vector u_i may be inaccurate. When $n < m$, then gap_n must be redefined as $\min(\min_{j \neq n}(|\sigma_n - \sigma_j|, \sigma_n))$. The gaps may be easily computed from the array of computed singular values using function SDISNA. The gaps computed by SDISNA are ensured not to be so small as to cause overflow when used as divisors. The same bound applies to the computed right singular vector $\hat{v}_i$ and a true vector v_i.

Let $\hat{\mathcal{S}}$ be the space spanned by a collection of computed left singular vectors $\{\hat{u}_i\,, i \in \mathcal{I}\}$, where $\mathcal{I}$ is a subset of the integers from 1 to n. Let $\mathcal{S}$ be the corresponding true space. Then

$$\theta(\hat{\mathcal{S}}, \mathcal{S}) \lesssim \frac{p(m,n)\epsilon\|A\|_2}{\text{gap}_{\mathcal{I}}}.$$

where

$$\text{gap}_{\mathcal{I}} = \min_{\substack{i \in \mathcal{I} \\ j \notin \mathcal{I}}} |\sigma_i - \sigma_j|$$

is the absolute gap between the singular values in $\mathcal{I}$ and the nearest other singular value. Thus, a cluster of close singular values which is far away from any other singular value may have a well determined space $\hat{\mathcal{S}}$ even if its individual singular vectors are ill-conditioned. The same bound applies to a set of right singular vectors $\{\hat{v}_i\,, i \in \mathcal{I}\}$[14].

There is a small possibility that PxGESVD will fail to achieve the above error bounds on a heterogeneous network of processors for reasons discussed in section 6.2. On a homogeneous network, PxGESVD is as robust as the corresponding LAPACK routine xGESVD. A future release will attempt to detect heterogeneity and warn the user to use an alternative algorithm.

In the special case of bidiagonal matrices, the singular values and singular vectors may be computed much more accurately. A bidiagonal matrix B has nonzero entries only on the main diagonal and the diagonal immediately above it (or immediately below it). PxGESVD computes the SVD of a

---

[14]These bounds are special cases of those in section 6.7 since the singular values and vectors of A are simply related to the eigenvalues and eigenvectors of the Hermitian matrix $\begin{pmatrix} 0 & A^H \\ A & 0 \end{pmatrix}$ [71, p. 427].

general matrix by first reducing it to bidiagonal form B, and then calling xBDSQR (subsection 3.3.6) to compute the SVD of B. Reduction of a dense matrix to bidiagonal form B can introduce additional errors, so the following bounds for the bidiagonal case do not apply to the dense case. For the error analysis of xBDSQR, see the LAPACK manual.

6.9   Error Bounds for the Generalized Symmetric Definite Eigenproblem

Three types of problems must be considered. In all cases A and B are real symmetric (or complex Hermitian) and B is positive definite. These decompositions are computed for real symmetric matrices by the driver routines PxSYGVX (see section 3.2.4). These decompositions are computed for complex Hermitian matrices by the driver routines PxHEGVX (see subsection 3.2.4). In each of the following three decompositions, Λ is real and diagonal with diagonal entries $\lambda_1 \leq \cdots \leq \lambda_n$, and the columns z_i of Z are linearly independent vectors. The λ_i are called **eigenvalues** and the z_i are **eigenvectors**.

1. $A - \lambda B$. The eigendecomposition may be written $Z^T A Z = \Lambda$ and $Z^T B Z = I$ (or $Z^H A Z = \Lambda$ and $Z^H B Z = I$ if A and B are complex). This may also be written $Az_i = \lambda_i B z_i$.

2. $AB - \lambda I$. The eigendecomposition may be written $Z^{-1} A Z^{-T} = \Lambda$ and $Z^T B Z = I$ ($Z^{-1} A Z^{-H} = \Lambda$ and $Z^H B Z = I$ if A and B are complex). This may also be written $ABz_i = \lambda_i z_i$.

3. $BA - \lambda I$. The eigendecomposition may be written $Z^T A Z = \Lambda$ and $Z^T B^{-1} Z = I$ ($Z^H A Z = \Lambda$ and $Z^H B^{-1} Z = I$ if A and B are complex). This may also be written $BAz_i = \lambda_i z_i$.

The approximate error bounds[11] for the computed eigenvalues $\hat{\lambda}_1 \leq \cdots \leq \hat{\lambda}_n$ are

$$|\hat{\lambda}_i - \lambda_i| \leq \text{EERRBD}(i) .$$

The approximate error bounds for the computed eigenvectors $\hat{z}_i$, which bound the acute angles between the computed eigenvectors and true eigenvectors z_i, are

$$\theta(\hat{z}_i, z_i) \leq \text{ZERRBD}(i) .$$

These bounds are computed differently, depending on which of the above three problems are to be solved. The following code fragments show how.

1. First we consider error bounds for problem 1.

```
        EPSMCH = PSLAMCH( ICTXT, 'E' )
        UNFL = PSLAMCH( ICTXT, 'U' )
*       Solve the eigenproblem A - lambda B (ITYPE = 1)
        ITYPE = 1
*       Compute the norms of A and B
```

```
              ANORM = PSLANSY( '1', UPLO, N, A, IA, JA, DESCA, WORK )
              BNORM = PSLANSY( '1', UPLO, N, B, IB, JB, DESCB, WORK )
*        The eigenvalues are returned in W
*        The eigenvectors are returned in A
              SUBROUTINE PSSYGVX( ITYPE, 'V', 'A', UPLO, N, A, IA, JA,
     $                           DESCA, B, IB, JB, DESCB, VL, VU, IL, IU,
     $                           UNFL, M, NZ, W, -1.0, Z, IZ, JZ, DESCZ,
     $                           WORK, LWORK, IWORK, LIWORK, IFAIL, ICLUSTR,
     $                           GAP, INFO )
         IF( INFO.GT.0 ) THEN
            PRINT *,'PSSYGVX did not converge, or B not positive definite'
         ELSE IF( N.GT.0 ) THEN
*           Get reciprocal condition number RCONDB of Cholesky factor of B
            CALL PSTRCON( '1', UPLO, 'N', N, B, IB, JB, DESCB, RCONDB,
     $                    WORK, LWORK, IWORK, LIWORK, INFO )
            RCONDB = MAX( RCONDB, EPSMCH )
            CALL SDISNA( 'Eigenvectors', N, N, W, RCONDZ, INFO )
            DO 10 I = 1, N
               EERRBD( I ) = ( EPSMCH / RCONDB**2 ) * ( ANORM / BNORM +
     $                       ABS( W( I ) ) )
               ZERRBD( I ) = ( EPSMCH / RCONDB**3 ) * ( ( ANORM / BNORM )
     $                       / RCONDZ( I ) + ( ABS( W( I ) ) ) /
     $                       RCONDZ( I ) ) * RCONDB )
10          CONTINUE
         END IF
```

For example, if $\text{PSLAMCH}(\text{ICTXT}, 'E') = 2^{-24} = 5.961 \cdot 10^{-8}$,

$$A = \begin{pmatrix} 100000 & 10099 & 2109 \\ 10099 & 100020 & 10012 \\ 2109 & 10112 & -48461 \end{pmatrix} \text{ and } B = \begin{pmatrix} 99 & 10 & 2 \\ 10 & 100 & 10 \\ 2 & 10 & 100 \end{pmatrix}$$

then ANORM = 120231, BNORM = 120, and RCONDB = .8326, and the approximate eigenvalues, approximate error bounds, and true errors are shown below.

i	λ_i	EERRBD(i)	true $|\hat{\lambda}_i - \lambda_i|$	ZERRBD(i)	true $\theta(\hat{v}_i, v_i)$
1	-500.0	$1.3 \cdot 10^{-4}$	$9.0 \cdot 10^{-6}$	$9.8 \cdot 10^{-8}$	$1.0 \cdot 10^{-9}$
2	$1000.$	$1.7 \cdot 10^{-4}$	$4.6 \cdot 10^{-5}$	$1.9 \cdot 10^{-5}$	$1.2 \cdot 10^{-7}$
3	$1010.$	$1.7 \cdot 10^{-4}$	$1.0 \cdot 10^{-4}$	$1.9 \cdot 10^{-5}$	$1.1 \cdot 10^{-7}$

2. Problem types 2 and 3 have the same error bounds. We illustrate only type 2.

```
         EPSMCH = PSLAMCH( ICTXT, 'E' )
*        Solve the eigenproblem A*B - lambda I (ITYPE = 2)
         ITYPE = 2
```

```
*       Compute the norms of A and B
        ANORM = PSLANSY( '1', UPLO, N, A, IA, JA, DESCA, WORK )
        BNORM = PSLANSY( '1', UPLO, N, B, IB, JB, DESCB, WORK )
*       The eigenvalues are returned in W
*       The eigenvectors are returned in A
        SUBROUTINE PSSYGVX( ITYPE, 'V', 'A', UPLO, N, A, IA, JA,
       $                    DESCA, B, IB, JB, DESCB, VL, VU, IL, IU,
       $                    UNFL, M, NZ, W, -1.0, Z, IZ, JZ, DESCZ,
       $                    WORK, LWORK, IWORK, LIWORK, IFAIL, ICLUSTR,
       $                    GAP, INFO )
        IF( INFO.GT.0 .AND. INFO.LE.N ) THEN
           PRINT *,'PSSYGVX did not converge'
        ELSE IF( INFO.GT.N ) THEN
           PRINT *,'B not positive definite'
        ELSE IF( N.GT.0 ) THEN
*          Get reciprocal condition number RCONDB of Cholesky factor of B
           CALL PSTRCON( '1', UPLO, 'N', N, B, IB, JB, DESCB, RCONDB,
       $                 WORK, LWORK, IWORK, LIWORK, INFO )
           RCONDB = MAX( RCONDB, EPSMCH )
           CALL SDISNA( 'Eigenvectors', N, N, W, RCONDZ, INFO )
           DO 10 I = 1, N
              EERRBD(I) = ( ANORM * BNORM ) * EPSMCH +
       $                  ( EPSMCH / RCONDB**2 ) * ABS( W( I ) )
              ZERRBD(I) = ( EPSMCH / RCONDB ) * ( ( ANORM * BNORM ) /
       $                  RCONDZ( I ) + 1.0 / RCONDB )
10         CONTINUE
        END IF
```

For the same A and B as above, the approximate eigenvalues, approximate error bounds, and true errors are shown below.

i	λ_i	EERRBD(i)	true $|\hat{\lambda}_i - \lambda_i|$	ZERRBD(i)	true $\theta(\hat{v}_i, v_i)$
1	$-4.817 \cdot 10^6$	1.3	$6.0 \cdot 10^{-3}$	$1.7 \cdot 10^{-7}$	$7.0 \cdot 10^{-9}$
2	$8.094 \cdot 10^6$	1.6	1.5	$3.4 \cdot 10^{-7}$	$3.3 \cdot 10^{-8}$
3	$1.219 \cdot 10^7$	1.9	4.5	$3.4 \cdot 10^{-7}$	$4.7 \cdot 10^{-8}$

Further Details: Error Bounds for the Generalized Symmetric Definite Eigenproblem

The error analysis of the driver routine PxSYGVX, or PxHEGVX in the complex case (see section 3.2.4), goes as follows. In all cases $\text{gap}_i = \min_{j \neq i} |\lambda_i - \lambda_j|$ is the **absolute gap** between λ_i and the nearest other eigenvalue.

1. $A - \lambda B$. The computed eigenvalues $\hat{\lambda}_i$ can differ from true eigenvalues λ_i by at most about

$$|\hat{\lambda}_i - \lambda_i| \lesssim p(n)\epsilon \cdot (\|B^{-1}\|_2 \|A\|_2 + \kappa_2(B) \cdot |\hat{\lambda}_i|) = \text{EERRBD}(i) \ .$$

The angular difference between the computed eigenvector $\hat{z}_i$ and a true eigenvector z_i is

$$\theta(\hat{z}_i, z_i) \lesssim p(n)\epsilon \frac{\|B^{-1}\|_2 \|A\|_2 (\kappa_2(B))^{1/2} + \kappa_2(B)|\hat{\lambda}_i|}{\text{gap}_i} = \texttt{ZERRBD}(i) \ .$$

2. $AB - \lambda I$ or $BA - \lambda I$. The computed eigenvalues $\hat{\lambda}_i$ can differ from true eigenvalues λ_i by at most about

$$|\hat{\lambda}_i - \lambda_i| \lesssim p(n)\epsilon \cdot (\|B\|_2 \|A\|_2 + \kappa_2(B) \cdot |\hat{\lambda}_i|) = \texttt{EERRBD}(i) \ .$$

The angular difference between the computed eigenvector $\hat{z}_i$ and a true eigenvector z_i is

$$\theta(\hat{z}_i, z_i) \lesssim p(n)\epsilon \left(\frac{\|B\|_2 \|A\|_2 (\kappa_2(B))^{1/2}}{\text{gap}_i} + \kappa_2(B) \right) = \texttt{ZERRBD}(i) \ .$$

The code fragments above replace $p(n)$ by 1 and make sure neither RCONDB nor RCONDZ is so small as to cause overflow when used as divisors in the expressions for error bounds.

These error bounds are large when B is ill-conditioned with respect to inversion ($\kappa_2(B)$ is large). Often, the eigenvalues and eigenvectors are much better conditioned than indicated here. We mention two ways to get tighter bounds. The first way is effective when the diagonal entries of B differ widely in magnitude:[15]

1. $A - \lambda B$. Let $D = \text{diag}(b_{11}^{-1/2}, \ldots, b_{nn}^{-1/2})$ be a diagonal matrix. Then replace B by DBD and A by DAD in the above bounds.

2. $AB - \lambda I$ or $BA - \lambda I$. Let $D = \text{diag}(b_{11}^{-1/2}, \ldots, b_{nn}^{-1/2})$ be a diagonal matrix. Then replace B by DBD and A by $D^{-1}AD^{-1}$ in the above bounds.

The second way to get tighter bounds does not actually supply guaranteed bounds, but its estimates are often better in practice. It is not guaranteed because it assumes the algorithm is backward stable, which is not necessarily true when B is ill-conditioned. It estimates the **chordal distance** between a true eigenvalue λ_i and a computed eigenvalue $\hat{\lambda}_i$:

$$\chi(\hat{\lambda}_i, \lambda_i) = \frac{|\hat{\lambda}_i - \lambda_i|}{\sqrt{1 + \hat{\lambda}_i^2} \cdot \sqrt{1 + \lambda_i^2}} .$$

To interpret this measure, we write $\lambda_i = \tan\theta$ and $\hat{\lambda}_i = \tan\hat{\theta}$. Then $\chi(\hat{\lambda}_i, \lambda_i) = |\sin(\hat{\theta} - \theta)|$. In other words, if $\hat{\lambda}_i$ represents the one-dimensional subspace $\hat{\mathcal{S}}$ consisting of the line through the origin with slope $\hat{\lambda}_i$, and λ_i represents the analogous subspace $\mathcal{S}$, then $\chi(\hat{\lambda}_i, \lambda_i)$ is the sine of the acute angle $\theta(\hat{\mathcal{S}}, \mathcal{S})$ between these subspaces. Thus, χ is bounded by one and is small when both arguments are large.[16] It applies only to the first problem, $A - \lambda B$:

---

[15]This bound is guaranteed only if the Level 3 BLAS are implemented in a conventional way, not in a fast way.

[16]Another interpretation of chordal distance is as half the usual Euclidean distance between the projections of $\hat{\lambda}_i$ and λ_i on the Riemann sphere, i.e., half the length of the chord connecting the projections.

Suppose a computed eigenvalue $\hat{\lambda}_i$ of $A - \lambda B$ is the exact eigenvalue of a perturbed problem $(A + E) - \lambda(B + F)$. Let x_i be the unit eigenvector ($\|x_i\|_2 = 1$) for the exact eigenvalue λ_i. Then if $\|E\|$ is small compared with $\|A\|$, and if $\|F\|$ is small compared with $\|B\|$, we have

$$\chi(\hat{\lambda}_i, \lambda_i) \lesssim \frac{\|E\| + \|F\|}{\sqrt{(x_i^H A x_i)^2 + (x_i^H B x_i)^2}} \quad .$$

Thus $1/\sqrt{(x_i^H A x_i)^2 + (x_i^H B x_i)^2}$ is a condition number for eigenvalue λ_i.

Chapter 7

Troubleshooting

Successful installation, testing, and use of ScaLAPACK rely heavily on the proper installation of its building blocks (PVM or MPI, BLACS, BLAS, and PBLAS). Frequently Asked Questions (FAQ) lists are maintained in the directories on *netlib* to answer some of the most common user questions. For the user's convenience, prebuilt ScaLAPACK and BLACS libraries are provided for a variety of computer architectures in the following URLs:

```
http://www.netlib.org/scalapack/archives/
http://www.netlib.org/blacs/archives/
```

Test suites are provided for PVM, the BLACS, the BLAS, and the PBLAS. It is highly recommended that each of these respective test suites be run prior to the execution of the ScaLAPACK test suite. Installation Guides are also provided for the BLACS and ScaLAPACK. Refer to the appropriate directory on *netlib* for further information.

We begin this chapter by discussing a set of first-step debugging hints to pinpoint where the problem is occurring. Following these debugging hints, we discuss the types of error messages that can be encountered during the execution of a ScaLAPACK routine: ScaLAPACK error messages, PBLAS error messages, BLACS error messages, and system-dependent error messages or failures.

If these suggestions do not help evaluate specific difficulties, we suggest that the user review the following "bug report checklist" and then feel free to contact the authors at scalapack@cs.utk.edu or blacs@cs.utk.edu, respectively. The user should tell us the type of machine on which the tests were run, the compiler and compiler options that were used, details of the BLACS library and message-passing library that were used and the BLAS library; also, the user should send a copy of the input file, if appropriate.

Bug Report Checklist

When the user sends e-mail to our mailing alias, some of the first questions we will ask are the following:

1. Have you run the BLAS, BLACS, PBLAS and ScaLAPACK test suites?

2. Have you checked the errata lists (`errata.scalapack` and `errata.blacs`) on *netlib*?

   ```
   http://www.netlib.org/scalapack/errata.scalapack
   http://www.netlib.org/blacs/errata.blacs
   ```

3. If you are using an optimized BLAS or BLACS library, have you tried using the reference implementations from *netlib*?

   ```
   http://www.netlib.org/blas/
   http://www.netlib.org/blacs/
   ```

4. If you are using an optimized MPI or PVM library, have you tried using the reference implementations from *netlib*?

   ```
   http://www.netlib.org/pvm3/
   http://www.netlib.org/mpi/
   ```

5. Have you attempted to replicate this error using the appropriate ScaLAPACK test code and/or one of the ScaLAPACK example routines?

7.1   Installation Debugging Hints

If the user encounters difficulty in the installation process, we suggest the following:

- Obtain prebuilt ScaLAPACK and BLACS libraries on *netlib* for a variety of architectures.

  ```
  http://www.netlib.org/scalapack/archives/
  http://www.netlib.org/blacs/archives/
  ```

- Obtain sample `SLmake.inc` files for a variety of architectures in the `SCALAPACK/INSTALL` directory in the scalapack distribution tar file. Sample `Bmake.inc` files are included in the `BLACS/BMAKES` directory in the blacs distribution file.

- Consult the ScaLAPACK FAQ list on *netlib*.

- Consult the `errata.scalapack` file in the *scalapack* directory on *netlib*, and/or the `errata.blacs` file in the *blacs* directory on *netlib*. These files contain a list of known difficulties that have been diagnosed and corrected (or will be corrected in the next release), or reported to the vendor as in the case of message-passing libraries or optimized BLAS.

- Always run the BLACS, BLAS, and PBLAS test suites to ensure that these libraries have been properly installed. (If PVM is the underlying message-passing layer, please also run the PVM test suite.) If a problem is detected in the BLAS or BLACS libraries, try linking to the reference implementations to be found in the respective *blas* or *blacs* directory on *netlib*.

- If ScaLAPACK is being tested on a heterogeneous cluster of computers, please ensure that all executables are linked with the same debug level of the BLACS. Otherwise, unpredictable results will occur because the debug level 1 BLACS (specified by BLACSDBGLVL=1 in Bmake.inc) perform error-checking and thus send more messages than the performance debug level 0 BLACS (specified by BLACSDBGLVL=0 in Bmake.inc).

7.2 Application Debugging Hints

We highly recommend the following as a list of debugging hints (and tools) for writing parallel application programs that call ScaLAPACK:

- Look at the ScaLAPACK example programs as a good starting point.

    http://www.netlib.org/scalapack/examples/

- Always check the value of INFO on exit from a ScaLAPACK routine.

- All routines in ScaLAPACK that require workspace also require the length of that workspace to be specified in the calling sequence. If in doubt about the amount of workspace to supply to a ScaLAPACK routine, supply $LWORK = -1$, and use the returned value in $WORK(1)$ as the correct value for $LWORK$. Refer to section 4.6.5 for further details on determining workspace requirements.

- If you are calling a ScaLAPACK routine that has an LAPACK equivalent, write a serial code calling LAPACK first. Code can be converted in pieces from LAPACK to ScaLAPACK by debugging on a one-process grid. When all of the LAPACK codes have been removed and the code has been fully parallelized, execute it on a multiple process grid.

- When writing a parallel program, first debug the code to work on one process, and then expand to more processes.

- When writing a parallel program, debug with small matrices.

- Use the TOOLS routine PxLAPRNT to print out each process's portion of a distributed matrix. A variety of utility routines are provided in the TOOLS directory and are commonly used as debugging aids in the development of the ScaLAPACK library.

- Sprinkle synchronization points via BLACS_BARRIER near suspected error.

- Link to the debug level 1 BLACS (specified by BLACSDBGLVL=1 in Bmake.inc) until the program is completely debugged.

- Specify the "Repeatability" flag in BLACS_SET.

- If running a heterogeneous application, please ensure that all executables are linked with the same debug level of the BLACS. Otherwise, unpredictable results will occur because the debug level 1 BLACS perform error-checking and thus send more messages than the performance debug level 0 BLACS.

- Always run the BLACS, BLAS, and PBLAS test suites to ensure that these libraries have been properly installed. (If PVM is the underlying message-passing layer, please also run the PVM test suite.) If a problem is detected in the BLAS or BLACS libraries, try linking to the reference implementations in the respective *blas* or *blacs* directory on *netlib*.

- Consult the `errata.scalapack` file in the *scalapack* directory on *netlib*, and/or the `errata.blacs` file in the *blacs* directory on *netlib*. These files contain a list of known difficulties that have been diagnosed and corrected (or will be corrected in the next release), or reported to the vendor as in the case of message-passing libraries or optimized BLAS.

- Refer to section 4.6.7 and the leading comments of the source code for the alignment restrictions currently needed in some of the ScaLAPACK routines.

7.3   Common Errors in Calling ScaLAPACK Routines

The user must read the leading comments of a ScaLAPACK routine before invoking the routine. The wording of the leading comments is explained in Chapter 4. Basic terminology is explained in the **Glossary** and **List of Notation**.

For the benefit of less experienced programmers, we provide a list of common programming errors in calling a ScaLAPACK routine. These errors may cause the ScaLAPACK routine to report a failure, as described in section 7.4; they may cause an error to be reported by the system; or they may lead to wrong results — see also section 7.5.

- Wrong number of arguments

- Arguments in the wrong order

- Argument of the wrong type (especially real and complex arguments of the wrong precision)

- Wrong dimensions for an array argument

- Insufficient space in a workspace argument

- Failure to assign a value to an input argument

- Routine designed for homogeneous computers was executed on a heterogeneous system (see section 6.2)

Some modern compilation systems, as well as software tools such as the Fortran 77 syntax checker `ftnchek` (freely available on *netlib*) and the portability checker in Toolpack [105], can check that arguments agree in number and type; and many compilation systems offer run-time detection of errors such as an array element out-of-bounds or use of an unassigned variable.

7.4   Failures Detected by ScaLAPACK Routines

A ScaLAPACK routine has two ways to report a failure to complete a computation successfully.

7.4.1 Invalid Arguments and PXERBLA

If an illegal value is supplied for one of the input arguments to a ScaLAPACK routine, it will call the error handler PXERBLA to write a message to the standard output unit of the form:

```
** On entry to PSGESV  parameter number  4 had an illegal value
```

This particular message could be caused by passing to PSGESV a value of NRHS that was less than zero, for example. The arguments are checked in order, beginning with the first. As mentioned in Chapter 4, if an error is detected in the jth entry of a descriptor array, which is the ith argument in the parameter list, the number passed to PXERBLA has been arbitrarily chosen to be $100 * i + j$. This allows the user to distinguish an error on a descriptor entry from an error on a scalar argument. Invalid arguments are often caused by the kind of error listed in section 7.3.

In the model implementation of PXERBLA that is supplied with ScaLAPACK, the only action that is performed is the printing of an error message to standard output. Program execution is *not* terminated. For the ScaLAPACK driver and computational routines, a RETURN statement is issued following the call to PXERBLA. Control returns to the higher-level calling routine, and it is left to the user to determine how the program should proceed. However, in the specialized low-level ScaLAPACK routines (auxiliary routines that are Level 2 equivalents of computational routines), the call to PXERBLA() is immediately followed by a call to BLACS_ABORT() to terminate program execution since recovery from an error at this level in the computation is not possible.

It is always good practice to check for a nonzero value of INFO on return from a ScaLAPACK routine.

7.4.2 Computational Failures and INFO > 0

A positive value of INFO on return from a ScaLAPACK routine indicates a failure in the course of the algorithm. Common causes are

- a matrix is singular (to working precision),

- a symmetric matrix is not positive definite, or

- an iterative algorithm for computing eigenvalues or eigenvectors fails to converge in the permitted number of iterations.

For example, if PSGESVX is called to solve a system of equations with a coefficient matrix that is approximately singular, it may detect exact singularity at the ith stage of the LU factorization, in which case it returns INFO $= i$; or (more probably) it may compute an estimate of the reciprocal condition number that is less than relative machine precision, in which case it returns INFO $= n+1$. Again, the documentation in Part II should be consulted for a description of the error.

When a failure with INFO > 0 occurs, control is *always* returned to the calling program; PXERBLA() is *not* called, and no error message is written. Thus, it is always good practice to check for a nonzero value of INFO on return from a ScaLAPACK routine.

A failure with INFO > 0 may indicate any of the following:

- An inappropriate routine was used:. For example, if a routine fails because a symmetric matrix turns out not to be positive definite, consider using a routine for symmetric indefinite matrices.

- A single-precision routine was used when double precision was needed. For example, if PS-GESVX reports approximate singularity (as illustrated above), the corresponding double precision routine PDGESVX may be able to solve the problem (but nevertheless the problem is ill-conditioned).

- A programming error occurred in generating the data supplied to a routine. For example, even though theoretically a matrix should be well-conditioned and positive-definite, a programming error in generating the matrix could easily destroy either of those properties.

- A programming error occurred in calling the routine, of the kind listed in section 7.3.

7.5   Wrong Results

Wrong results from ScaLAPACK routines are most often caused by incorrect usage. It is also possible that wrong results are caused by a bug outside of ScaLAPACK, in the compiler or in one of the library routines, such as the BLAS, the BLACS, or the underlying message-passing layer, that are linked with ScaLAPACK. Test suites are available for ScaLAPACK, the PBLAS, the BLACS, and the BLAS. The ScaLAPACK installation guide [24] or the BLACS installation guide should be consulted for descriptions of the tests and for advice on resolving problems.

A list of known problems, compiler errors, and bugs in ScaLAPACK routines is maintained on *netlib*; see Chapter 1.

Users who suspect they have found a new bug in a ScaLAPACK routine are encouraged to report it promptly to the developers as directed in Chapter 1. The bug report should include a test case, a description of the problem and expected results, and the actions, if any, that the user has already taken to fix the bug.

7.6   Error Handling in the PBLAS

If a PBLAS routine is called with an invalid value for any of its arguments, it must report the fact and terminate the execution of the program. In the model implementation, each routine, on detecting an error, calls a common error-handling PBLAS routine, passing to it the current BLACS context, the name of the routine and the number of the first argument that is in error. If an error is detected in the jth entry of a descriptor array, which is the ith argument in the parameter list, the number passed to the PBLAS error-handler routine has been arbitrarily chosen to be $100 * i + j$. This allows the user to distinguish an error on a descriptor entry from an error on a scalar argument. For efficiency purposes, the PBLAS routines performs only a local validity check of their argument list. If an error is detected in at least one process of the current context, the program execution is stopped.

A global validity check of the input arguments passed to a PBLAS routine must be performed in the higher-level calling procedure. To understand the need and cost of global checking, as well as the reason why this type of checking is not performed in the PBLAS, consider the following example. The value of a global input argument is legal but differs from one process to another. The results are unpredictable. In order to detect this kind of error situation, a synchronization point would be necessary, which may result in a significant performance degradation. Since every process must call the same routine to perform the desired operation successfully, it is natural and safe to restrict somewhat the amount of checking operations performed in the PBLAS routines.

Specialized implementations may call system-specific exception-handling facilities, either via an auxiliary routine for error handling or directly from the routine. In addition, the testing programs can take advantage of this exception handling mechanism by simulating specific erroneous input argument lists and then verifying that particular errors are correctly detected.

For complete details on the specification of all routines, please refer to [26]. Appendix D.2 contains the Quick Reference Guide to the PBLAS. An *html* version of this Quick Reference Guide, along with the leading comments from each of the routines, is available on the ScaLAPACK homepage.

7.7 Error Handling in the BLACS

This section describes the BLACS error-handling features. The BLACS error-handling behavior may be changed at compile time by using the C preprocessor macro `BlacsDebugLvl`. a call to BLACS_GET (see [54] for details) will help determine what debug level the BLACS are using.

If the BLACS are compiled with a BLACS debug level of 0, little error checking is performed. A few critical items will be checked (for instance, BLACS_GRIDINIT will still not allow the user to allocate a process grid with more processes than there are available), but for performance reasons, the BLACS will not check most of the parameters.

It is therefore highly recommended that users link their code to a BLACS library compiled with debug level 1 while debugging their code. BLACS debug level 1 mainly does parameter checking. A few other services are also provided. For instance, users will be warned if a process sends a message to itself. Having a process send to itself is legal, but it displays poor performance and requires enough buffer space that it can occasionally cause hangs for large messages. The BLACS therefore issue a warning when this behavior is detected.

Many times, the debug level 0 code will simply hang, leaving the developer without any clue as to what has gone wrong. This may be caused, for instance, by trying to receive from a process that is not in the current context. The debug level 1 BLACS can detect this type of user error, and issue a (we hope helpful) message.

The BLACS issue three types of messages:

1. *BLACS warning*: BLACS detect risky behavior, but attempt to correct or ignore. Warning

message is printed, and execution proceeds.

2. *BLACS error*: BLACS detects an error, prints an error message, and kills the machine via a call to BLACS_ABORT.

3. *System error*: The BLACS receive an error message from the underlying system, which is then passed on to the user, and the BLACS kills the machine.

7.7.1   BLACS Warning and Error Messages

All BLACS warning messages are printed by the internal routine `BlacsWarn`, and all BLACS error messages are printed by the internal routine `BlacsErr`. The only real difference between `BlacsWarn` and `BlacsErr` is that `BlacsErr` calls BLACS_ABORT after the message is printed.

With these central routines handling BLACS error messages, it should be relatively easy for the programmer to modify error handling if the default routines are not adequate for his needs. One particularly annoying problem is that on many systems a print to the screen takes a long time to finish. `BlacsErr` may then kill the machine before the print reaches the screen, and the error message is lost. In this case, the user may wish to make `BlacsErr` wait before killing, or not kill at all, for instance.

BLACS warning messages have the following form:

```
BLACS WARNING '<explanation string>'
from {<p>,<q>}, pnum=<pnum>, Contxt=<ictxt>, on line <#> of file '<fname>'.
```

BLACS error messages have the following form:

```
BLACS ERROR '<explanation string>'
from {<p>,<q>}, pnum=<pnum>, Contxt=<ictxt>, on line <#> of file '<fname>'.
```

The meaning of these parameters are as follows:

- `explanation string` The message that should help the user determine what is wrong. For example, on an incorrect call to BLACS_GRIDINIT, the user might get:
  `Process 0 had 2 x 4 grid; correct is 1 x 4.`

- {p, q}: The row and column process grid coordinates of the process issuing the warning/error.

- pnum: The process number returned in the first argument of BLACS_PINFO.

- ictxt: The integer context handle. Please note that this value is not the same across all processes. For instance, process {0, 0} may have `ictxt` = 0 and process {0, 1} have `ictxt` = 1 for the same context. However, the pnum and ictxt together provide an unambiguous process/context identifier.

- #: The line number within the file fname that issued the warning.

- fname: The file name where the routine that issued the warning/error is located.

Not all of this information may be available at the time an error or warning is issued. For instance, if the error occurs before the creation of the grid, the process grid coordinates will be unavailable. For any value that the BLACS cannot figure out, a -1 is printed to indicate that the value is unknown.

Examples of these BLACS error messages can be found on the BLACS homepage (http://www.netlib.org/blacs/index.html) or in "A User's Guide to the BLACS" [54].

7.8 System Error Messages

Occasionally, ScaLAPACK will receive an error message from the underlying system. At this time, the BLACS will print the system error message, and exit. Since these error messages come from the underlying system, their form will necessarily vary depending on which BLACS version is being used. The user may need to obtain vendor documentation or on-line manpages describing system error messages in order to understand the message. For example, if the PVM BLACS are being used, a PVM error number will be returned. The PVM quick reference guide, for example, could then be consulted to translate the error number into an understandable error message.

Examples of system error messages can be found on the BLACS homepage (http://www.netlib.org/blacs/index.html) or in "A User's Guide to the BLACS" [54].

7.9 Poor Performance

ScaLAPACK ultimately relies on an efficient implementation of the BLAS and the data distribution for load balance. Refer to Chapter 5.

To avoid poor performance from ScaLAPACK routines, note the following recommendations:

BLAS: One should use machine-specific optimized BLAS if they are available. Many manufacturers and research institutions have developed, or are developing, efficient versions of the BLAS for particular machines. The BLAS enable LAPACK and ScaLAPACK routines to achieve high performance with transportable software. Users are urged to determine whether such an implementation of the BLAS exists for their platform. When such an optimized implementation of the BLAS is available, it should be used to ensure optimal performance. If such a machine-specific implementation of the BLAS does not exist for a particular platform, one should consider installing a publicly available set of BLAS that requires only an efficient implementation of the matrix-matrix multiply BLAS routine xGEMM. Examples of such implementations are [35, 90]. A machine-specific and efficient implementation of the routine GEMM can be automatically generated by publicly available software such as [16]. Although a reference implementation of the Fortran77 BLAS is available from the *blas* directory on *netlib*, these routines are not expected to perform as well as a specially tuned implementation on most high-performance computers – on some machines it may give much worse performance – but it allows users to run LAPACK and ScaLAPACK software on machines that do not offer any other implementation of the BLAS.

BLACS: With the few exceptions mentioned in section 5.2.3, the performance achieved by the BLACS should be close to the one of the underlying message-passing library it is calling. Since publicly available implementations of the BLACS exist for a range of native message-passing libraries such as NX for the Intel supercomputers and MPL for the IBM SP series, as well as more generic interfaces such as PVM and MPI, users should select the BLACS implementation that is based on the most efficient message-passing library available. Some vendors, such as Cray and IBM, supply an optimized implementation of the BLACS for their systems. Users are urged to rely on these BLACS libraries whenever possible.

LWORK $\geq$ WORK(1): In some ScaLAPACK eigenvalue routines, such as the symmetric eigenproblems (PxSYEV and PxSYEVX/PxHEEVX) and the generalized symmetric eigenproblem (PxSYGVX/PxHEGVX), a larger value of $LWORK$ can guarantee the orthogonality of the returned eigenvectors at the risk of potentially degraded performance of the algorithm. The minimum amount of workspace required is returned in the first element of the work array, but a larger amount of workspace can allow for additional orthogonalization if desired by the user. Refer to section 5.3.6 and the leading comments of the source code for complete details.

Appendix A

Index of ScaLAPACK Routines

A separate index is provided for each of the following classifications of routines: driver and computational routines, auxiliary routines, and matrix redistribution/copy routines.

A.1 Index of Driver and Computational Routines

Notes

1. This index lists related pairs of real and complex routines together, for example, PSGETRF and PCGETRF.

2. Driver routines are listed in bold type, for example, **PSGESV** and **PCGESV**.

3. Routines are listed in alphanumeric order of the real (single precision) routine name (which always begins with PS-). (See section 3.1.3 for details of the ScaLAPACK naming scheme.)

4. Double precision routines are not listed here; they have names beginning with PD- instead of PS-, or PZ- instead of PC-.

5. This index gives only a brief description of the purpose of each routine. For a precise description, consult the Specifications in Part II, where the routines appear in the same order as here.

6. The text of the descriptions applies to both real and complex routines, except where alternative words or phrases are indicated, for example, "symmetric/Hermitian", "orthogonal/unitary", or "quasi-triangular/triangular". For the real routines A^H is equivalent to A^T. (The same convention is used in Part II.)

7. A few routines for real matrices have no complex equivalent (for example, PSSTEBZ).

159

Routine		Description
Real	Complex	
PSDBSV	**PCDBSV**	Solves a general banded system of linear equations $AX = B$ without pivoting.
PSDBTRF	PCDBTRF	Computes an LU factorization of a general band matrix without pivoting.
PSDBTRS	PCDBTRS	Solves a general banded system of linear equations $AX = B$, $A^T X = B$ or $A^H X = B$, using the LU factorization computed by PSDBTRF/PCDBTRF.
PSDTSV	**PCDTSV**	Solves a general tridiagonal system of linear equations $AX = B$ without pivoting.
PSDTTRF	PCDTTRF	Computes an LU factorization of a general tridiagonal matrix without pivoting.
PSDTTRS	PCDTTRS	Solves a general tridiagonal system of linear equations $AX = B$, $A^T X = B$ or $A^H X = B$, using the LU factorization computed by PSDTTRF/PCDTTRF.
PSGBSV	**PCGBSV**	Solves a general banded system of linear equations $AX = B$ using partial pivoting with local row interchanges.
PSGBTRF	PCGBTRF	Computes an LU factorization of a general band matrix, using partial pivoting with local row interchanges.
PSGBTRS	PCGBTRS	Solves a general banded system of linear equations $AX = B$, $A^T X = B$ or $A^H X = B$, using the LU factorization computed by PSGBTRF/PCGBTRF.
PSGEBRD	PCGEBRD	Reduces a general rectangular matrix to real bidiagonal form by orthogonal/unitary transformations.
PSGECON	PCGECON	Estimates the reciprocal of the condition number of a general matrix, in either the 1-norm or the infinity-norm, using the LU factorization computed by PSGETRF/PCGETRF.
PSGEEQU	PCGEEQU	Computes row and column scalings to equilibrate a general rectangular matrix and reduce its condition number.
PSGEHRD	PCGEHRD	Reduces a general matrix to upper Hessenberg form by an orthogonal/unitary similarity transformation.
PSGELQF	PCGELQF	Computes an LQ factorization of a general rectangular matrix.
PSGELS	**PCGELS**	Computes the least squares solution to an overdetermined system of linear equations, $AX = B$ or $A^H X = B$, or the minimum norm solution of an underdetermined system, where A is a general rectangular matrix of full rank, using a QR or LQ factorization of A.
PSGEQLF	PCGEQLF	Computes a QL factorization of a general rectangular matrix.
PSGEQPF	PCGEQPF	Computes a QR factorization with column pivoting of a general rectangular matrix.

Routine		Description
Real	Complex	
PSGEQRF	PCGEQRF	Computes a QR factorization of a general rectangular matrix.
PSGERFS	PCGERFS	Improves the computed solution to a general system of linear equations $AX = B$, $A^T X = B$ or $A^H X = B$, and provides forward and backward error bounds for the solution.
PSGERQF	PCGERQF	Computes an RQ factorization of a general rectangular matrix.
PSGESV	**PCGESV**	Solves a general system of linear equations $AX = B$.
PSGESVD		Computes the singular value decomposition (SVD) of a general rectangular matrix.
PSGESVX	**PCGESVX**	Solves a general system of linear equations $AX = B$, $A^T X = B$ or $A^H X = B$, and provides an estimate of the condition number and error bounds on the solution.
PSGETRF	PCGETRF	Computes an LU factorization of a general matrix, using partial pivoting with row interchanges.
PSGETRI	PCGETRI	Computes the inverse of a general matrix, using the LU factorization computed by PSGETRF/PCGETRF.
PSGETRS	PCGETRS	Solves a general system of linear equations $AX = B$, $A^T X = B$ or $A^H X = B$, using the LU factorization computed by PSGETRF/PCGETRF.
PSGGQRF	PCGGQRF	Computes a generalized QR factorization of a pair of matrices.
PSGGRQF	PCGGRQF	Computes a generalized RQ factorization of a pair of matrices.
PSORGLQ	PCUNGLQ	Generates all or part of the orthogonal/unitary matrix Q from an LQ factorization determined by PSGELQF/PCGELQF.
PSORGQL	PCUNGQL	Generates all or part of the orthogonal/unitary matrix Q from a QL factorization determined by PSGEQLF/PCGEQLF.
PSORGQR	PCUNGQR	Generates all or part of the orthogonal/unitary matrix Q from a QR factorization determined by PSGEQRF/PCGEQRF.
PSORGRQ	PCUNGRQ	Generates all or part of the orthogonal/unitary matrix Q from an RQ factorization determined by PSGERQF/PCGERQF.
PSORMBR	PCUNMBR	Multiplies a general matrix by one of the orthogonal/unitary transformation matrices from a reduction to bidiagonal form determined by PSGEBRD/PCGEBRD.
PSORMHR	PCUNMHR	Multiplies a general matrix by the orthogonal/unitary transformation matrix from a reduction to Hessenberg form determined by PSGEHRD/PCGEHRD.
PSORMLQ	PCUNMLQ	Multiplies a general matrix by the orthogonal/unitary matrix from an LQ factorization determined by PSGELQF/PCGELQF.
PSORMQL	PCUNMQL	Multiplies a general matrix by the orthogonal/unitary matrix from a QL factorization determined by PSGEQLF/PCGEQLF.

Routine		Description
Real	Complex	
PSORMQR	PCUNMQR	Multiplies a general matrix by the orthogonal/unitary matrix from a QR factorization determined by PSGEQRF/PCGEQRF.
PSORMRQ	PCUNMRQ	Multiplies a general matrix by the orthogonal/unitary matrix from an RQ factorization determined by PSGERQF/PCGERQF.
PSORMRZ	PCUNMRZ	Multiplies a general matrix by the orthogonal/unitary matrix from an RZ factorization determined by PSTZRZF/PCTZRZF.
PSORMTR	PCUNMTR	Multiplies a general matrix by the orthogonal/unitary transformation matrix from a reduction to tridiagonal form determined by PSSYTRD/PCHETRD.
PSPBSV	**PCPBSV**	Solves a symmetric/Hermitian positive definite banded system of linear equations $AX = B$.
PSPBTRF	PCPBTRF	Computes the Cholesky factorization of a symmetric/Hermitian positive definite band matrix.
PSPBTRS	PCPBTRS	Solves a symmetric/Hermitian positive definite banded system of linear equations $AX = B$, using the Cholesky factorization computed by PSPBTRF/PCPBTRF.
PSPOCON	PCPOCON	Estimates the reciprocal of the condition number of a symmetric/Hermitian positive definite matrix, using the Cholesky factorization computed by PSPOTRF/PCPOTRF.
PSPOEQU	PCPOEQU	Computes row and column scalings to equilibrate a symmetric/Hermitian positive definite matrix and reduce its condition number.
PSPORFS	PCPORFS	Improves the computed solution to a symmetric/Hermitian positive definite system of linear equations $AX = B$, and provides forward and backward error bounds for the solution.
PSPOSV	**PCPOSV**	Solves a symmetric/Hermitian positive definite system of linear equations $AX = B$.
PSPOSVX	**PCPOSVX**	Solves a symmetric/Hermitian positive definite system of linear equations $AX = B$, and provides an estimate of the condition number and error bounds on the solution.
PSPOTRF	PCPOTRF	Computes the Cholesky factorization of a symmetric/Hermitian positive definite matrix.
PSPOTRI	PCPOTRI	Computes the inverse of a symmetric/Hermitian positive definite matrix, using the Cholesky factorization computed by PSPOTRF/PCPOTRF.
PSPOTRS	PCPOTRS	Solves a symmetric/Hermitian positive definite system of linear equations $AX = B$, using the Cholesky factorization computed by PSPOTRF/PCPOTRF.

| Routine | | Description |
Real	Complex	
PSPTSV	**PCPTSV**	Solves a symmetric/Hermitian positive definite tridiagonal system of linear equations $AX = B$.
PSPTTRF	PCPTTRF	Computes the LDL^H factorization of a symmetric/Hermitian positive definite tridiagonal matrix.
PSPTTRS	PCPTTRS	Solves a symmetric/Hermitian positive definite tridiagonal system of linear equations, using the LDL^H factorization computed by PSPTTRF/PCPTTRF.
PSSTEBZ		Computes selected eigenvalues of a real symmetric tridiagonal matrix by bisection.
PSSTEIN	PCSTEIN	Computes selected eigenvectors of a real symmetric tridiagonal matrix by inverse iteration.
PSSYEV		Computes all eigenvalues and, optionally, eigenvectors of a symmetric/Hermitian matrix.
PSSYEVX	**PCHEEVX**	Computes selected eigenvalues and eigenvectors of a symmetric/Hermitian matrix.
PSSYGST	PCHEGST	Reduces a symmetric/Hermitian definite generalized eigenproblem $Ax = \lambda Bx$, $ABx = \lambda x$, or $BAx = \lambda x$, to standard form, where B has been factorized by PSPOTRF/PCPOTRF.
PSSYGVX	**PCHEGVX**	Computes all or selected eigenvalues and the eigenvectors of a generalized symmetric/Hermitian definite generalized eigenproblem, $Ax = \lambda Bx$, $ABx = \lambda x$, or $BAx = \lambda x$.
PSSYTRD	PCHETRD	Reduces a symmetric/Hermitian matrix to real symmetric tridiagonal form by an orthogonal/unitary similarity transformation.
PSTRCON	PCTRCON	Estimates the reciprocal of the condition number of a triangular matrix, in either the 1-norm or the infinity-norm.
PSTRRFS	PCTRRFS	Provides forward and backward error bounds for the solution of a triangular system of linear equations $AX = B$, $A^T X = B$ or $A^H X = B$.
PSTRTRI	PCTRTRI	Computes the inverse of a triangular matrix.
PSTRTRS	PCTRTRS	Solves a triangular system of linear equations $AX = B$, $A^T X = B$ or $A^H X = B$.
PSTZRZF	PCTZRZF	Reduces an upper trapezoidal matrix to upper triangular form using orthogonal/unitary transformations.

A.2 Index of Auxiliary Routines

Notes

1. This index lists related pairs of real and complex routines together.

2. Routines are listed in alphanumeric order of the real (single precision) routine name (which always begins with PS-). (See section 3.1.3 for details of the ScaLAPACK naming scheme.)

3. A few complex routines have no real equivalents, and they are listed first.

4. Double-precision routines are not listed here; they have names beginning with PD- instead of PS-, or PZ- instead of PC-. The only exceptions to this simple rule are that the double-precision versions of PCMAX1, PSCSUM1, and PCSRSCL are named PZMAX1, PDZSUM1, and PZDRSCL.

5. A few routines in the list have names that are independent of data type: PXERBLA.

6. This index gives only a brief description of the purpose of each routine. For a precise description, consult the leading comments in the code, which have been written in the same style as for the driver and computational routines.

Routine		Description
Real	Complex	
	PCLACGV	Conjugates a complex vector.
	PCMAX1	Finds the index of the element whose real part has maximum absolute value (similar to the Level 1 PBLAS PCAMAX, but using the absolute value of the real part).
	PSCSUM1	Forms the 1-norm of a complex vector (similar to the Level 1 PBLAS PSCASUM, but using the true absolute value).
PSDBTRSV	PCDBTRSV	Is called by PSDBTRS/PCDBTRS.
PSDTTRSV	PCDTTRSV	Is called by PSDTTRS/PCDTTRS.
PSGEBD2	PCGEBD2	Reduces a general rectangular matrix to real bidiagonal form by an orthogonal/unitary transformation (unblocked algorithm).
PSGEHD2	PCGEHD2	Reduces a general matrix to upper Hessenberg form by an orthogonal/unitary similarity transformation (unblocked algorithm).
PSGELQ2	PCGELQ2	Computes an LQ factorization of a general rectangular matrix (unblocked algorithm).
PSGEQL2	PCGEQL2	Computes a QL factorization of a general rectangular matrix (unblocked algorithm).
PSGEQR2	PCGEQR2	Computes a QR factorization of a general rectangular matrix (unblocked algorithm).
PSGERQ2	PCGERQ2	Computes an RQ factorization of a general rectangular matrix (unblocked algorithm).
PSGETF2	PCGETF2	Computes an LU factorization of a general matrix, using partial pivoting with row interchanges (local unblocked algorithm).
PSLABAD		Returns the square root of the underflow and overflow thresholds if the exponent-range is very large.
PSLABRD	PCLABRD	Reduces the first nb rows and columns of a general rectangular matrix A to real bidiagonal form by an orthogonal/unitary transformation, and returns auxiliary matrices that are needed to apply the transformation to the unreduced part of A.
PSLACHKIEE		(real) performs a simple check for the features of the IEEE standard
PSLACON	PCLACON	Estimates the 1-norm of a square matrix, using reverse communication for evaluating matrix-vector products.
PSLACONSB		(real) looks for two consecutive small subdiagonal elements
PSLACP2	PCLACP2	copies all or part of a distributed matrix to another distributed matrix
PSLACP3		(real) copies from a global parallel array into a local replicated array or vice versa.
PSLACPY	PCLACPY	Copies all or part of one two-dimensional array to another.
PSLAEVSWP	PCLAEVSWP	Moves the eigenvectors from where they are computed, to a ScaLAPACK standard block cyclic array.

Routine		Description
Real	Complex	
PSLAHQR		Computes the eigenvalues and Schur factorization of an upper Hessenberg matrix, using the double-shift/single-shift QR algorithm.
PSLAHRD	PCLAHRD	Reduces the first nb columns of a general rectangular matrix A so that elements below the kth subdiagonal are zero, by an orthogonal/unitary transformation, and returns auxiliary matrices which are needed to apply the transformation to the unreduced part of A.
PSLAIECT		Exploits IEEE arithmetic to accelerate the computations of eigenvalues.
PSLAMCH		Determines machine parameters for floating-point arithmetic.
PSLANGE	PCLANGE	Returns the value of the 1-norm, Frobenius norm, infinity-norm, or the largest absolute value of any element, of a general rectangular matrix.
PSLANHS	PCLANHS	Returns the value of the 1-norm, Frobenius norm, infinity-norm, or the largest absolute value of any element, of an upper Hessenberg matrix.
PSLANSY	PCLANSY PCLANHE	Returns the value of the 1-norm, Frobenius norm, infinity-norm, or the largest absolute value of any element, of a real symmetric/complex symmetric/complex Hermitian matrix.
PSLANTR	PCLANTR	Returns the value of the 1-norm, Frobenius norm, infinity-norm, or the largest absolute value of any element, of a triangular matrix.
PSLAPIV	PCLAPIV	Applies a permutation matrix to a general distributed matrix, resulting in row or column pivoting.
PSLAQGE	PCLAQGE	Scales a general rectangular matrix, using row and column scaling factors computed by PSGEEQU/PCGEEQU.
PSLAQSY	PCLAQSY	Scales a symmetric/Hermitian matrix, using scaling factors computed by PSPOEQU/PCPOEQU.
PSLARED1D		Redistributes an array assuming that the input array, BYCOL, is distributed across rows and that all process columns contain the same copy of BYCOL.
PSLARED2D		Redistributes an array assuming that the input array, BYROW, is distributed across columns and that all process rows contain the same copy of BYROW. The output array, BYALL, will be identical on all processes.
PSLARF	PCLARF	Applies an elementary reflector to a general rectangular matrix.
PSLARFB	PCLARFB	Applies a block reflector or its transpose/conjugate-transpose to a general rectangular matrix.
	PCLARFC	(complex) Applies (multiplies by) the conjugate transpose of an elementary reflector to a general matrix.
PSLARFG	PCLARFG	Generates an elementary reflector (Householder matrix).
PSLARFT	PCLARFT	Forms the triangular factor T of a block reflector $H = I - VTV^H$.

Routine		Description
Real	Complex	
PSLARZ	PCLARZ	Applies an elementary reflector as returned by PSTZRZF/PCTZRZF to a general matrix.
PSLARZB	PCLARZB	Applies a block reflector or its transpose/conjugate-transpose as returned by PSTZRZF/PCTZRZF to a general matrix.
	PCLARZC	(complex) Applies (multiplies by) the conjugate transpose of an elementary reflector as returned by PSTZRZF/PCTZRZF to a general matrix.
PSLARZT	PCLARZT	Forms the triangular factor T of a block reflector $H = I - VTV^H$ as returned by PSTZRZF/PCTZRZF.
PSLASCL	PCLASCL	Multiplies a general rectangular matrix by a real scalar defined as c_{to}/c_{from}.
PSLASET	PCLASET	Initializes the off-diagonal elements of a matrix to α and the diagonal elements to β.
PSLASMSUB		(real) Looks for a small subdiagonal element from the bottom of the matrix that it can safely set to zero.
PSLASNBT		Computes the position of the sign bit of a floating-point number.
PSLASSQ	PCLASSQ	Updates a sum of squares represented in scaled form.
PSLASWP	PCLASWP	Performs a sequence of row interchanges on a general rectangular matrix.
PSLATRA	PCLATRA	Computes the trace of a general square distributed matrix.
PSLATRD	PCLATRD	Reduces the first nb rows and columns of a symmetric/Hermitian matrix A to real tridiagonal form by an orthogonal/unitary similarity transformation, and returns auxiliary matrices that are needed to apply the transformation to the unreduced part of A.
PSLATRS	PCLATRS	Solves a triangular system of equations $Ax = \sigma b$, $A^T x = \sigma b$, or $A^H x = \sigma b$, where σ is a scale factor set to prevent overflow.
PSLATRZ	PCLATRZ	Reduces the M-by-N real/complex upper trapezoidal matrix to upper triangular form by orthogonal/unitary transformations.
PSLAUU2	PCLAUU2	Computes the product UU^H or $L^H L$, where U and L are upper or lower triangular matrices (local unblocked algorithm).
PSLAUUM	PCLAUUM	Computes the product UU^H or $L^H L$, where U and L are upper or lower triangular matrices.
PSLAWIL		(real) Forms the Wilkinson transform.
PSORG2L	PCUNG2L	Generates all or part of the orthogonal/unitary matrix Q from a QL factorization determined by PSGEQLF/PCGEQLF (unblocked algorithm).
PSORG2R	PCUNG2R	Generates all or part of the orthogonal/unitary matrix Q from a QR factorization determined by PSGEQRF/PCGEQRF (unblocked algorithm).
PSORGL2	PCUNGL2	Generates all or part of the orthogonal/unitary matrix Q from an LQ factorization determined by PSGELQF/PCGELQF (unblocked algorithm).

Routine		Description
Real	Complex	
PSORGR2	PCUNGR2	Generates all or part of the orthogonal/unitary matrix Q from an RQ factorization determined by PSGERQF/PCGERQF (unblocked algorithm).
PSORM2L	PCUNM2L	Multiplies a general matrix by the orthogonal/unitary matrix from a QL factorization determined by PSGEQLF/PCGEQLF (unblocked algorithm).
PSORM2R	PCUNM2R	Multiplies a general matrix by the orthogonal/unitary matrix from a QR factorization determined by PSGEQRF/PCGEQRF (unblocked algorithm).
PSORML2	PCUNML2	Multiplies a general matrix by the orthogonal/unitary matrix from an LQ factorization determined by PSGELQF/PCGELQF (unblocked algorithm).
PSORMR2	PCUNMR2	Multiplies a general matrix by the orthogonal/unitary matrix from an RQ factorization determined by PSGERQF/PCGERQF (unblocked algorithm).
PSPBTRSV	PCPBTRSV	Solves a single triangular linear system via frontsolve or backsolve where the triangular matrix is a factor of a banded matrix computed by PSPBTRF/PCPBTRF.
PSPTTRSV	PCPTTRSV	Solves a single triangular linear system via frontsolve or backsolve where the triangular matrix is a factor of a tridiagonal matrix computed by PSPTTRF/PCPTTRF.
PSPOTF2	PCPOTF2	Computes the Cholesky factorization of a symmetric/Hermitian positive definite matrix (local unblocked algorithm).
PSRSCL	PCSRSCL	Multiplies a vector by the reciprocal of a real scalar.
PSSYGS2	PCHEGS2	Reduces a symmetric/Hermitian definite generalized eigenproblem $Ax = \lambda Bx$, $ABx = \lambda x$, or $BAx = \lambda x$, to standard form, where B has been factorized by PSPOTRF/PCPOTRF (local unblocked algorithm).
PSSYTD2	PCHETD2	Reduces a symmetric/Hermitian matrix to real symmetric tridiagonal form by an orthogonal/unitary similarity transformation (local unblocked algorithm).
PSTRTI2	PCTRTI2	Computes the inverse of a triangular matrix (local unblocked algorithm).
PXERBLA		Error-handling routine called by ScaLAPACK routines if an input parameter has an invalid value.
SDBTF2	CDBTF2	Computes an LU factorization of a general band matrix with no pivoting (local unblocked algorithm).
SDBTRF	CDBTRF	Computes an LU factorization of a general band matrix with no pivoting (local blocked algorithm).

Routine		Description
Real	Complex	
SDTTRF	CDTTRF	Computes an *LU* factorization of a general tridiagonal matrix with no pivoting (local blocked algorithm).
SDTTRSV	CDTTRSV	Solves a general tridiagonal system of linear equations $AX = B$, $A^T X = B$ or $A^H X = B$, using the *LU* factorization computed by SDTTRF/CDTTRF.
SLAMSH		Sends multiple shifts through a small (single node) matrix to see how consecutive small subdiagonal elements are modified by subsequent shifts in an effort to maximize the number of bulges that can be sent through.
SLAREF		Applies one or several Householder reflectors of size 3 to one or two matrices (if column is specified) on either their rows or columns.
SLASORTE		Sorts eigenpairs so that real eigenpairs are together and complex are together. This way one can employ 2-by-2 shifts easily since every 2nd subdiagonal is guaranteed to be zero.
SLASRT2		Modified LAPACK routine SLASRT, which sorts numbers in increasing or decreasing order using Quick Sort, reverting to Insertion sort on arrays of size ≤ 20.
SPTTRSV	CPTTRSV	Solves a symmetric/Hermitian positive definite tridiagonal system of linear equations, using the LDL^H factorization computed by SPTTRF/CPTTRF.
SSTEIN2		Modified LAPACK routine SSTEIN, which computes the eigenvectors of a real symmetric tridiagonal matrix T corresponding to specified eigenvalues, using inverse iteration.
SSTEQR2		Modified LAPACK routine SSTEQR, which computes all eigenvalues and, optionally, eigenvectors of a symmetric tridiagonal matrix using the implicit QL or QR method.

A.3   Matrix Redistribution/Copy Routines

ScaLAPACK provides two matrix redistribution/copy routines for each data type [107, 49, 106]. These routines provide a truly general copy from any block cyclicly distributed (sub)matrix to any other block cyclicly distributed (sub)matrix. These routines are the only ones in the entire ScaLAPACK library which provide *inter-context* operations. By this we mean that they can take a (sub)matrix in context A (distributed over process grid A) and copy it to a (sub)matrix in context B.

There need be no relation between the two operand (sub)matrices other than their global size and the fact that they are both legal block cyclicly distributed (sub)matrices. This means that they may be distributed across different process grids, have varying block sizes, and differing matrix starting points, be contained in different size distributed matrices, etc.

Because of the generality of these routines, they may be used for many operations not usually associated with copy routines. For instance, they may be used to a take a matrix on one process and distribute it across a process grid, or the reverse. If a supercomputer is grouped into a virtual parallel machine with a workstation, for instance, this routine can be used to move the matrix from the workstation to the supercomputer and back. In ScaLAPACK, these routines are called to copy matrices from a two-dimensional process grid to a one-dimensional process grid. They can be used to redistribute matrices so that distributions providing maximal performance can be used by various component libraries, as well. This list of uses is hardly exhaustive, but it gives an idea of the power of a general copy in parallel computing.

The two routine classifications are as follows:

- P_GEMR2D copies between general, rectangular matrices.

- P_TRMR2D copies between trapezoidal matrices.

All routines are available in integer, single precision real, double precision real, single precision complex, and double precision complex. In the following sections, we describe only the singe precision routines for each data type. Double precision routines are the same as their single precision counterparts, but they have names beginning with PD- instead of PS-, or PZ- instead of PC-.

Note that these routines require an array descriptor of type DESC_(DTYPE_)=1.

A.3.1   Fortran Interface

PxGEMR2D

```
    SUBROUTINE PSGEMR2D( M, N, A, IA, JA, DESCA, B, IB, JB, DESCB,
   $                     ICTXT )
    INTEGER          IA, IB, ICTXT, JA, JB, M, N
    INTEGER          DESCA( * ), DESCB( * )
    REAL             A( * ), B( * )
```

```
      SUBROUTINE PCGEMR2D( M, N, A, IA, JA, DESCA, B, IB, JB, DESCB,
   $                       ICTXT )
      INTEGER          IA, IB, ICTXT, JA, JB, M, N
      INTEGER          DESCA( * ), DESCB( * )
      COMPLEX          A( * ), B( * )


      SUBROUTINE PIGEMR2D( M, N, A, IA, JA, DESCA, B, IB, JB, DESCB,
   $                       ICTXT )
      INTEGER          IA, IB, ICTXT, JA, JB, M, N
      INTEGER          DESCA( * ), DESCB( * )
      INTEGER          A( * ), B( * )
```

Purpose

PxGEMR2D copies the indicated (sub)matrix of A to the indicated (sub)matrix of B. A and B can have arbitrary block-cyclic distributions: they can be distributed across different process grids, have different blocking factors, etc.

Particular care must be taken when the process grid over which matrix A is distributed (call this context A) is disjoint from the process grid over which matrix B is distributed (call this context B). The general rules for which parameters need to be set are:

- All calling processes must have the correct M and N.

- Processes in context A must correctly define all parameters describing A.

- Processes in context B must correctly define all parameters describing B.

- Processes which are not members of context A must pass DESCA(CTXT_) = −1 and need not set other parameters describing A.

- Processes which are not members of context B must pass DESCB(CTXT_) = −1 and need not set other parameters describing B.

Arguments

M   (global input) INTEGER
On entry, M specifies the number of rows of the (sub)matrix A to be copied. $M \geq 0$.

N   (global input) INTEGER
On entry, N specifies the number of columns of the (sub)matrix A to be copied. $N \geq$ zero.

A   (local input) REAL/COMPLEX/INTEGER array, dimension (LLD_A,LOCc(JA+N−1))
On entry, the source matrix.

IA (global input) INTEGER
On entry,the global row index of the beginning of the (sub)matrix of A to copy.
$1 \le IA \le M\_A - M + 1$.

JA (global input) INTEGER
On entry,the global column index of the beginning of the (sub)matrix of A to copy.
$1 \le JA \le N\_A - N + 1$.

DESCA (global and local input) INTEGER array, dimension (DLEN_)
The array descriptor for the distributed matrix A.
Only DESCA(DTYPE_)=1 is supported, and thus DLEN_ = 9.
If the calling process is not part of the context of A, then DESCA(CTXT_) must be equal to
−1.

B (local output) REAL/COMPLEX/INTEGER array, dimension (LLD_B,LOCc(JB+N−1))
On exit, the defined (sub)matrix is overwritten by the indicated (sub)matrix from A.

IB (global input) INTEGER
On entry, the global row index of the beginning of the (sub)matrix of B that will be overwritten.
$1 \le IB \le M\_B - M + 1$.

JB (global input) INTEGER
On entry, the global column index of the beginning of the submatrix of B that will be overwritten.
$1 \le JB \le N\_B - N + 1$.

DESCB (global and local input) INTEGER array, dimension (DLEN_)
The array descriptor for the distributed matrix B.
Only DESCB(DTYPE_)=1 is supported, and thus DLEN_ = 9.
If the calling process is not part of the context of B, then DESCB(CTXT_) must be equal to
−1.

ICTXT (global input) INTEGER
The context encompassing at least the union of all processes in context A and context B. All
processes in the context ICTXT must call this routine, even if they do not own a piece of
either matrix.

PxTRMR2D

```
SUBROUTINE PSTRMR2D( UPLO, DIAG, M, N, A, IA, JA, DESCA, B, IB,
$                    JB, DESCB, ICTXT )
CHARACTER          DIAG, UPLO
INTEGER            IA, IB, ICTXT, JA, JB, M, N
INTEGER            DESCA( * ), DESCB( * )
REAL               A( * ), B( * )
```

```
      SUBROUTINE PCTRMR2D( UPLO, DIAG, M, N, A, IA, JA, DESCA, B, IB,
     $                    JB, DESCB, ICTXT )
      CHARACTER          DIAG, UPLO
      INTEGER            IA, IB, ICTXT, JA, JB, M, N
      INTEGER            DESCA( * ), DESCB( * )
      COMPLEX            A( * ), B( * )


      SUBROUTINE PITRMR2D( UPLO, DIAG, M, N, A, IA, JA, DESCA, B, IB,
     $                    JB, DESCB, ICTXT )
      CHARACTER          DIAG, UPLO
      INTEGER            IA, IB, ICTXT, JA, JB, M, N
      INTEGER            DESCA( * ), DESCB( * )
      INTEGER            A( * ), B( * )
```

Purpose

PxTRMR2D copies the indicated (sub)matrix of A to the indicated (sub)matrix of B. A and B can have arbitrary block-cyclic distributions: they can be distributed across different process grids, have different blocking factors, etc.

The (sub)matrix to be copied is assumed to be trapezoidal. So only the upper or the lower part will be copied. The other part is unchanged.

Particular care must be taken when the process grid over which matrix A is distributed (call this context A) is disjoint from the process grid over which matrix B is distributed (call this context B). The general rules for which parameters need to be set are as follows:

- All calling processes must have the correct M and N.

- Processes in context A must correctly define all parameters describing A.

- Processes in context B must correctly define all parameters describing B.

- Processes that are not members of context A must pass DESCA(CTXT_) = −1 and need not set other parameters describing A.

- Processes that are not members of context B must pass DESCB(CTXT_) = −1 and need not set other parameters describing B.

Arguments

UPLO (global input) CHARACTER*1
   On entry, UPLO specifies whether we should copy the upper part or the lower part of the indicated (sub)matrix:

   = 'U': Copy the upper triangular part.

= 'L': Copy the lower triangular part.

DIAG (global input) CHARACTER*1
   On entry, DIAG specifies whether we should copy the diagonal.

   = 'U': Do NOT copy the diagonal of the (sub)matrix.

   = 'N': Do copy the diagonal of the (sub)matrix.

M (global input) INTEGER
   On entry, M specifies the number of rows of the (sub)matrix to be copied. $M \geq 0$.

N (global input) INTEGER
   On entry, N specifies the number of columns of the (sub)matrix to be copied. $N \geq 0$.

A (local input) REAL/COMPLEX/INTEGER array, dimension (LLD_A,LOCc(JA+N−1))
   On entry, the source matrix.

IA (global input) INTEGER
   On entry,the global row index of the beginning of the (sub)matrix of A to copy.
   $1 \leq IA \leq M\_A - M + 1$.

JA (global input) INTEGER
   On entry,the global column index of the beginning of the (sub)matrix of A to copy.
   $1 \leq JA \leq N\_A - N + 1$.

DESCA (global and local input) INTEGER array, dimension (DLEN_)
   The array descriptor for the distributed matrix A.
   Only DESCA(DTYPE_)=1 is supported, and thus DLEN_ = 9.
   If the current process is not part of the context of A, then DESCA(CTXT_) must be equal
   to −1.

B (local output) REAL/COMPLEX/INTEGER array, dimension (LLD_B,LOCc(JB+N−1))
   On exit, the defined (sub)matrix is overwritten by the indicated (sub)matrix from A.

IB (global input) INTEGER
   On entry, the global row index of the beginning of the (sub)matrix of B that will be overwritten.
   $1 \leq IB \leq M\_B - M + 1$.

JB (global input) INTEGER
   On entry, the global column index of the beginning of the submatrix of B that will be overwritten.
   $1 \leq JB \leq N\_B - N + 1$.

DESCB (global and local input) INTEGER array, dimension (DLEN_)
   The array descriptor for the distributed matrix B.
   Only DESCB(DTYPE_)=1 is supported, and thus DLEN_ = 9.
   If the calling process is not part of the context of B, then DESCB(CTXT_) must be equal to
   −1.

ICTXT (global input) INTEGER

   The context encompassing at least the union of all processes in context A and context B. All
   processes in the context ICTXT must call this routine, even if they do not own a piece of
   either matrix.

A.3.2 C Interface

```
void Cp_gemr2d(int m, int n, TYPE *A, int IA, int JA, int *descA,
           TYPE *B, int IB, int JB, int *descB, int gcontext);

void Cp_trmr2d(char *uplo, char *diag, int m, int n, TYPE *A, int IA,
           int JA, int *descA, TYPE *B, int IB, int JB, int *descB,
           int gcontext);
```

where _ and TYPE are as defined below:

_	Data and Precision	Type
s	single-precision real	float
d	double-precision real	double
c	single-precision complex	float
z	double-precision complex	double
i	integer	int

A.3.3 Code Fragment Calling C Interface Cpdgemr2d

```
/* scatter of the matrix A from 1 processor to a P*Q grid */
   Cpdgemr2d(m, n,
           Aseq, ia, ja, &descA_1x1,
           Apar, ib, jb, &descA_PxQ, gcontext);

/* computation of the system solution */
   Cpdgesv( m, n,
           Apar , 1, 1, &descA_PxQ, ipiv ,
           Cpar, 1, 1, &descC_PxQ, &info);

/* gather of the solution matrix C on 1 processor */
   Cpdgemr2d(m, n,
           Cpar, ia, ja, &descC_PxQ,
           Cseq, ib, jb, &descC_1x1, gcontext);
```

Appendix B

Call Conversion: LAPACK to ScaLAPACK and BLAS to PBLAS

This section is designed to assist people in converting serial programs based on calls to the BLAS and LAPACK to parallel programs using the PBLAS and ScaLAPACK.

B.1 Translating BLAS-based programs to the PBLAS

With a concrete understanding of array descriptors (Chapter 4), it is relatively simple to translate the serial version of a BLAS call into its parallel equivalent. Translating BLAS calls to PBLAS calls primarily consists of the following steps:

- a 'P' has to be inserted in front of the routine name,

- the leading dimensions should be replaced by the global *array descriptors*, and

- the global indices into the distributed matrices should be inserted as separate parameters in the calling sequence.

An example of translating a DGEMM call to a PDGEMM call is given below.

```
    CALL DGEMM( 'No transpose', 'No transpose', M-J-JB+1, N-J-JB+1,
   $           JB, -ONE, A( J+JB, J ), LDA, A( J, J+JB ), LDA, ONE,
   $           A( J+JB, J+JB ), LDA )

                                  ↓

    CALL PDGEMM( 'No transpose', 'No transpose', M-J-JB+JA, N-J-JB+JA,
   $           JB, -ONE, A, J+JB, J, DESCA, A, J, J+JB, DESCA, ONE,
   $           A, J+JB, J+JB, DESCA )
```

This simple translation process considerably simplifies the implementation phase of linear algebra codes built on top of the BLAS.

The steps necessary to write a program to call a PBLAS routine are analogous to the steps presented in section 2.4.

B.2 Translating LAPACK-based programs to ScaLAPACK

This section demonstrates how sequential LAPACK-based programs are parallelized and converted to ScaLAPACK.

As with the BLAS conversion, it is relatively simple to translate the serial version of an LAPACK call into its parallel equivalent. Translating LAPACK calls to ScaLAPACK calls primarily consists of the following steps:

- a 'P' has to be inserted in front of the routine name,

- the leading dimensions should be replaced by the global *array descriptors*, and

- the global indices into the distributed matrices should be inserted as separate parameters in the calling sequence.

As an example of this translation process, let us consider the parallelization of the LAPACK driver routine DGESV, which solves a general system of linear equations. The calling sequence comparison for DGESV versus its ScaLAPACK equivalent, PDGESV, is presented below.

```
CALL DGESV( N, NRHS, A( I, J ), LDA, IPIV, B( I, 1 ), LDB, INFO )
```

$$\downarrow$$

```
CALL PDGESV( N, NRHS, A, I, J, DESCA, IPIV, B, I, 1, DESCB, INFO )
```

For a more complete example, let us consider parallelizing a serial LU factorization code, as demonstrated in sections B.2.1 and B.2.2. Note that the parallel routine assumes the existence of the auxiliary routines PDGETF2 (unblocked *LU*) and PDLASWP (parallel swap routine) in addition to the PBLAS. With this in mind, the serial and parallel versions are very similar since most of the details of the parallel implementation such as communication and synchronization are hidden at lower levels of the software.

B.2.1   Sequential LU Factorization

```
      SUBROUTINE DGETRF( M, N, A, LDA, IPIV, INFO )
*
*  LU factorization of a M-by-N matrix A using partial pivoting with
*  row interchanges.
*
      INTEGER           INFO, LDA, M, N, IPIV( * )
      DOUBLE PRECISION  A( LDA, * )
*
      INTEGER           I, IINFO, J, JB, NB
      PARAMETER         ( NB = 64 )
      EXTERNAL          DGEMM, DGETF2, DLASWP, DTRSM
      INTRINSIC         MIN
*
      DO 20 J = 1, MIN(M,N), NB
         JB = MIN( MIN(M,N)-J+1, NB )
*
*        Factor diagonal block and test for exact singularity.
*
         CALL DGETF2( M-J+1, JB, A(J,J), LDA, IPIV(J), IINFO )
*
*        Adjust INFO and the pivot indices.
*
         IF( INFO.EQ.0 .AND. IINFO.GT.0 ) INFO = IINFO + J - 1
         DO 10 I = J, MIN(M,J+JB-1)
            IPIV(I) = J - 1 + IPIV(I)
   10    CONTINUE
*
*        Apply interchanges to columns 1:J-1 and J+JB:N.
*
         CALL DLASWP( J-1, A, LDA, J, J+JB-1, IPIV, 1 )
         IF( J+JB.LE.N ) THEN
            CALL DLASWP( N-J-JB+1, A(1,J+JB), LDA, J, J+JB-1, IPIV, 1 )
*
*        Compute block row of U and update trailing submatrix.
*
         CALL DTRSM( 'Left', 'Lower', 'No transpose', 'Unit', JB,
     $               N-J-JB+1, 1.0D+0, A(J,J), LDA, A(J,J+JB), LDA )
         IF( J+JB.LE.M )
     $      CALL DGEMM( 'No transpose', 'No transpose', M-J-JB+1,
     $                  N-J-JB+1, JB, -1.0D+0, A(J+JB,J), LDA,
     $                  A(J,J+JB), LDA, 1.0D+0, A(J+JB,J+JB), LDA )
         END IF
   20 CONTINUE
      RETURN
*
      END
```

B.2.2 Parallel LU Factorization

```
      SUBROUTINE PDGETRF( M, N, A, IA, JA, DESCA, IPIV, INFO )
*
      INTEGER           IA, INFO, JA, M, N, DESCA( * ), IPIV( * )
      DOUBLE PRECISION  A( * )
*
* LU factorization of a M-by-N distributed matrix A(IA:IA+M-1,JA:JA+N-1)
* using partial pivoting with row interchanges.
*
      INTEGER           I, IINFO, J, JB
      EXTERNAL          IGAMN2D, PDGEMM, PDGETF2, PDLASWP, PDTRSM
      INTRINSIC         MIN
*
      DO 10 J = JA, JA+MIN(M,N)-1, DESCA( NB_ )
         JB = MIN( MIN(M,N)-J+JA, DESCA( NB_ ) )
         I = IA + J - JA
*
*        Factor diagonal block and test for exact singularity.
*
         CALL PDGETF2( M-J+JA, JB, A, I, J, DESCA, IPIV, IINFO )
         IF( INFO.EQ.0 .AND. IINFO.GT.0 ) INFO = IINFO + J - JA
*
*        Apply interchanges to columns JA:J-JA and J+JB:JA+N-1.
*
         CALL PDLASWP( 'Forward', 'Rows', J-JA, A, IA, JA, DESCA,
     $                 I, I+JB-1, IPIV )
         IF( J-JA+JB+1.LE.N ) THEN
            CALL PDLASWP( 'Forward', 'Rows', N-J-JB+JA, A, IA, J+JB,
     $                    DESCA, I, I+JB-1, IPIV )
*
*        Compute block row of U and update trailing submatrix.
*
            CALL PDTRSM( 'Left', 'Lower', 'No transpose', 'Unit', JB,
     $                   N-J-JB+JA, 1.0D+0, A, I, J, DESCA, A, I, J+JB,
     $                   DESCA )
            IF( J-JA+JB+1.LE.M ) THEN
     $         CALL PDGEMM( 'No transpose', 'No transpose', M-J-JB+JA,
     $                      N-J-JB+JA, JB, -1.0D+0, A, I+JB, J, DESCA, A,
     $                      I, J+JB, DESCA, 1.0D+0, A, I+JB, J+JB, DESCA )
         END IF
   10 CONTINUE
      IF( INFO.EQ.0 ) INFO = MIN(M,N) + 1
      CALL IGAMN2D( ICTXT, 'Row', ' ', 1, 1, INFO, 1, I, J, -1, -1, MYCOL )
      IF( INFO.EQ.MIN(M,N)+1 ) INFO = 0
*
      RETURN
*
      END
```

The required steps to call a ScaLAPACK routine from a parallel program are demonstrated in Example Program #1 in Chapter 2 and explained in section 2.3.

Appendix C

Example Programs

This Appendix provides additional example programs. Section C.1 presents a more memory-efficient program calling the ScaLAPACK driver routine **PDGESV**. Section C.2 presents an HPF program calling the HPF interface to ScaLAPACK. These example programs are available from the respective URLs:

```
http://www.netlib.org/scalapack/examples/scaex.shar
http://www.netlib.org/scalapack/examples/sample_hpf_gesv.f
```

C.1   Example Program #2

In Chapter 2, we presented an example program using ScaLAPACK. Here we present a second example—a more flexible and memory efficient program to solve a system of linear equations using the ScaLAPACK driver routine **PDGESV**. In this example we will read the input matrices from a file, distribute these matrices to the processes in the grid. After calling the ScaLAPACK routine, we will write the output solution matrix to a file. The input data files for the program are **SCAEX.dat, SCAEXMAT.dat, and SCAEXRHS.dat.**

This program is also available in the scalapack directory on netlib (http://www.netlib.org/scalapack/examples/scaex.shar).

SCAEX.dat is:

```
'ScaLAPACK Example Program 2'
'May 1997'
'SCAEX.out'              output file name (if any)
6                       device out
6                       value of N
1                       value of NRHS
2                       values of NB
2                       values of NPROW
2                       values of NPCOL
```

SCAEXMAT.dat is:

181

```
6 6
 6.0000D+0
 3.0000D+0
 0.0000D+0
 0.0000D+0
 3.0000D+0
 0.0000D+0
 0.0000D+0
-3.0000D+0
-1.0000D+0
 1.0000D+0
 1.0000D+0
 0.0000D+0
-1.0000D+0
 0.0000D+0
11.0000D+0
 0.0000D+0
 0.0000D+0
10.0000D+0
 0.0000D+0
 0.0000D+0
 0.0000D+0
-11.0000D+0
 0.0000D+0
 0.0000D+0
 0.0000D+0
 0.0000D+0
 0.0000D+0
 2.0000D+0
-4.0000D+0
 0.0000D+0
 0.0000D+0
 0.0000D+0
 0.0000D+0
 8.0000D+0
 0.0000D+0
-10.0000D+0
```

SCAEXRHS.dat is:

```
6  1
  72.000000000000000000D+00
   0.000000000000000000D+00
 160.000000000000000000D+00
   0.000000000000000000D+00
   0.000000000000000000D+00
   0.000000000000000000D+00
```

```
      PROGRAM PDSCAEX
*
*  -- ScaLAPACK example code --
*     University of Tennessee, Knoxville, Oak Ridge National Laboratory,
*     and University of California, Berkeley.
*
*     Written by Antoine Petitet, (petitet@cs.utk.edu)
*
*     This program solves a linear system by calling the ScaLAPACK
*     routine PDGESV. The input matrix and right-and-sides are
*     read from a file. The solution is written to a file.
*
*     .. Parameters ..
      INTEGER           DBLESZ, INTGSZ, MEMSIZ, TOTMEM
      PARAMETER         ( DBLESZ = 8, INTGSZ = 4, TOTMEM = 2000000,
     $                    MEMSIZ = TOTMEM / DBLESZ )
      INTEGER           BLOCK_CYCLIC_2D, CSRC_, CTXT_, DLEN_, DTYPE_,
     $                  LLD_, MB_, M_, NB_, N_, RSRC_
      PARAMETER         ( BLOCK_CYCLIC_2D = 1, DLEN_ = 9, DTYPE_ = 1,
     $                    CTXT_ = 2, M_ = 3, N_ = 4, MB_ = 5, NB_ = 6,
     $                    RSRC_ = 7, CSRC_ = 8, LLD_ = 9 )
      DOUBLE PRECISION  ONE
      PARAMETER         ( ONE = 1.0D+0 )
*     ..
*     .. Local Scalars ..
      CHARACTER*80      OUTFILE
      INTEGER           IAM, ICTXT, INFO, IPA, IPACPY, IPB, IPPIV, IPX,
     $                  IPW, LIPIV, MYCOL, MYROW, N, NB, NOUT, NPCOL,
     $                  NPROCS, NPROW, NP, NQ, NQRHS, NRHS, WORKSIZ
      DOUBLE PRECISION  ANORM, BNORM, EPS, XNORM, RESID
*     ..
*     .. Local Arrays ..
      INTEGER           DESCA( DLEN_ ), DESCB( DLEN_ ), DESCX( DLEN_ )
      DOUBLE PRECISION  MEM( MEMSIZ )
*     ..
*     .. External Subroutines ..
      EXTERNAL          BLACS_EXIT, BLACS_GET, BLACS_GRIDEXIT,
     $                  BLACS_GRIDINFO, BLACS_GRIDINIT, BLACS_PINFO,
     $                  DESCINIT, IGSUM2D, PDSCAEXINFO, PDGESV,
     $                  PDGEMM, PDLACPY, PDLAPRNT, PDLAREAD, PDLAWRITE
*     ..
*     .. External Functions ..
      INTEGER           ICEIL, NUMROC
      DOUBLE PRECISION  PDLAMCH, PDLANGE
      EXTERNAL          ICEIL, NUMROC, PDLAMCH, PDLANGE
*     ..
*     .. Intrinsic Functions ..
      INTRINSIC         DBLE, MAX
*     ..
*     .. Executable Statements ..
*
*     Get starting information
*
      CALL BLACS_PINFO( IAM, NPROCS )
      CALL PDSCAEXINFO( OUTFILE, NOUT, N, NRHS, NB, NPROW, NPCOL, MEM,
     $                  IAM, NPROCS )
*
*     Define process grid
*
      CALL BLACS_GET( -1, 0, ICTXT )
      CALL BLACS_GRIDINIT( ICTXT, 'Row-major', NPROW, NPCOL )
      CALL BLACS_GRIDINFO( ICTXT, NPROW, NPCOL, MYROW, MYCOL )
*
*     Go to bottom of process grid loop if this case doesn't use my
```

```
*     process
*
      IF( MYROW.GE.NPROW .OR. MYCOL.GE.NPCOL )
     $   GO TO 20
*
      NP    = NUMROC( N, NB, MYROW, 0, NPROW )
      NQ    = NUMROC( N, NB, MYCOL, 0, NPCOL )
      NQRHS = NUMROC( NRHS, NB, MYCOL, 0, NPCOL )
*
*     Initialize the array descriptor for the matrix A and B
*
      CALL DESCINIT( DESCA, N, N, NB, NB, 0, 0, ICTXT, MAX( 1, NP ),
     $              INFO )
      CALL DESCINIT( DESCB, N, NRHS, NB, NB, 0, 0, ICTXT, MAX( 1, NP ),
     $              INFO )
      CALL DESCINIT( DESCX, N, NRHS, NB, NB, 0, 0, ICTXT, MAX( 1, NP ),
     $              INFO )
*
*     Assign pointers into MEM for SCALAPACK arrays, A is
*     allocated starting at position MEM( 1 )
*
      IPA = 1
      IPACPY = IPA + DESCA( LLD_ )*NQ
      IPB = IPACPY + DESCA( LLD_ )*NQ
      IPX = IPB + DESCB( LLD_ )*NQRHS
      IPPIV = IPX + DESCB( LLD_ )*NQRHS
      LIPIV = ICEIL( INTGSZ*( NP+NB ), DBLESZ )
      IPW = IPPIV + MAX( NP, LIPIV )
*
      WORKSIZ = MAX( NB, NP )
*
*     Check for adequate memory for problem size
*
      INFO = 0
      IF( IPW+WORKSIZ.GT.MEMSIZ ) THEN
         IF( IAM.EQ.0 )
     $      WRITE( NOUT, FMT = 9998 ) 'test', ( IPW+WORKSIZ )*DBLESZ
         INFO = 1
      END IF
*
*     Check all processes for an error
*
      CALL IGSUM2D( ICTXT, 'All', ' ', 1, 1, INFO, 1, -1, 0 )
      IF( INFO.GT.0 ) THEN
         IF( IAM.EQ.0 )
     $      WRITE( NOUT, FMT = 9999 ) 'MEMORY'
         GO TO 10
      END IF
*
*     Read from file and distribute matrices A and B
*
      CALL PDLAREAD( 'SCAEXMAT.dat', MEM( IPA ), DESCA, 0, 0,
     $              MEM( IPW ) )
      CALL PDLAREAD( 'SCAEXRHS.dat', MEM( IPB ), DESCB, 0, 0,
     $              MEM( IPW ) )
*
*     Make a copy of A and the rhs for checking purposes
*
      CALL PDLACPY( 'All', N, N, MEM( IPA ), 1, 1, DESCA,
     $              MEM( IPACPY ), 1, 1, DESCA )
      CALL PDLACPY( 'All', N, NRHS, MEM( IPB ), 1, 1, DESCB,
     $              MEM( IPX ), 1, 1, DESCX )
*
*********************************************************************
```

```
*     Call ScaLAPACK PDGESV routine
****************************************************************************
*
      IF( IAM.EQ.0 ) THEN
         WRITE( NOUT, FMT = * )
         WRITE( NOUT, FMT = * )
     $         '************************************************'
         WRITE( NOUT, FMT = * )
     $         'Example of ScaLAPACK routine call: (PDGESV)'
         WRITE( NOUT, FMT = * )
     $         '************************************************'
         WRITE( NOUT, FMT = * )
         WRITE( NOUT, FMT = * ) 'A * X = B, Matrix A:'
         WRITE( NOUT, FMT = * )
      END IF
      CALL PDLAPRNT( N, N, MEM( IPA ), 1, 1, DESCA, 0, 0,
     $              'A', NOUT, MEM( IPW ) )
      IF( IAM.EQ.0 ) THEN
         WRITE( NOUT, FMT = * )
         WRITE( NOUT, FMT = * ) 'Matrix B:'
         WRITE( NOUT, FMT = * )
      END IF
      CALL PDLAPRNT( N, NRHS, MEM( IPB ), 1, 1, DESCB, 0, 0,
     $              'B', NOUT, MEM( IPW ) )
*
      CALL PDGESV( N, NRHS, MEM( IPA ), 1, 1, DESCA, MEM( IPPIV ),
     $            MEM( IPB ), 1, 1, DESCB, INFO )
*
      IF( MYROW.EQ.0 .AND. MYCOL.EQ.0 ) THEN
         WRITE( NOUT, FMT = * )
         WRITE( NOUT, FMT = * ) 'INFO code returned by PDGESV = ', INFO
         WRITE( NOUT, FMT = * )
         WRITE( NOUT, FMT = * ) 'Matrix X = A^{-1} * B'
         WRITE( NOUT, FMT = * )
      END IF
      CALL PDLAPRNT( N, NRHS, MEM( IPB ), 1, 1, DESCB, 0, 0, 'X', NOUT,
     $              MEM( IPW ) )
      CALL PDLAWRITE( 'SCAEXSOL.dat', N, NRHS, MEM( IPB ), 1, 1, DESCB,
     $               0, 0, MEM( IPW ) )
*
*     Compute residual ||A * X  - B|| / ( ||X|| * ||A|| * eps * N )
*
      EPS = PDLAMCH( ICTXT, 'Epsilon' )
      ANORM = PDLANGE( 'I', N, N, MEM( IPA ), 1, 1, DESCA, MEM( IPW ) )
      BNORM = PDLANGE( 'I', N, NRHS, MEM( IPB ), 1, 1, DESCB,
     $                MEM( IPW ) )
      CALL PDGEMM( 'No transpose', 'No transpose', N, NRHS, N, ONE,
     $            MEM( IPACPY ), 1, 1, DESCA, MEM( IPB ), 1, 1, DESCB,
     $            -ONE, MEM( IPX ), 1, 1, DESCX )
      XNORM = PDLANGE( 'I', N, NRHS, MEM( IPX ), 1, 1, DESCX,
     $                MEM( IPW ) )
      RESID = XNORM / ( ANORM * BNORM * EPS * DBLE( N ) )
*
      IF( MYROW.EQ.0 .AND. MYCOL.EQ.0 ) THEN
         WRITE( NOUT, FMT = * )
         WRITE( NOUT, FMT = * )
     $      '||A * X  - B|| / ( ||X|| * ||A|| * eps * N ) = ', RESID
         WRITE( NOUT, FMT = * )
         IF( RESID.LT.10.0D+0 ) THEN
            WRITE( NOUT, FMT = * ) 'The answer is correct.'
         ELSE
            WRITE( NOUT, FMT = * ) 'The answer is suspicious.'
         END IF
      END IF
```

```
*
   10 CONTINUE
*
      CALL BLACS_GRIDEXIT( ICTXT )
*
   20 CONTINUE
*
*     Print ending messages and close output file
*
      IF( IAM.EQ.0 ) THEN
         WRITE( NOUT, FMT = * )
         WRITE( NOUT, FMT = * )
         WRITE( NOUT, FMT = 9997 )
         WRITE( NOUT, FMT = * )
         IF( NOUT.NE.6 .AND. NOUT.NE.0 )
     $      CLOSE ( NOUT )
      END IF
*
      CALL BLACS_EXIT( 0 )
*
 9999 FORMAT( 'Bad ', A6, ' parameters: going on to next test case.' )
 9998 FORMAT( 'Unable to perform ', A, ': need TOTMEM of at least',
     $         I11 )
 9997 FORMAT( 'END OF TESTS.' )
*
      STOP
*
*     End of PDSCAEX
*
      END
```

```
      SUBROUTINE PDSCAEXINFO( SUMMRY, NOUT, N, NRHS, NB, NPROW, NPCOL,
     $                        WORK, IAM, NPROCS )
*
*  -- ScaLAPACK example code --
*     University of Tennessee, Knoxville, Oak Ridge National Laboratory,
*     and University of California, Berkeley.
*
*     Written by Antoine Petitet, (petitet@cs.utk.edu)
*
*     This program solves a linear system by calling the ScaLAPACK
*     routine PDGESV. The input matrix and right-and-sides are
*     read from a file. The solution is written to a file.
*
*     .. Scalar Arguments ..
      CHARACTER*( * )    SUMMRY
      INTEGER            IAM, N, NRHS, NB, NOUT, NPCOL, NPROCS, NPROW
*     ..
*     .. Array Arguments ..
      INTEGER            WORK( * )
*     ..
*
* =======================================================================
*
*     .. Parameters ..
      INTEGER            NIN
      PARAMETER          ( NIN = 11 )
*     ..
*     .. Local Scalars ..
      CHARACTER*79       USRINFO
      INTEGER            ICTXT
*     ..
*     .. External Subroutines ..
      EXTERNAL           BLACS_ABORT, BLACS_GET, BLACS_GRIDEXIT,
     $                   BLACS_GRIDINIT, BLACS_SETUP, IGEBR2D, IGEBS2D
*     ..
*     .. Intrinsic Functions ..
      INTRINSIC          MAX, MIN
*     ..
*     .. Executable Statements ..
*
*     Process 0 reads the input data, broadcasts to other processes and
*     writes needed information to NOUT
*
      IF( IAM.EQ.0 ) THEN
*
*        Open file and skip data file header
*
         OPEN( NIN, FILE='SCAEX.dat', STATUS='OLD' )
         READ( NIN, FMT = * ) SUMMRY
         SUMMRY = ' '
*
*        Read in user-supplied info about machine type, compiler, etc.
*
         READ( NIN, FMT = 9999 ) USRINFO
*
*        Read name and unit number for summary output file
*
         READ( NIN, FMT = * ) SUMMRY
         READ( NIN, FMT = * ) NOUT
         IF( NOUT.NE.0 .AND. NOUT.NE.6 )
     $      OPEN( NOUT, FILE = SUMMRY, STATUS = 'UNKNOWN' )
*
*        Read and check the parameter values for the tests.
*
```

```
*         Get matrix dimensions
*
          READ( NIN, FMT = * ) N
          READ( NIN, FMT = * ) NRHS
*
*         Get value of NB
*
          READ( NIN, FMT = * ) NB
*
*         Get grid shape
*
          READ( NIN, FMT = * ) NPROW
          READ( NIN, FMT = * ) NPCOL
*
*         Close input file
*
          CLOSE( NIN )
*
*         If underlying system needs additional set up, do it now
*
          IF( NPROCS.LT.1 ) THEN
             NPROCS = NPROW * NPCOL
             CALL BLACS_SETUP( IAM, NPROCS )
          END IF
*
*         Temporarily define blacs grid to include all processes so
*         information can be broadcast to all processes
*
          CALL BLACS_GET( -1, 0, ICTXT )
          CALL BLACS_GRIDINIT( ICTXT, 'Row-major', 1, NPROCS )
*
*         Pack information arrays and broadcast
*
          WORK( 1 ) = N
          WORK( 2 ) = NRHS
          WORK( 3 ) = NB
          WORK( 4 ) = NPROW
          WORK( 5 ) = NPCOL
          CALL IGEBS2D( ICTXT, 'All', ' ', 5, 1, WORK, 5 )
*
*         regurgitate input
*
          WRITE( NOUT, FMT = 9999 )
     $                  'SCALAPACK example driver.'
          WRITE( NOUT, FMT = 9999 ) USRINFO
          WRITE( NOUT, FMT = * )
          WRITE( NOUT, FMT = 9999 )
     $                  'The matrices A and B are read from '//
     $                  'a file.'
          WRITE( NOUT, FMT = * )
          WRITE( NOUT, FMT = 9999 )
     $                  'An explanation of the input/output '//
     $                  'parameters follows:'
*
          WRITE( NOUT, FMT = 9999 )
     $                  'N        : The order of the matrix A.'
          WRITE( NOUT, FMT = 9999 )
     $                  'NRHS     : The number of right and sides.'
          WRITE( NOUT, FMT = 9999 )
     $                  'NB       : The size of the square blocks the'//
     $                  ' matrices A and B are split into.'
          WRITE( NOUT, FMT = 9999 )
     $                  'P        : The number of process rows.'
          WRITE( NOUT, FMT = 9999 )
```

```
     $                'Q      : The number of process columns.'
           WRITE( NOUT, FMT = * )
           WRITE( NOUT, FMT = 9999 )
     $                'The following parameter values will be used:'
           WRITE( NOUT, FMT = 9998 ) 'N     ', N
           WRITE( NOUT, FMT = 9998 ) 'NRHS ', NRHS
           WRITE( NOUT, FMT = 9998 ) 'NB   ', NB
           WRITE( NOUT, FMT = 9998 ) 'P    ', NPROW
           WRITE( NOUT, FMT = 9998 ) 'Q    ', NPCOL
           WRITE( NOUT, FMT = * )
*
      ELSE
*
*         If underlying system needs additional set up, do it now
*
           IF( NPROCS.LT.1 )
     $        CALL BLACS_SETUP( IAM, NPROCS )
*
*         Temporarily define blacs grid to include all processes so
*         information can be broadcast to all processes
*
           CALL BLACS_GET( -1, 0, ICTXT )
           CALL BLACS_GRIDINIT( ICTXT, 'Row-major', 1, NPROCS )
*
           CALL IGEBR2D( ICTXT, 'All', ' ', 5, 1, WORK, 5, 0, 0 )
           N     = WORK( 1 )
           NRHS  = WORK( 2 )
           NB    = WORK( 3 )
           NPROW = WORK( 4 )
           NPCOL = WORK( 5 )
*
      END IF
*
      CALL BLACS_GRIDEXIT( ICTXT )
*
      RETURN
*
   20 WRITE( NOUT, FMT = 9997 )
      CLOSE( NIN )
      IF( NOUT.NE.6 .AND. NOUT.NE.0 )
     $   CLOSE( NOUT )
      CALL BLACS_ABORT( ICTXT, 1 )
*
      STOP
*
 9999 FORMAT( A )
 9998 FORMAT( 2X, A5, '  :          ', I6 )
 9997 FORMAT( ' Illegal input in file ',40A,'.  Aborting run.' )
*
*     End of PDSCAEXINFO
*
      END
```

```
      SUBROUTINE PDLAREAD( FILNAM, A, DESCA, IRREAD, ICREAD, WORK )
*
*  -- ScaLAPACK example --
*     University of Tennessee, Knoxville, Oak Ridge National Laboratory,
*     and University of California, Berkeley.
*
*     written by Antoine Petitet, (petitet@cs.utk.edu)
*
*     .. Scalar Arguments ..
      INTEGER            ICREAD, IRREAD
*     ..
*     .. Array Arguments ..
      CHARACTER*(*)      FILNAM
      INTEGER            DESCA( * )
      DOUBLE PRECISION   A( * ), WORK( * )
*     ..
*
*  Purpose
*  =======
*
*  PDLAREAD reads from a file named FILNAM a matrix and distribute
*  it to the process grid.
*
*  Only the process of coordinates {IRREAD, ICREAD} read the file.
*
*  WORK must be of size >= MB_ = DESCA( MB_ ).
*
*  =====================================================================
*
*     .. Parameters ..
      INTEGER            NIN
      PARAMETER          ( NIN = 11 )
      INTEGER            BLOCK_CYCLIC_2D, CSRC_, CTXT_, DLEN_, DTYPE_,
     $                   LLD_, MB_, M_, NB_, N_, RSRC_
      PARAMETER          ( BLOCK_CYCLIC_2D = 1, DLEN_ = 9, DTYPE_ = 1,
     $                     CTXT_ = 2, M_ = 3, N_ = 4, MB_ = 5, NB_ = 6,
     $                     RSRC_ = 7, CSRC_ = 8, LLD_ = 9 )
*     ..
*     .. Local Scalars ..
      INTEGER            H, I, IB, ICTXT, ICURCOL, ICURROW, II, J, JB,
     $                   JJ, K, LDA, M, MYCOL, MYROW, N, NPCOL, NPROW
*     ..
*     .. Local Arrays ..
      INTEGER            IWORK( 2 )
*     ..
*     .. External Subroutines ..
      EXTERNAL           BLACS_GRIDINFO, INFOG2L, DGERV2D, DGESD2D,
     $                   IGEBS2D, IGEBR2D
*     ..
*     .. External Functions ..
      INTEGER            ICEIL
      EXTERNAL           ICEIL
*     ..
*     .. Intrinsic Functions ..
      INTRINSIC          MIN
*     ..
*     .. Executable Statements ..
*
*     Get grid parameters
*
      ICTXT = DESCA( CTXT_ )
      CALL BLACS_GRIDINFO( ICTXT, NPROW, NPCOL, MYROW, MYCOL )
*
      IF( MYROW.EQ.IRREAD .AND. MYCOL.EQ.ICREAD ) THEN
```

```
              OPEN( NIN, FILE=FILNAM, STATUS='OLD' )
              READ( NIN, FMT = * ) ( IWORK( I ), I = 1, 2 )
              CALL IGEBS2D( ICTXT, 'All', ' ', 2, 1, IWORK, 2 )
          ELSE
              CALL IGEBR2D( ICTXT, 'All', ' ', 2, 1, IWORK, 2, IRREAD,
     $                     ICREAD )
          END IF
          M = IWORK( 1 )
          N = IWORK( 2 )
*
          IF( M.LE.0 .OR. N.LE.0 )
     $      RETURN
*
          IF( M.GT.DESCA( M_ ).OR. N.GT.DESCA( N_ ) ) THEN
              IF( MYROW.EQ.0 .AND. MYCOL.EQ.0 ) THEN
                  WRITE( *, FMT = * ) 'PDLAREAD: Matrix too big to fit in'
                  WRITE( *, FMT = * ) 'Abort ...'
              END IF
              CALL BLACS_ABORT( ICTXT, 0 )
          END IF
*
          II = 1
          JJ = 1
          ICURROW = DESCA( RSRC_ )
          ICURCOL = DESCA( CSRC_ )
          LDA = DESCA( LLD_ )
*
*         Loop over column blocks
*
          DO 50 J = 1, N, DESCA( NB_ )
             JB = MIN( DESCA( NB_ ), N-J+1 )
             DO 40 H = 0, JB-1
*
*               Loop over block of rows
*
                DO 30 I = 1, M, DESCA( MB_ )
                   IB = MIN( DESCA( MB_ ), M-I+1 )
                   IF( ICURROW.EQ.IRREAD .AND. ICURCOL.EQ.ICREAD ) THEN
                      IF( MYROW.EQ.IRREAD .AND. MYCOL.EQ.ICREAD ) THEN
                         DO 10 K = 0, IB-1
                            READ( NIN, FMT = * ) A( II+K+(JJ+H-1)*LDA )
 10                      CONTINUE
                      END IF
                   ELSE
                      IF( MYROW.EQ.ICURROW .AND. MYCOL.EQ.ICURCOL ) THEN
                         CALL DGERV2D( ICTXT, IB, 1, A( II+(JJ+H-1)*LDA ),
     $                                LDA, IRREAD, ICREAD )
                      ELSE IF( MYROW.EQ.IRREAD .AND. MYCOL.EQ.ICREAD ) THEN
                         DO 20 K = 1, IB
                            READ( NIN, FMT = * ) WORK( K )
 20                      CONTINUE
                         CALL DGESD2D( ICTXT, IB, 1, WORK, DESCA( MB_ ),
     $                                ICURROW, ICURCOL )
                      END IF
                   END IF
                   IF( MYROW.EQ.ICURROW )
     $                II = II + IB
                   ICURROW = MOD( ICURROW+1, NPROW )
 30             CONTINUE
*
                II = 1
                ICURROW = DESCA( RSRC_ )
 40          CONTINUE
*
```

```
         IF( MYCOL.EQ.ICURCOL )
   $        JJ = JJ + JB
         ICURCOL = MOD( ICURCOL+1, NPCOL )
*
   50 CONTINUE
*
      IF( MYROW.EQ.IRREAD .AND. MYCOL.EQ.ICREAD ) THEN
         CLOSE( NIN )
      END IF
*
      RETURN
*
*     End of PDLAREAD
*
      END
```

```
      SUBROUTINE PDLAWRITE( FILNAM, M, N, A, IA, JA, DESCA, IRWRIT,
     $                      ICWRIT, WORK )
*
*  -- ScaLAPACK example --
*     University of Tennessee, Knoxville, Oak Ridge National Laboratory,
*     and University of California, Berkeley.
*
*     written by Antoine Petitet, (petitet@cs.utk.edu)
*
*     .. Scalar Arguments ..
      INTEGER            IA, ICWRIT, IRWRIT, JA, M, N
*     ..
*     .. Array Arguments ..
      CHARACTER*(*)      FILNAM
      INTEGER            DESCA( * )
      DOUBLE PRECISION   A( * ), WORK( * )
*     ..
*
*  Purpose
*  =======
*
*  PDLAWRITE writes to a file named FILNAMa distributed matrix sub( A )
*  denoting A(IA:IA+M-1,JA:JA+N-1). The local pieces are sent to and
*  written by the process of coordinates (IRWWRITE, ICWRIT).
*
*  WORK must be of size >= MB_ = DESCA( MB_ ).
*
*  =====================================================================
*
*     .. Parameters ..
      INTEGER            NOUT
      PARAMETER          ( NOUT = 13 )
      INTEGER            BLOCK_CYCLIC_2D, CSRC_, CTXT_, DLEN_, DTYPE_,
     $                   LLD_, MB_, M_, NB_, N_, RSRC_
      PARAMETER          ( BLOCK_CYCLIC_2D = 1, DLEN_ = 9, DTYPE_ = 1,
     $                     CTXT_ = 2, M_ = 3, N_ = 4, MB_ = 5, NB_ = 6,
     $                     RSRC_ = 7, CSRC_ = 8, LLD_ = 9 )
*     ..
*     .. Local Scalars ..
      INTEGER            H, I, IACOL, IAROW, IB, ICTXT, ICURCOL,
     $                   ICURROW, II, IIA, IN, J, JB, JJ, JJA, JN, K,
     $                   LDA, MYCOL, MYROW, NPCOL, NPROW
*     ..
*     .. External Subroutines ..
      EXTERNAL           BLACS_BARRIER, BLACS_GRIDINFO, INFOG2L,
     $                   DGERV2D, DGESD2D
*     ..
*     .. External Functions ..
      INTEGER            ICEIL
      EXTERNAL           ICEIL
*     ..
*     .. Intrinsic Functions ..
      INTRINSIC          MIN
*     ..
*     .. Executable Statements ..
*
*     Get grid parameters
*
      ICTXT = DESCA( CTXT_ )
      CALL BLACS_GRIDINFO( ICTXT, NPROW, NPCOL, MYROW, MYCOL )
*
      IF( MYROW.EQ.IRWRIT .AND. MYCOL.EQ.ICWRIT ) THEN
         OPEN( NOUT, FILE=FILNAM, STATUS='UNKNOWN' )
         WRITE( NOUT, FMT = * ) M, N
```

```fortran
      END IF
*
      CALL INFOG2L( IA, JA, DESCA, NPROW, NPCOL, MYROW, MYCOL,
     $              IIA, JJA, IAROW, IACOL )
      ICURROW = IAROW
      ICURCOL = IACOL
      II = IIA
      JJ = JJA
      LDA = DESCA( LLD_ )
*
*     Handle the first block of column separately
*
      JN = MIN( ICEIL( JA, DESCA( NB_ ) ) * DESCA( NB_ ), JA+N-1 )
      JB = JN-JA+1
      DO 60 H = 0, JB-1
         IN = MIN( ICEIL( IA, DESCA( MB_ ) ) * DESCA( MB_ ), IA+M-1 )
         IB = IN-IA+1
         IF( ICURROW.EQ.IRWRIT .AND. ICURCOL.EQ.ICWRIT ) THEN
            IF( MYROW.EQ.IRWRIT .AND. MYCOL.EQ.ICWRIT ) THEN
               DO 10 K = 0, IB-1
                  WRITE( NOUT, FMT = 9999 ) A( II+K+(JJ+H-1)*LDA )
   10          CONTINUE
            END IF
         ELSE
            IF( MYROW.EQ.ICURROW .AND. MYCOL.EQ.ICURCOL ) THEN
               CALL DGESD2D( ICTXT, IB, 1, A( II+(JJ+H-1)*LDA ), LDA,
     $                       IRWRIT, ICWRIT )
            ELSE IF( MYROW.EQ.IRWRIT .AND. MYCOL.EQ.ICWRIT ) THEN
               CALL DGERV2D( ICTXT, IB, 1, WORK, DESCA( MB_ ),
     $                       ICURROW, ICURCOL )
               DO 20 K = 1, IB
                  WRITE( NOUT, FMT = 9999 ) WORK( K )
   20          CONTINUE
            END IF
         END IF
         IF( MYROW.EQ.ICURROW )
     $      II = II + IB
         ICURROW = MOD( ICURROW+1, NPROW )
         CALL BLACS_BARRIER( ICTXT, 'All' )
*
*        Loop over remaining block of rows
*
         DO 50 I = IN+1, IA+M-1, DESCA( MB_ )
            IB = MIN( DESCA( MB_ ), IA+M-I )
            IF( ICURROW.EQ.IRWRIT .AND. ICURCOL.EQ.ICWRIT ) THEN
               IF( MYROW.EQ.IRWRIT .AND. MYCOL.EQ.ICWRIT ) THEN
                  DO 30 K = 0, IB-1
                     WRITE( NOUT, FMT = 9999 ) A( II+K+(JJ+H-1)*LDA )
   30             CONTINUE
               END IF
            ELSE
               IF( MYROW.EQ.ICURROW .AND. MYCOL.EQ.ICURCOL ) THEN
                  CALL DGESD2D( ICTXT, IB, 1, A( II+(JJ+H-1)*LDA ),
     $                          LDA, IRWRIT, ICWRIT )
               ELSE IF( MYROW.EQ.IRWRIT .AND. MYCOL.EQ.ICWRIT ) THEN
                  CALL DGERV2D( ICTXT, IB, 1, WORK, DESCA( MB_ ),
     $                          ICURROW, ICURCOL )
                  DO 40 K = 1, IB
                     WRITE( NOUT, FMT = 9999 ) WORK( K )
   40             CONTINUE
               END IF
            END IF
            IF( MYROW.EQ.ICURROW )
     $         II = II + IB
```

```
               ICURROW = MOD( ICURROW+1, NPROW )
               CALL BLACS_BARRIER( ICTXT, 'All' )
   50      CONTINUE
*
         II = IIA
         ICURROW = IAROW
   60 CONTINUE
*
      IF( MYCOL.EQ.ICURCOL )
     $   JJ = JJ + JB
      ICURCOL = MOD( ICURCOL+1, NPCOL )
      CALL BLACS_BARRIER( ICTXT, 'All' )
*
*     Loop over remaining column blocks
*
      DO 130 J = JN+1, JA+N-1, DESCA( NB_ )
         JB = MIN( DESCA( NB_ ), JA+N-J )
         DO 120 H = 0, JB-1
            IN = MIN( ICEIL( IA, DESCA( MB_ ) ) * DESCA( MB_ ), IA+M-1 )
            IB = IN-IA+1
            IF( ICURROW.EQ.IRWRIT .AND. ICURCOL.EQ.ICWRIT ) THEN
               IF( MYROW.EQ.IRWRIT .AND. MYCOL.EQ.ICWRIT ) THEN
                  DO 70 K = 0, IB-1
                     WRITE( NOUT, FMT = 9999 ) A( II+K+(JJ+H-1)*LDA )
   70             CONTINUE
               END IF
            ELSE
               IF( MYROW.EQ.ICURROW .AND. MYCOL.EQ.ICURCOL ) THEN
                  CALL DGESD2D( ICTXT, IB, 1, A( II+(JJ+H-1)*LDA ),
     $                          LDA, IRWRIT, ICWRIT )
               ELSE IF( MYROW.EQ.IRWRIT .AND. MYCOL.EQ.ICWRIT ) THEN
                  CALL DGERV2D( ICTXT, IB, 1, WORK, DESCA( MB_ ),
     $                          ICURROW, ICURCOL )
                  DO 80 K = 1, IB
                     WRITE( NOUT, FMT = 9999 ) WORK( K )
   80             CONTINUE
               END IF
            END IF
            IF( MYROW.EQ.ICURROW )
     $         II = II + IB
            ICURROW = MOD( ICURROW+1, NPROW )
            CALL BLACS_BARRIER( ICTXT, 'All' )
*
*           Loop over remaining block of rows
*
            DO 110 I = IN+1, IA+M-1, DESCA( MB_ )
               IB = MIN( DESCA( MB_ ), IA+M-I )
               IF( ICURROW.EQ.IRWRIT .AND. ICURCOL.EQ.ICWRIT ) THEN
                  IF( MYROW.EQ.IRWRIT .AND. MYCOL.EQ.ICWRIT ) THEN
                     DO 90 K = 0, IB-1
                        WRITE( NOUT, FMT = 9999 ) A( II+K+(JJ+H-1)*LDA )
   90                CONTINUE
                  END IF
               ELSE
                  IF( MYROW.EQ.ICURROW .AND. MYCOL.EQ.ICURCOL ) THEN
                     CALL DGESD2D( ICTXT, IB, 1, A( II+(JJ+H-1)*LDA ),
     $                             LDA, IRWRIT, ICWRIT )
                   ELSE IF( MYROW.EQ.IRWRIT .AND. MYCOL.EQ.ICWRIT ) THEN
                     CALL DGERV2D( ICTXT, IB, 1, WORK, DESCA( MB_ ),
     $                             ICURROW, ICURCOL )
                     DO 100 K = 1, IB
                        WRITE( NOUT, FMT = 9999 ) WORK( K )
  100                CONTINUE
                  END IF
```

```
                  END IF
                  IF( MYROW.EQ.ICURROW )
     $                II = II + IB
                  ICURROW = MOD( ICURROW+1, NPROW )
                  CALL BLACS_BARRIER( ICTXT, 'All' )
  110        CONTINUE
*
             II = IIA
             ICURROW = IAROW
  120     CONTINUE
*
          IF( MYCOL.EQ.ICURCOL )
     $       JJ = JJ + JB
          ICURCOL = MOD( ICURCOL+1, NPCOL )
          CALL BLACS_BARRIER( ICTXT, 'All' )
*
  130 CONTINUE
*
      IF( MYROW.EQ.IRWRIT .AND. MYCOL.EQ.ICWRIT ) THEN
         CLOSE( NOUT )
      END IF
*
 9999 FORMAT( D30.18 )
*
      RETURN
*
*     End of PDLAWRITE
*
      END
```

C.2 HPF Interface to ScaLAPACK

We are investigating issues related to interfacing ScaLAPACK with High Performance Fortran (HPF) [91]. As a part of this effort, we have provided prototype interfaces to some of the ScaLA-PACK routines. We are collecting user feedback on these codes, as well as allowing additional time for compiler maturation, before providing a more complete interface.

Initially, interfaces are provided for the following ScaLAPACK routines: the general and symmetric positive definite linear equation solvers (PxGESV and PxPOSV), the linear least squares solver (PxGELS), and the PBLAS matrix multiply routine (PxGEMM).

```
LA_GESV(A, B, IPIV, INFO)
   TYPE, intent(inout), dimension(:,:) :: A, B
   integer, optional, intent(out) :: IPIV(:), INFO

LA_POSV(A, B, UPLO, INFO)
   TYPE, intent(inout), dimension(:,:) :: A, B
   character(LEN=1), optional, intent(in) :: UPLO
   integer, optional, intent(out) :: INFO

LA_GELS(A, B, TRANS, INFO)
   TYPE, intent(inout), dimension(:,:) :: A, B
   character(LEN=1), optional, intent(in) :: TRANS
   integer, optional, intent(out) :: INFO

LA_GEMM(A, B, C, transA, transB, alpha, beta)
   TYPE, intent(in), dimension(:,:) :: A, B
   TYPE, intent(inout), dimension(:,:) :: C
   character(LEN=1), optional, intent(in) :: transA, transB
   TYPE, optional, intent(in) :: alpha, beta
```

With this interface, all matrices are inherited, and query functions are used to determine the distribution of the matrices. Only when ScaLAPACK cannot handle the user's distribution are the matrices redistributed. In such a case, it is done transparently to the user, and only performance will show that it has occurred.

The prototype interfaces can be downloaded from *netlib* at the following URL:

```
http://www.netlib.org/scalapack/prototypes/slhpf.tar.gz
```

Questions or comments on these routines may be mailed to scalapack@cs.utk.edu.

The following example code is a complete HPF code calling and testing the ScaLAPACK LU factorization/solve in HPF.

This program is also available in the scalapack directory on netlib (http://www.netlib.org/scalapack/examples/sample_hpf_gesv.f).

```
      program simplegesv
      use HPF_LAPACK
      integer, parameter :: N=500, NRHS=20, NB=64, NBRHS=64, P=1, Q=3
      integer, parameter :: DP=kind(0.0D0)
      integer :: IPIV(N)
      real(DP) :: A(N, N), X(N, NRHS), B(N, NRHS)
!HPF$ PROCESSORS PROC(P,Q)
!HPF$ DISTRIBUTE A(cyclic(NB), cyclic(NB)) ONTO PROC
!HPF$ DISTRIBUTE (cyclic(NB), cyclic(NBRHS)) ONTO PROC :: B, X


!
!     Randomly generate the coefficient matrix A and the solution
!     matrix X.  Set the right hand side matrix B such that B = A * X.
!
      call random_number(A)
      call random_number(X)
      B = matmul(A, X)
!
!     Solve the linear system; the computed solution overwrites B
!
      call la_gesv(A, B, IPIV)
!
!     As a simple test, print the largest difference (in absolute value)
!     between the computed solution (B) and the generated solution (X).
!
      print*,'MAX( ABS(X~ - X) ) = ',maxval( abs(B - X) )
!
!     Shutdown the ScaLAPACK system, I'm done
!
      call SLhpf_exit()

      stop
      end
```

Appendix D

Quick Reference Guides

Quick References Guides are provided for ScaLAPACK, the PBLAS, and the BLACS. In addition, quick reference guides are also available for LAPACK and the BLAS.

```
http://www.netlib.org/scalapack/scalapackqref.ps
http://www.netlib.org/scalapack/pblasqref.ps
http://www.netlib.org/blacs/cblacsqref.ps
http://www.netlib.org/blacs/f77blacsqref.ps
http://www.netlib.org/blas/blasqr.ps
```

D.1   ScaLAPACK Quick Reference Guide

A *postscript* version of this Quick Reference Guide is available on the ScaLAPACK homepage.

```
http://www.netlib.org/scalapack/scalapackqref.ps
```

Simple Drivers

Simple Driver Routines for Linear Equations

Matrix Type	Routine
General	PSGESV(N, NRHS, A, IA, JA, IPIV, B, IB, JB, INFO)
	PCGESV(N, NRHS, A, IA, JA, IPIV, B, IB, JB, INFO)
General Band (no pivoting)	PSDBSV(N, BWL, BWU, NRHS, A, JA, DESCA, B, IB, DESCB, WORK,LWORK, INFO)
	PCDBSV(N, BWL, BWU, NRHS, A, JA, DESCA, B, IB, DESCB, WORK, LWORK, INFO)
General Band (partial pivoting)	PSGBSV(N, BWL, BWU, NRHS, A, JA, DESCA, IPIV, B, IB, DESCB, WORK, LWORK, INFO)
	PCGBSV(N, BWL, BWU, NRHS, A, JA, DESCA, IPIV, B, IB, DESCB, WORK, LWORK, INFO)
General Tridiagonal (no pivoting)	PSDTSV(N, NRHS, DL, D, DU, JA, DESCA, B, IB, DESCB, WORK, LWORK, INFO)
	PCDTSV(N, NRHS, DL, D, DU, JA, DESCA, B, IB, DESCB, WORK, LWORK, INFO)
Symmetric/Hermitian Positive Definite	PSPOSV(UPLO, N, NRHS, A, IA, JA, DESCA, B, IB, JB, DESCB, INFO)
	PCPOSV(UPLO, N, NRHS, A, IA, JA, DESCA, B, IB, JB, DESCB, INFO)
Symmetric/Hermitian Positive Definite Band	PSPBSV(UPLO, N, BW, NRHS, A, JA, DESCA, B, IB, DESCB, WORK, LWORK, INFO)
	PCPBSV(UPLO, N, BW, NRHS, A, JA, DESCA, B, IB, DESCB, WORK, LWORK, INFO)
Symmetric/Hermitian Positive Definite Tridiagonal	PSPTSV(N, NRHS, D, E, JA, DESCA, B, IB, DESCB, WORK, LWORK, INFO)
	PCPTSV(N, NRHS, D, E, JA, DESCA, B, IB, DESCB, WORK, LWORK, INFO)

Simple Driver Routines for Standard and Generalized Linear Least Squares Problems

Problem Type	Routine
Solve Using Orthogonal Factor, Assuming Full Rank	PSGELS(TRANS, M, N, NRHS, A, IA, JA, DESCA, B, IB, JB, DESCB, WORK, LWORK, INFO)
	PCGELS(TRANS, M, N, NRHS, A, IA, JA, DESCA, B, IB, JB, DESCB, WORK, LWORK, INFO)

Simple Driver Routines for Standard Eigenvalue and Singular Value Problems

Matrix/Problem Type	Routine
Symmetric/Hermitian Eigenvalues/vectors	PSSYEV(JOBZ, UPLO, N, A, IA, JA, DESCA, W, Z, IZ, JZ, DESCZ, WORK, LWORK, INFO)
General Singular Values/Vectors	PSGESVD(JOBU, JOBVT, M, N, A, IA, JA, DESCA, S, U, IU, JU, DESCU, VT, IVT, JVT, DESCVT, WORK, LWORK, INFO)

Expert Drivers

Expert Driver Routines for Linear Equations

Matrix Type	Routine

General	PSGESVX(FACT, TRANS, N, NRHS, A, IA, JA, DESCA, AF, IAF, JAF, DESCAF,
	    IPIV, EQUED, R, C, B, IB, JB, DESCB, X, IX, JX, DESCX,
	    RCOND, FERR, BERR, WORK, LWORK, IWORK, LIWORK, INFO)
	PCGESVX(FACT, TRANS, N, NRHS, A, IA, JA, DESCA, AF, IAF, JAF, DESCAF,
	    IPIV, EQUED, R, C, B, IB, JB, DESCB, X, IX, JX, DESCX,
	    RCOND, FERR, BERR, WORK, LWORK, RWORK, LRWORK, INFO)

Symmetric/Hermitian
Positive Definite
	PSPOSVX(FACT, UPLO, N, NRHS, A, IA, JA, DESCA, AF, IAF, JAF, DESCAF,
	    EQUED, S, B, IB, JB, DESCB, X, IX, JX, DESCX,
	    RCOND, FERR, BERR, WORK, LWORK, IWORK, LIWORK, INFO)
	PCPOSVX(FACT, UPLO, N, NRHS, A, IA, JA, DESCA, AF, IAF, JAF, DESCAF,
	    EQUED, S, B, IB, JB, DESCB, X, IX, JX, DESCX,
	    RCOND, FERR, BERR, WORK, LWORK, RWORK, LRWORK, INFO)

Expert Driver Routines for Standard and Generalized Symmetric Eigenvalue Problems

Matrix/Problem Type	Routine

Symmetric
Eigenvalues/vectors
	PSSYEVX(JOBZ, RANGE, UPLO, N, A, IA, JA, DESCA, VL, VU, IL, IU,
	    ABSTOL, M, NZ, W, ORFAC, Z, IZ, JZ, DESCZ, WORK, LWORK,
	    IWORK, LIWORK, IFAIL, ICLUSTR, GAP, INFO)

Hermitian
Eigenvalues/vectors
	PCHEEVX(JOBZ, RANGE, UPLO, N, A, IA, JA, DESCA, VL, VU, IL, IU,
	    ABSTOL, M, NZ, W, ORFAC, Z, IZ, JZ, DESCZ, WORK, LWORK,
	    RWORK, LRWORK, IWORK, LIWORK, IFAIL, ICLUSTR, GAP, INFO)

Symmetric
Eigenvalues/vectors
	PSSYGVX(IBTYPE, JOBZ, RANGE, UPLO, N, A, IA, JA, DESCA, B, IB, JB, DESCB,
	    VL, VU, IL, IU, ABSTOL, M, NZ, W, ORFAC, Z, IZ, JZ, DESCZ,
	    WORK, LWORK, IWORK, LIWORK, IFAIL, ICLUSTR, GAP, INFO)

Hermitian
Eigenvalues/vectors
	PCHEGVX(IBTYPE, JOBZ, RANGE, UPLO, N, A, IA, JA, DESCA, B, IB, JB, DESCB,
	    VL, VU, IL, IU, ABSTOL, M, NZ, W, ORFAC, Z, IZ, JZ, DESCZ,
	    WORK, LWORK, RWORK, LRWORK, IWORK, LIWORK,
	    IFAIL, ICLUSTR, GAP, INFO)

Meaning of prefixes

Routines beginning with ''PS'' are available in:

PS - REAL
PD - DOUBLE PRECISION

Routines beginning with ''PC'' are available in:

PC - COMPLEX
PZ - COMPLEX*16

Note: COMPLEX*16 may not be supported by all machines

D.2   Quick Reference Guide to the PBLAS

An *html* version of this Quick Reference Guide, along with the leading comments from each of the routines, is available via the ScaLAPACK homepage.

    http://www.netlib.org/scalapack/index.html

At the lowest level, the efficiency of the PBLAS is determined by the local performance of the BLAS and the BLACS. In addition, depending on the shape of its input and output distributed matrices, the PBLAS select the best algorithm in terms of data transfer across the process grid. Transparent to the user, this relatively simple selection process ensures high efficiency independent of the actual computation performed.

Level 1 PBLAS

```
           dim scalar      vector               vector
P_SWAP ( N,               X, IX, JX, DESCX, INCX, Y, IY, JY, DESCY, INCY )
P_SCAL ( N, ALPHA,        X, IX, JX, DESCX, INCX )
P_COPY ( N,               X, IX, JX, DESCX, INCX, Y, IY, JY, DESCY, INCY )
P_AXPY ( N, ALPHA,        X, IX, JX, DESCX, INCX, Y, IY, JY, DESCY, INCY )
P_DOT  ( N, DOT,          X, IX, JX, DESCX, INCX, Y, IY, JY, DESCY, INCY )
P_DOTU ( N, DOTU,         X, IX, JX, DESCX, INCX, Y, IY, JY, DESCY, INCY )
P_DOTC ( N, DOTC,         X, IX, JX, DESCX, INCX, Y, IY, JY, DESCY, INCY )
P_NRM2 ( N, NORM2,        X, IX, JX, DESCX, INCX )
P_ASUM ( N, ASUM,         X, IX, JX, DESCX, INCX )
P_AMAX ( N, AMAX, INDX, X, IX, JX, DESCX, INCX )
```

Level 2 PBLAS

```
           options          dim  scalar matrix              vector             scalar vector
P_GEMV (        TRANS,     M, N, ALPHA, A, IA, JA, DESCA, X, IX, JX, DESCX, INCX, BETA, Y, IY, JY, DESCY, INCY )
P_HEMV ( UPLO,                N, ALPHA, A, IA, JA, DESCA, X, IX, JX, DESCX, INCX, BETA, Y, IY, JY, DESCY, INCY )
P_SYMV ( UPLO,                N, ALPHA, A, IA, JA, DESCA, X, IX, JX, DESCX, INCX, BETA, Y, IY, JY, DESCY, INCY )
P_TRMV ( UPLO, TRANS, DIAG,   N,        A, IA, JA, DESCA, X, IX, JX, DESCX, INCX )
P_TRSV ( UPLO, TRANS, DIAG,   N,        A, IA, JA, DESCA, X, IX, JX, DESCX, INCX )
```

```
           options          dim  scalar vector                   vector                   matrix
P_GER  (                  M, N, ALPHA, X, IX, JX, DESCX, INCX, Y, IY, JY, DESCY, INCY, A, IA, JA, DESCA )
P_GERU (                  M, N, ALPHA, X, IX, JX, DESCX, INCX, Y, IY, JY, DESCY, INCY, A, IA, JA, DESCA )
P_GERC (                  M, N, ALPHA, X, IX, JX, DESCX, INCX, Y, IY, JY, DESCY, INCY, A, IA, JA, DESCA )
P_HER  ( UPLO,               N, ALPHA, X, IX, JX, DESCX, INCX,                          A, IA, JA, DESCA )
P_HER2 ( UPLO,               N, ALPHA, X, IX, JX, DESCX, INCX, Y, IY, JY, DESCY, INCY, A, IA, JA, DESCA )
P_SYR  ( UPLO,               N, ALPHA, X, IX, JX, DESCX, INCX,                          A, IA, JA, DESCA )
P_SYR2 ( UPLO,               N, ALPHA, X, IX, JX, DESCX, INCX, Y, IY, JY, DESCY, INCY, A, IA, JA, DESCA )
```

Level 3 PBLAS

```
           options                    dim    scalar matrix           matrix           scalar matrix
P_GEMM (          TRANSA, TRANSB,   M, N, K, ALPHA, A, IA, JA, DESCA, B, IB, JB, DESCB, BETA, C, IC, JC, DESCC )
P_SYMM ( SIDE, UPLO,                M, N,    ALPHA, A, IA, JA, DESCA, B, IB, JB, DESCB, BETA, C, IC, JC, DESCC )
P_HEMM ( SIDE, UPLO,                M, N,    ALPHA, A, IA, JA, DESCA, B, IB, JB, DESCB, BETA, C, IC, JC, DESCC )
P_SYRK (       UPLO, TRANS,            N, K, ALPHA, A, IA, JA, DESCA,                    BETA, C, IC, JC, DESCC )
P_HERK (       UPLO, TRANS,            N, K, ALPHA, A, IA, JA, DESCA,                    BETA, C, IC, JC, DESCC )
P_SYR2K(       UPLO, TRANS,            N, K, ALPHA, A, IA, JA, DESCA, B, IB, JB, DESCB, BETA, C, IC, JC, DESCC )
P_HER2K(       UPLO, TRANS,            N, K, ALPHA, A, IA, JA, DESCA, B, IB, JB, DESCB, BETA, C, IC, JC, DESCC )
P_TRAN (                            M, N,    ALPHA, A, IA, JA, DESCA,                    BETA, C, IC, JC, DESCC )
P_TRANU(                            M, N,    ALPHA, A, IA, JA, DESCA,                    BETA, C, IC, JC, DESCC )
P_TRANC(                            M, N,    ALPHA, A, IA, JA, DESCA,                    BETA, C, IC, JC, DESCC )
P_TRMM ( SIDE, UPLO, TRANSA,   DIAG, M, N,   ALPHA, A, IA, JA, DESCA, B, IB, JB, DESCB )
P_TRSM ( SIDE, UPLO, TRANSA,   DIAG, M, N,   ALPHA, A, IA, JA, DESCA, B, IB, JB, DESCB )
```

Name	Operation	Prefixes
P_SWAP	$x \leftrightarrow y$	S, D, C, Z
P_SCAL	$x \leftarrow \alpha x$	S, D, C, Z, CS, ZD
P_COPY	$y \leftarrow x$	S, D, C, Z
P_AXPY	$y \leftarrow \alpha x + y$	S, D, C, Z
P_DOT	$dot \leftarrow x^T y$	S, D
P_DOTU	$dotu \leftarrow x^T y$	C, Z
P_DOTC	$dotc \leftarrow x^H y$	C, Z
P_NRM2	$norm2 \leftarrow \|x\|_2$	S, D, SC, DZ
P_ASUM	$asum \leftarrow \|re(x)\|_1 + \|im(x)\|_1$	S, D, SC, DZ
P_AMAX	$indx \leftarrow 1^{st}\ k \ni \|Re(x_k)\| + \|Im(x_k)\|$ $= max(\|Re(x_i)\| + \|Im(x_i)\|) = amax$	S, D, C, Z
P_GEMV	$y \leftarrow \alpha op(A)x + \beta y, op(A) = A, A^T, A^H, A - m \times n$	S, D, C, Z
P_HEMV	$y \leftarrow \alpha A x + \beta y$	C, Z
P_SYMV	$y \leftarrow \alpha A x + \beta y$	S, D
P_TRMV	$x \leftarrow Ax, x \leftarrow A^T x, x \leftarrow A^H x,$	S, D, C, Z
P_TRSV	$x \leftarrow \alpha A^{-1}x, x \leftarrow \alpha A^{-T}x, x \leftarrow \alpha A^{-H}x,$	S, D, C, Z
P_GER	$A \leftarrow \alpha xy^T + A, A - m \times n$	S, D
P_GERU	$A \leftarrow \alpha xy^T + A, A - m \times n$	C, Z
P_GERC	$A \leftarrow \alpha xy^H + A, A - m \times n$	C, Z
P_HER	$A \leftarrow \alpha xx^H + A$	C, Z
P_HER2	$A \leftarrow \alpha xy^H + y(\alpha x)^H + A$	C, Z
P_SYR	$A \leftarrow \alpha xx^T + A$	S, D
P_SYR2	$A \leftarrow \alpha xy^T + \alpha yx^T + A$	S, D
P_GEMM	$C \leftarrow \alpha op(A)op(B) + \beta C, op(X) = X, X^T, X^H, C - m \times n$	S, D, C, Z
P_SYMM	$C \leftarrow \alpha AB + \beta C, C \leftarrow \alpha BA + \beta C, C - m \times n, A = A^T$	S, D, C, Z
P_HEMM	$C \leftarrow \alpha AB + \beta C, C \leftarrow \alpha BA + \beta C, C - m \times n, A = A^H$	C, Z
P_SYRK	$C \leftarrow \alpha AA^T + \beta C, C \leftarrow \alpha A^T A + \beta C, C - n \times n$	S, D, C, Z
P_HERK	$C \leftarrow \alpha AA^H + \beta C, C \leftarrow \alpha A^H A + \beta C, C - n \times n$	C, Z
P_SYR2K	$C \leftarrow \alpha AB^T + \alpha BA^T + \beta C, C \leftarrow \alpha A^T B + \alpha B^T A + \beta C, C - n \times n$	S, D, C, Z
P_HER2K	$C \leftarrow \alpha AB^H + \bar{\alpha} BA^H + \beta C, C \leftarrow \alpha A^H B + \bar{\alpha} B^H A + \beta C, C - n \times n$	C, Z
P_TRAN	$C \leftarrow \beta C + \alpha A^T, A - n \times m, C - m \times n$	S, D
P_TRANU	$C \leftarrow \beta C + \alpha A^T, A - n \times m, C - m \times n$	C, Z
P_TRANC	$C \leftarrow \beta C + \alpha A^H, A - n \times m, C - m \times n$	C, Z
P_TRMM	$B \leftarrow \alpha op(A)B, B \leftarrow \alpha Bop(A), op(A) = A, A^T, A^H, B - m \times n$	S, D, C, Z
P_TRSM	$B \leftarrow \alpha op(A^{-1})B, B \leftarrow \alpha Bop(A^{-1}), op(A) = A, A^T, A^H, B - m \times n$	S, D, C, Z

Meaning of prefixes

S - REAL   C - COMPLEX
D - DOUBLE PRECISION   Z - COMPLEX*16
                (may not be supported by all machines)

Level 2 and Level 3 PBLAS
Matrix Types

GE - GEneral
SY - SYmmetric
HE - HErmitian
TR - TRiangular

Level 2 and Level 3 PBLAS
Options

Dummy options arguments are declared as CHARACTER*1 and may be passed as character strings.
TRANS_ = 'No transpose', 'Transpose', 'Conjugate transpose', (X, X^T, X^H)
UPLO = 'Upper triangular', 'Lower Triangular'
DIAG = 'Non-unit triangular', 'Unit triangular'
SIDE = 'Left' or 'Right' (A or op(A) on the left, or A or op(A) on the right)

For real matrices, TRANS_ = 'T' and TRANS_ = 'C' have the same meaning.

For Hermitian matrices, TRANS_ ='T' is not allowed.

For complex symmetric matrices, TRANS_ ='C' is not allowed.

Array Descriptor, Increment

The array descriptor $DESCA$ is an integer array of dimension 9. It describes the two-dimensional block-cyclic mapping of the matrix A.

The first two entries are the descriptor type and the BLACS context. The third and fourth entries are the dimensions of the matrix (row, column). The fifth and sixth entries are the row- and column block sizes used to distribute the matrix. The seventh and eighth are the coordinates of the process containing the first entry of the matrix. The last entry contains the leading dimension of the local array containing the matrix elements.

The increment specified for vectors is always global. So far only 1 and DESCA(M_) are supported.

D.3   Quick Reference Guide to the BLACS

An *html* version of this Quick Reference Guide, along with the leading comments from each of the routines, is available on the BLACS homepage.

```
http://www.netlib.org/blacs/index.html
```

Fortran Interface

<u>Initialization</u>
```
BLACS_PINFO   ( MYPNUM, NPROCS )
BLACS_SETUP   ( MYPNUM, NPROCS )
BLACS_GET     ( ICTXT, WHAT, VAL )
BLACS_SET     ( ICTXT, WHAT, VAL )
BLACS_GRIDINIT( ICTXT, ORDER,              NPROW, NPCOL )
BLACS_GRIDMAP ( ICTXT, USERMAP, LDUMAP, NPROW, NPCOL )
```

<u>Destruction</u>
```
BLACS_FREEBUFF( ICTXT, WAIT )
BLACS_GRIDEXIT( ICTXT )
BLACS_ABORT   ( ICTXT, ERRORNUM )
BLACS_EXIT    ( DONEFLAG )
```

<u>Sending</u>
```
□GESD2D( ICTXT,                      M, N, A, LDA, RDEST, CDEST )

□GEBS2D( ICTXT, SCOPE, TOP,          M, N, A, LDA                )

□TRSD2D( ICTXT,          UPLO, DIAG, M, N, A, LDA, RDEST, CDEST )

□TRBS2D( ICTXT, SCOPE, TOP, UPLO, DIAG, M, N, A, LDA            )
```

<u>Receiving</u>
```
□GERV2D( ICTXT,                      M, N, A, LDA, RSRC, CSRC )

□GEBR2D( ICTXT, SCOPE, TOP,          M, N, A, LDA, RSRC, CSRC )

□TRRV2D( ICTXT,          UPLO, DIAG, M, N, A, LDA, RSRC, CSRC )

□TRBR2D( ICTXT, SCOPE, TOP, UPLO, DIAG, M, N, A, LDA, RSRC, CSRC )
```

<u>Combine Operations</u>
```
□GAMX2D( ICTXT, SCOPE, TOP, M, N, A, LDA, RA, CA, RCFLAG, RDEST, CDEST )

□GAMN2D( ICTXT, SCOPE, TOP, M, N, A, LDA, RA, CA, RCFLAG, RDEST, CDEST )

□GSUM2D( ICTXT, SCOPE, TOP, M, N, A, LDA,                 RDEST, CDEST )
```

All routines preceded by a □ have the following prefixes: S, D, C, Z, I.

Informational and Miscellaneous

```
BLACS_GRIDINFO( ICTXT, NPROW, NPCOL, MYROW, MYCOL )
BLACS_PNUM    ( ICTXT,              PROW,  PCOL )
BLACS_PCOORD  ( ICTXT, PNUM,        PROW,  PCOL )
BLACS_BARRIER ( ICTXT, SCOPE )
```

Non-standard

```
SETPVMTIDS ( NTASKS, TIDS )
DCPUTIME00 ( )
DWALLTIME00( )
KSENDID    ( ICTXT,        RDEST, CDEST )
KRECVID    ( ICTXT,        RSRC,  CSRC )
KBSID      ( ICTXT, SCOPE               )
KBRID      ( ICTXT, SCOPE, RSRC,  CSRC  )
```

Declarations

```
CHARACTER           DIAG, ORDER, SCOPE, TOP, UPLO
INTEGER             BLACS_PNUM, CDEST, ICTXT, CSRC, DONEFLAG
INTEGER             ERRORNUM, LDA, RCFLAG, M, MAXID, MINID, N
INTEGER             NBRANCHES, NPCOL, NPROW
INTEGER             PCOL, PNUM, PROW, RDEST, RSRC, WAIT
INTEGER             CA( * ), RA( * )
DOUBLE PRECISION    DCPUTIME00, DWALLTIME00
REAL/DOUBLE         A( LDA, * )
COMPLEX/COMPLEX*16  A( LDA, * )
 or
INTEGER             A( LDA, * )
```

Options

```
UPLO  = 'Upper triangular', 'Lower triangular';
DIAG  = 'Non-unit triangular', 'Unit triangular';
SCOPE = 'All', 'row', 'column';
TOP   = (SEE DESCRIPTION BELOW).
```

Broadcast Topologies

```
TOP   = ' ' : System dependent default topology;
      = 'I' : increasing ring;
      = 'D' : decreasing ring;
      = 'H' : hypercube (minimum spanning tree);
      = 'S' : split-ring;
      = 'F' : fully connected;
      = 'M' : nodes divided into I increasing
              rings, where I is set with call
              to BLACS_SET;
      = 'T' : tree broadcast with NBRANCHES = I,
              where I is set with call to
              BLACS_SET;
      = '1' : tree broadcast with NBRANCHES = 1;
      = '2' : tree broadcast with NBRANCHES = 2;
        :
        :
      = '9' : tree broadcast with NBRANCHES = 9.
```

Global Topologies

```
TOP   = ' '  : System dependent default topology;
      = '1'  : tree gather with NBRANCHES = 1;
      = '2'  : tree gather with NBRANCHES = 2;
          .
          .
      = '9'  : tree gather with NBRANCHES = 9;
      = 'T'  : tree gather with NBRANCHES = I,
               where I is set with call to
               BLACS_SET;
      = 'F'  : Fully connected;
      = 'H'  : if RDEST = -1, a specialized
               "leave on all" hypercube topology
               called bidirectional exchange is used.
               Otherwise, TOP = '1' is substituted.
```

Notation

Underlined parameters are output arguments. If a routine is underlined, it is a function that returns a value. The prefix **P** usually stands for process. Other standard notations are

```
GE - GENERAL      TR - TRAPEZOIDAL
SD - SEND         BS - BROADCAST/SEND
RV - RECEIVE      BR - BROADCAST/RECEIVE
GAMX - General element-wise Absolute value MAXIMUM
GAMN - General element-wise Absolute value MINIMUM
GSUM - General element-wise SUMMATION
```

Key Ideas:

A BLACS context is created via a call to either **BLACS_GRIDINIT** or **BLACS_GRIDMAP**. No routine requiring a context may be used until one of these routines has been called. Multiple calls to **BLACS_GRIDINIT** or **BLACS_GRIDMAP** result in the creation of new contexts. To preserve resources, the user should free unused contexts by calling **BLACS_GRIDEXIT**. When all BLACS operations are done, a call to **BLACS_EXIT** frees any remaining contexts and shuts down all BLACS operations.

Please note that **BLACS_GRIDINIT** and **BLACS_GRIDMAP** accept system contexts as input. A default system context encompassing all available processes may be obtained by a call to **BLACS_GET**.

BLACS_SET can only be used to change the message ID range before the creation of the first context. Subsequent calls will be ignored.

Topology Hints

Topologies allow the user to optimize communication patterns for a particular operation. If the user does not have a communication pipe to maintain, the default TOP = ' ' is recommended.

C Interface

Initialization

```
void Cblacs_pinfo   ( int *mypnum, int *nprocs )
void Cblacs_setup   ( int *mypnum, int *nprocs )
void Cblacs_get     ( int ictxt, int what, int *val )
void Cblacs_set     ( int ictxt, int what, int *val )
void Cblacs_gridinit( int *ictxt, char *order,          int nprow, int npcol )
void Cblacs_gridmap ( int *ictxt, int *pmap, int ldpmap, int nprow, int npcol )
```

Destruction

```
void Cblacs_freebuff( int ictxt, int wait )
void Cblacs_gridexit( int ictxt )
void Cblacs_abort   ( int ictxt, int errornum )
void Cblacs_exit    ( int doneflag )
```

Sending

```
void C□gesd2d( int ictxt,                                       int m, int n, TYPE *A, int lda,
               int rdest, int cdest )

void C□gebs2d( int ictxt, char *scope, char *top,              int m, int n, TYPE *A, int lda )

void C□trsd2d( int ictxt, char *uplo, char *diag,              int m, int n, TYPE *A, int lda,
               int rdest, int cdest )

void C□trbs2d( int ictxt, char *scope, char *top, char *uplo, char *diag, int m, int n, TYPE *A, int lda )
```

Receiving

```
void C□gerv2d( int ictxt,                                       int m, int n, TYPE *A, int lda,
               int rsrc, int csrc )

void C□gebr2d( int ictxt, char *scope, char *top,              int m, int n, TYPE *A, int lda,
               int rsrc, int csrc )

void C□trrv2d( int ictxt, char *uplo, char *diag,              int m, int n, TYPE *A, int lda,
               int rsrc, int csrc )

void C□trbr2d( int ictxt, char *scope, char *top, char *uplo, char *diag, int m, int n, TYPE *A, int lda,
               int rsrc, int csrc )
```

Combine Operations

```
void C□gamx2d( int ictxt, char *scope, char *top, int m, int n, TYPE *A, int lda, int *RA, int *CA,
               int RCflag, int rdest, int cdest )

void C□gamn2d( int ictxt, char *scope, char *top, int m, int n, TYPE *A, int lda, int *RA, int *CA,
               int RCflag, int rdest, int cdest )

void C□gsum2d( int ictxt, char *scope, char *top, int m, int n, TYPE *A, int lda,
               int rdest, int cdest )
```

Definition of □

□is	Data operated on is	TYPE is
s	single precision real	float
d	double precision real	double
c	single precision complex	float
z	double precision complex	double
i	integer	int

Informational and Miscellaneous

```
void Cblacs_gridinfo( int ictxt, int *nprow, int *npcol, int *myprow, int *mypcol )
int   Cblacs_pnum   ( int ictxt,               int  prow, int  pcol )
void Cblacs_pcoord  ( int ictxt, int pnum, int *prow, int *pcol )
void Cblacs_barrier( int ictxt, char *scope )
```

Non-standard

```
void    Csetpvmtids ( int ntasks, int *tids )
double Cdcputime00 ( )
double Cdwalltime00( )
int    Cksendid     ( int ictxt,               int rdest, int cdest )
int    Ckrecvid     ( int ictxt,               int rsrc,  int csrc  )
int    Ckbsid       ( int ictxt, char *scope                        )
int    Ckbrid       ( int ictxt, char *scope, int rsrc,  int csrc  )
```

Options

```
UPLO  = "Upper triangular", "Lower triangular";
DIAG  = "Non-unit triangular", "Unit triangular";
SCOPE = "All", "row", "column";
TOP   = (SEE DESCRIPTION BELOW).
```

Broadcast Topologies

```
TOP   = " " : System dependent default topology;
      = "I" : increasing ring;
      = "D" : decreasing ring;
      = "H" : hypercube (minimum spanning tree);
      = "S" : split-ring;
      = "F" : fully connected;
      = "M" : nodes divided into I increasing
              rings, where I is set with call
              to Cblacs_set;
      = "T" : tree broadcast with NBRANCHES = I,
              where I is set with call to
              Cblacs_set;
      = "1" : tree broadcast with NBRANCHES = 1;
      = "2" : tree broadcast with NBRANCHES = 2;
          ⋮
      = "9" : tree broadcast with NBRANCHES = 9.
```

Global Topologies

```
TOP   = " " : System dependent default topology;
      = "1" : tree gather with NBRANCHES = 1;
      = "2" : tree gather with NBRANCHES = 2;
         .
         .
      = "9" : tree gather with NBRANCHES = 9;
      = "T" : tree gather with NBRANCHES = I,
              where I is set with call to
              Cblacs_set;
      = "F" : Fully connected;
      = "H" : if rdest = -1, a specialized
              "leave on all" hypercube topology
              called bidirectional exchange is used.
              Otherwise, TOP = "1" is substituted.
```

Notation

Underlined parameters are output arguments. If a routine is underlined it is a function that returns a value. The prefix **p** usually stands for process. Other standard notations are:

```
GE - GENERAL      TR - TRAPEZOIDAL
SD - SEND         BS - BROADCAST/SEND
RV - RECEIVE      BR - BROADCAST/RECEIVE
GAMX - General element-wise Absolute value MAXIMUM
GAMN - General element-wise Absolute value MINIMUM
GSUM - General element-wise SUMMATION
```

Key Ideas:

A BLACS context is created via a call to either `Cblacs_gridinit` or `Cblacs_gridmap`. No routine requiring a context may be used until one of these routines has been called. Multiple calls to `Cblacs_gridinit` or `Cblacs_gridmap` result in the creation of new contexts. To preserve resources, the user should free unused contexts by calling `Cblacs_gridexit`. When all BLACS operations are done, a call to `Cblacs_exit` frees any remaining contexts, and shuts down all BLACS operations.

Please note that `Cblacs_gridinit` and `Cblacs_gridmap` accept system contexts as input. A default system context encompassing all available processes may be obtained by a call to `Cblacs_get`.

`Cblacs_set` can only be used to set the BLACS' message ID range before the creation of the first context. Subsequent calls will be ignored.

Topology Hints

Topologies allow the user to optimize communication patterns for a particular operation. If the user does not have a communication pipe to maintain, the default TOP = " " is recommended.

Glossary

The following is a glossary of terms and notation used throughout this users guide and the leading comments of the source code. The first time notation from this glossary appears in the text, it will be *italicized*.

- **Array descriptor**: Contains the information required to establish the mapping between a global matrix entry and its corresponding process and memory location.

  The notations x_ used in the entries of the array descriptor denote the attributes of a global matrix. For example, M_ denotes the number of rows, and M_A specifically denotes the number of rows in global matrix A. See sections 4.2, 4.3.3, 4.4.5, 4.4.6, and 4.5.1 for complete details.

- **BLACS**: Basic Linear Algebra Communication Subprograms, a message-passing library designed for linear algebra. They provide.a portability layer for communication between ScaLA-PACK and message-passing systems such as MPI and PVM, as well as native message-passing libraries such as NX and MPL. See section 1.3.4.

- **BLAS**: Basic Linear Algebra Subprograms [57, 59, 93], a standard for subroutines for common linear algebra computations such as dot-products, matrix-vector multiplication, and matrix-matrix multiplication. They provide a portability layer for computation. See section 1.3.2.

- **Block size**: The number of contiguous rows or columns of a global matrix to be distributed consecutively to each of the processes in the process grid. The block size is quantified by the notation $MB \times NB$, where MB is the row block size and NB is the column block size.

  The distribution block size can be square, $MB = NB$, or rectangular, $MB \neq NB$. **Block size** is also referred to as the **partitioning unit** or **blocking factor**.

- **Distributed memory computer**: A term used in two senses:

  - A computer marketed as a distributed memory computer (such as the Cray T3 computers, the IBM SP computers, or the Intel Paragon), including one or more message-passing libraries.

  - A distributed shared-memory computer (e.g., the Origin 2000) or network of workstations (e.g., the Berkeley NOW) with message passing.

ScaLAPACK delivers high performance on these computers provided that they include certain key features such as an efficient message-passing system, a one-to-one mapping of processes to processors, a gang scheduler and a well-connected communication network.

- **Distribution**: Method by which the entries of a global matrix are allocated among the processes, also commonly referred to as **decomposition** or **data layout**. Examples of distributions used by ScaLAPACK include block and block-cyclic distributions and these will be illustrated and explained in detail later.

  Data distribution in ScaLAPACK is controlled primarily by the process grid and the block size.

- **Global**: A term "global" used in two ways:

  - To define the mathematical matrix, e.g. the global matrix A.

  - To identify arguments that must have the same value on all processes.

- $LOC_c(\mathbf{K_-})$: Number of columns that a process receives if K_- columns of a matrix are distributed over c columns of its process row.

  To be consistent in notation, we have used a "modifying character" subscript on LOC to denote the dimension of the process grid to which we are referring. The subscript "r" indicates "row" whenever it is appended to LOC; likewise, the subscript "c" indicates "column" when it is appended to LOC.

  The value of $LOC_c()$ may differ from process to process within the process grid. For example, in figure 4.6 (section 4.3.4), we can see that for process $(0,0)$ $LOC_c(\text{N}_-)= 4$; however, for process $(0,1)$ $LOC_c(\text{N}_-) = 3$.

- $LOC_r(\mathbf{K_-})$: Number of rows that a process would receive if K_- rows of a matrix are distributed over r rows of its process column.

  To be consistent in notation, we have used a "modifying character" subscript on LOC to denote the dimension of the process grid to which we are referring. The subscript "r" indicates "row" whenever it is appended to LOC; likewise, the subscript "c" indicates "column" when it is appended to LOC.

  The value of $LOC_r()$ may differ from process to process within the process grid. For example, in figure 4.6 (section 4.3.4), we can see that for process $(0,0)$ $LOC_r(\text{M}_-)= 5$; however, for process $(1,0)$ $LOC_r(\text{M}_-) = 4$.

- **Local**: A term used in two ways:

  - To express the array elements or blocks stored on each process, e.g., the local part of the global matrix A, also referred to as the **local array**. The size of the local array may differ from process to process. See section 2.3 for further details.

  - To identify arguments that may have different values on different processes.

- **Local leading dimension** of a local array: Specification of entry size for local array. When a global array is distributed among the processes in the process grid, locally the entries are stored in a two-dimensional array, the size of which may vary from process to process. Thus, a leading dimension needs to be specified for each local array. For example, in Figure 2.2 in section 2.3, we can see that for process $(0,0)$ the local leading dimension of the local array A (denoted LLD_A) is 5, whereas for process $(1,0)$ the local leading dimension of local array A is 4.

- *MYCOL*: The calling process's column coordinate in the process grid. Each process within the process grid is uniquely identified by its process coordinates $(MYROW, MYCOL)$.

- *MYROW*: The calling process's row coordinate in the process grid. Each process within the process grid is uniquely identified by its process coordinates $(MYROW, MYCOL)$.

- P: The total number of processes in the process grid, i.e., $P = P_r \times P_c$.

  In terms of notation for process grids, we have used a "modifying character" subscript on P to denote the dimension of the process grid to which we are referring. The subscript "r" indicates "row" whenever it is appended to P, and thus P_r is the number of process rows in the process grid. Likewise, the subscript "c" indicates "column" when it is appended to P, and thus P_c is the number of process columns in the process grid.

- P_c: The number of process columns in the process grid (i.e., the second dimension of the two-dimensional process grid).

- P_r: The number of process rows in the process grid (i.e., the first dimension of the two-dimensional process grid).

- **PBLAS**: A distributed-memory version of the BLAS (Basic Linear Algebra Subprograms), also referred to as the **Parallel BLAS** or **Parallel Basic Linear Algebra Subprograms**. Refer to section 1.3.3 for further details.

- **Process**: Basic unit or thread of execution that minimally includes a stack, registers, and memory. Multiple processes may share a physical processor. The term processor refers to the actual hardware.

  In ScaLAPACK, each process is treated as if it were a processor: the process must exist for the lifetime of the ScaLAPACK run, and its execution should affect other processes' execution only through the use of message-passing calls. With this in mind, we use the term process in all sections of this users guide except those dealing with timings. When discussing timings, we specify processors as our unit of execution, since speedup will be determined largely by actual hardware resources.

  In ScaLAPACK, algorithms are presented in terms of *processes*, rather than physical processors. In general there may be several processes on a processor, in which case we assume that the runtime system handles the scheduling of processes. In the absence of such a runtime system, ScaLAPACK assumes one process per processor.

- **Process column**: A specific column of processes within the two-dimensional process grid. For further details, consult the definition of **process grid**.

- **Process grid**: The way we logically view a parallel machine as a one- or two-dimensional rectangular grid of processes.

For two-dimensional process grids, the variable P_r is used to indicate the number of rows in the process grid (i.e., the first dimension of the two-dimensional process grid). The variable P_c is used to indicate the number of columns in the process grid (i.e., the second dimension of the two-dimensional process grid). The collection of processes need not physically be connected in the two-dimensional process grid.

For example, the following figure shows six processes mapped to a 2×3 grid, where $P_r = 2$ and $P_c = 3$.

	0	1	2
0	0	1	2
1	3	4	5

A user may perform an operation within a **process row** or **process column** of the process grid. A **process row** refers to a specific row of processes within the process grid, and a **process column** refers to a specific column of processes with the process grid. In the example, **process row 0** contains the processes with natural ordering **0**, **1**, and **2**, and **process column 0** contains the processes with natural ordering **0** and **3**.

For further details, please refer to section 4.1.1.

- **Process row**: A specific row of processes within the two-dimensional process grid. For further details, consult the definition of **process grid**.

- **Scope**: A term used in two ways:

  - The portion of the process grid within which an operation is defined. For example, in the Level 1 PBLAS, the resultant output array or scalar will be global or local within a process column or row of the process grid, and undefined elsewhere.

    Equivalently, in Appendix D.3, scope indicates the processes that participate in the broadcast or global combine operations. Scope can equal "all", "row", or "column".

  - The portion of the parallel program within which the definition of an argument remains unchanged. When the scope of an argument is defined as global, the argument must have the same value on all processes. When the scope of an argument is defined as local, the argument may have different values on different processes.

Refer to section 4.1.3 for further details.

Part II

Specifications of Routines

Notes

1. The specifications that follow give the calling sequence, purpose, and descriptions of the arguments of each ScaLAPACK driver and computational routine (but not of auxiliary routines).

2. Specifications of pairs of real and complex routines have been merged (for example PSGETRF/PCGETRF).

3. Specifications are given only for *single-precision* routines. To adapt them for the double precision version of the software, simply interpret REAL as DOUBLE PRECISION, COMPLEX as COMPLEX*16 (or DOUBLE COMPLEX), and the initial letters PS- and PC- of ScaLAPACK routine names as PD- and PZ-.

4. Specifications are arranged in alphabetical order of the real routine name.

5. The text of the specifications has been derived from the leading comments in the source-text of the routines. It makes only limited use of mathematical typesetting facilities. To eliminate redundancy, A^H has been used throughout the specifications. Thus, the reader should note that A^H is equivalent to A^T in the real case.

6. If there is a discrepancy between the specifications listed in this section and the actual source code, the source code should be regarded as the most up to date.

Included in the leading comments of each subroutine (immediately preceding the Argument section) is a brief note describing the **array descriptor** and some commonly used expressions in calculating workspace. For brevity, we have listed this information below and not included it in the specifications of the routines.

```
* Notes
* =====
*
* Each global data object is described by an associated description
* vector.  This vector stores the information required to establish
* the mapping between an object element and its corresponding process
* and memory location.
*
* Let A be a generic term for any 2D block cyclicly distributed array.
* Such a global array has an associated description vector DESCA.
* In the following comments, the character _ should be read as
* "of the global array".
*
* NOTATION        STORED IN      EXPLANATION
* --------------- -------------- --------------------------------------
* DTYPE_A(global) DESCA( DTYPE_ )The descriptor type.  In this case,
*                                DTYPE_A = 1.
* CTXT_A (global) DESCA( CTXT_ ) The BLACS context handle, indicating
```

```
*                                     the BLACS process grid A is distribu-
*                                     ted over. The context itself is glo-
*                                     bal, but the handle (the integer
*                                     value) may vary.
*  M_A    (global) DESCA( M_ )        The number of rows in the global
*                                     array A.
*  N_A    (global) DESCA( N_ )        The number of columns in the global
*                                     array A.
*  MB_A   (global) DESCA( MB_ )       The blocking factor used to distribute
*                                     the rows of the array.
*  NB_A   (global) DESCA( NB_ )       The blocking factor used to distribute
*                                     the columns of the array.
*  RSRC_A (global) DESCA( RSRC_ )     The process row over which the first
*                                     row of the array A is distributed.
*  CSRC_A (global) DESCA( CSRC_ )     The process column over which the
*                                     first column of the array A is
*                                     distributed.
*  LLD_A  (local)  DESCA( LLD_ )      The leading dimension of the local
*                                     array.  LLD_A >= MAX(1,LOCr(M_A)).
*
*  Let K be the number of rows or columns of a distributed matrix,
*  and assume that its process grid has dimension p x q.
*  LOCr( K ) denotes the number of elements of K that a process
*  would receive if K were distributed over the p processes of its
*  process column.
*  Similarly, LOCc( K ) denotes the number of elements of K that a
*  process would receive if K were distributed over the q processes of
*  its process row.
*  The values of LOCr() and LOCc() may be determined via a call to the
*  ScaLAPACK tool function, NUMROC:
*          LOCr( M ) = NUMROC( M, MB_A, MYROW, RSRC_A, NPROW ),
*          LOCc( N ) = NUMROC( N, NB_A, MYCOL, CSRC_A, NPCOL ).
*  An upper bound for these quantities may be computed by:
*          LOCr( M ) <= ceil( ceil(M/MB_A)/NPROW )*MB_A
*          LOCc( N ) <= ceil( ceil(N/NB_A)/NPCOL )*NB_A
```

PSDBSV/PCDBSV

```
SUBROUTINE PSDBSV( N, BWL, BWU, NRHS, A, JA, DESCA, B, IB, DESCB,
$                  WORK, LWORK, INFO )

    INTEGER        BWL, BWU, IB, INFO, JA, LWORK, N, NRHS
    INTEGER        DESCA( * ), DESCB( * )
    REAL           A( * ), B( * ), WORK( * )

SUBROUTINE PCDBSV( N, BWL, BWU, NRHS, A, JA, DESCA, B, IB, DESCB,
$                  WORK, LWORK, INFO )

    INTEGER        BWL, BWU, IB, INFO, JA, LWORK, N, NRHS
    INTEGER        DESCA( * ), DESCB( * )
    COMPLEX        A( * ), B( * ), WORK( * )
```

Purpose

PSDBSV/PCDBSV solves a system of linear equations

$$A(1{:}N, JA{:}JA{+}N{-}1) * X = B(IB{:}IB{+}N{-}1, 1{:}NRHS),$$

where $A(1{:}N, JA{:}JA{+}N{-}1)$ is an n-by-n real/complex diagonally dominant-like banded distributed matrix with bwl subdiagonals and bwu superdiagonals, and X and B are n-by-nrhs distributed matrices.

The LU decomposition without pivoting is used to factor A as $A = P*L*U*P^T$, where P is a permutation matrix and L and U are banded lower and upper triangular matrices respectively. The permutations are performed for the sake of parallelism.

Arguments

N   (global input) INTEGER
The number of rows and columns to be operated on, i.e. the order of the distributed submatrix $A(1{:}N, JA{:}JA{+}N{-}1)$. $N \geq 0$.

BWL   (global input) INTEGER
The number of subdiagonals within the band of A. $0 \leq BWL \leq N{-}1$.

BWU   (global input) INTEGER
The number of superdiagonals within the band of A. $0 \leq BWU \leq N{-}1$.

NRHS   (global input) INTEGER
The number of right hand sides, i.e., the number of columns of the distributed submatrix $B(IB{:}IB{+}N{-}1, 1{:}NRHS)$. $NRHS \geq 0$.

A   (local input/local output) REAL/COMPLEX pointer into the local memory to an array of dimension (LLD_A, LOCc(JA+N−1))
On entry, this array contains the local pieces of the global array A.
On exit, details of the factorization. Note, the resulting factorization is *not* the same factorization as returned from LAPACK. Additional permutations are performed on the matrix for the sake of parallelism.

JA   (global input) INTEGER
The index in the global array A that points to the start of the matrix to be operated on (which may be either all of A or a submatrix of A).

DESCA   (global and local input) INTEGER array, dimension (DLEN_)
The array descriptor for the distributed matrix A.
If DESCA(DTYPE_)=501 then DLEN_$\geq$7,
else if DESCA(DTYPE_)=1 then DLEN_$\geq$9.

B   (local input/local output) REAL/COMPLEX pointer into the local memory to an array of dimension (LLD_B, LOCc(NRHS))
On entry, this array contains the local pieces of the right hand sides B(IB:IB+N−1, 1:NRHS).
On exit, this array contains the local pieces of the distributed solution matrix X.

IB   (global input) INTEGER
The row index in the global array B that points to the first row of the matrix to be operated on (which may be either all of B or a submatrix of B).

DESCB   (global and local input) INTEGER array, dimension (DLEN_)
The array descriptor for the distributed matrix B.
If DESCB(DTYPE_)=502 then DLEN_$\geq$7,
else if DESCB(DTYPE_)=1 then DLEN_$\geq$9.

WORK   (local workspace/local output) REAL/COMPLEX array, dimension (LWORK)
On exit, WORK(1) returns the minimal and optimal LWORK.

LWORK   (local or global input) INTEGER
The dimension of the array WORK.
LWORK is local input and must be at least
LWORK $\geq$ NB*(BWL+BWU)+6*max(BWL,BWU)*max(BWL,BWU)
+max((max(BWL,BWU)*NRHS), max(BWL,BWU)*max(BWL,BWU)).
If LWORK = −1, then LWORK is global input and a workspace query is assumed; the routine only calculates the minimum and optimal size for all work arrays. Each of these values is returned in the first entry of the corresponding work array, and no error message is issued by PXERBLA.

INFO   (global output) INTEGER
= 0:   successful exit
< 0:   If the ith argument is an array and the j-entry had an illegal value, then INFO = −(i*100+j), if the ith argument is a scalar and had an illegal value, then INFO = −i.
> 0:   If INFO = K $\leq$ NPROCS, the submatrix stored on processor INFO−NPROCS and factored locally was not diagonally dominant-like, and the factorization was not completed.
If INFO = K > NPROCS, the submatrix stored on processor INFO−NPROCS representing interactions with other processors was not nonsingular, and the factorization was not completed.

PSDBTRF/PCDBTRF

```
SUBROUTINE PSDBTRF( N, BWL, BWU, A, JA, DESCA, AF, LAF, WORK,
$                   LWORK, INFO )
INTEGER             BWL, BWU, INFO, JA, LAF, LWORK, N
INTEGER             DESCA( * )
REAL                A( * ), AF( * ), WORK( * )

SUBROUTINE PCDBTRF( N, BWL, BWU, A, JA, DESCA, AF, LAF, WORK,
$                   LWORK, INFO )
INTEGER             BWL, BWU, INFO, JA, LAF, LWORK, N
INTEGER             DESCA( * )
COMPLEX             A( * ), AF( * ), WORK( * )
```

Purpose

PSDBTRF/PCDBTRF computes an LU factorization of a real/complex n-by-n diagonally dominant-like banded distributed matrix A(1:N, JA:JA+N−1) without pivoting.

The resulting factorization is *not* the same factorization as returned from LAPACK. Additional permutations are performed on the matrix for the sake of parallelism.

Arguments

N   (global input) INTEGER
        The number of rows and columns to be operated on, i.e. the order of
        the distributed submatrix A(1:N, JA:JA+N−1). N ≥ 0.

BWL   (global input) INTEGER
        The number of subdiagonals within the band of A. 0 ≤ BWL ≤ N−1.

BWU   (global input) INTEGER
        The number of superdiagonals within the band of A. 0 ≤ BWU ≤ N−1.

A   (local input/local output) REAL/COMPLEX pointer into the local
        memory to an array of dimension (LLD_A, LOCc(JA+N−1))
        On entry, this array contains the local pieces of the N-by-N general
        banded distributed matrix A(1:N, JA:JA+N−1) to be factored.
        On exit, details of the factorization. Note, the resulting factorization
        is *not* the same factorization as returned from LAPACK. Additional
        permutations are performed on the matrix for the sake of parallelism.

JA   (global input) INTEGER
        The index in the global array A that points to the start of the matrix
        to be operated on (which may be either all of A or a submatrix of A).

DESCA   (global and local input) INTEGER array, dimension (DLEN_)
        The array descriptor for the distributed matrix A.
        If DESCA(DTYPE_)=501 then DLEN_≥7,
        else if DESCA(DTYPE_)=1 then DLEN_≥9.

AF   (local output) REAL/COMPLEX array of dimension (LAF)
        Auxiliary Fillin space. Fillin is created during the factorization routine
        PSDBTRF/PCDBTRF and this is stored in AF. If a linear system is to
        be solved using PSDBTRS/PCDBTRS after the factorization routine,
        AF must not be altered.

LAF   (local input) INTEGER
        The dimension of the array AF.
        LAF ≥ NB*(BWL+BWU)+6*max(BWL,BWU)*max(BWL,BWU).
        If LAF is not large enough, an error code will be returned and the
        minimum acceptable size will be returned in AF(1).

WORK   (local workspace/local output) REAL/COMPLEX array of dimension
        (LWORK)
        On exit, WORK(1) returns the minimal and optimal LWORK.

LWORK   (local or global input) INTEGER
        The dimension of the array WORK.
        LWORK is local input and must be at least
        LWORK ≥ max(BWL,BWU)*max(BWL,BWU).
        If LWORK = −1, then LWORK is global input and a workspace query is
        assumed; the routine only calculates the minimum and optimal size for
        all work arrays. Each of these values is returned in the first entry of the
        corresponding work array, and no error message is issued by PXERBLA.

INFO   (global output) INTEGER
        = 0: successful exit
        < 0: If the ith argument is an array and the j-entry had an illegal
              value, then INFO = −(i*100+j), if the ith argument is a scalar
              and had an illegal value, then INFO = −i.
        > 0: If INFO = K ≤ NPROCS, the submatrix stored on proces-
              sor INFO−NPROCS and factored locally was not diagonally
              dominant-like, and the factorization was not completed.
              If INFO = K > NPROCS, the submatrix stored on processor
              INFO−NPROCS representing interactions with other processors
              was not nonsingular, and the factorization was not completed.

PSDBTRS/PCDBTRS

```
SUBROUTINE PSDBTRS( TRANS, N, BWL, BWU, NRHS, A, JA, DESCA, B, IB,
$                   DESCB, AF, LAF, WORK, LWORK, INFO )
CHARACTER           TRANS
INTEGER             BWL, BWU, IB, INFO, JA, LAF, LWORK, N, NRHS
INTEGER             DESCA( * ), DESCB( * )
REAL                A( * ), AF( * ), B( * ), WORK( * )
```

```
      SUBROUTINE PCDBTRS( TRANS, N, BWL, BWU, NRHS, A, JA, DESCA, B, IB,
     $                    DESCB, AF, LAF, WORK, LWORK, INFO )
      CHARACTER           TRANS
      INTEGER             BWL, BWU, IB, INFO, JA, LAF, LWORK, N, NRHS
      COMPLEX             DESCA( * ), DESCB( * )
      COMPLEX             A( * ), AF( * ), B( * ), WORK( * )
```

Purpose

PSDBTRS/PCDBTRS solves a system of linear equations

$A(1:N, JA:JA+N-1) * X = B(IB:IB+N-1, 1:NRHS)$, or
$A(1:N, JA:JA+N-1)^T * X = B(IB:IB+N-1, 1:NRHS)$, or
$A(1:N, JA:JA+N-1)^H * X = B(IB:IB+N-1, 1:NRHS)$,

with a diagonally dominant-like banded distributed matrix A using the LU factorization computed by PSDBTRF/PCDBTRF.

Arguments

TRANS   (global input) CHARACTER*1
        = 'N': Solve with $A(1:N, JA:JA+N-1)$;
        = 'T': Solve with $A(1:N, JA:JA+N-1)^T$;
        = 'C': Solve with $A(1:N, JA:JA+N-1)^H$;

N   (global input) INTEGER
        The number of rows and columns to be operated on, i.e. the order of
        the distributed submatrix $A(1:N, JA:JA+N-1)$. $N \geq 0$.

BWL   (global input) INTEGER
        The number of subdiagonals within the band of A. $0 \leq BWL \leq N-1$.

BWU   (global input) INTEGER
        The number of superdiagonals within the band of A. $0 \leq BWU \leq N-1$.

A   (local input) REAL/COMPLEX pointer into the local memory to an
        array of dimension (LLD_A, LOCc(JA+N-1))
        On entry, this array contains details of the LU factorization of the band
        matrix A, as computed by PSDBTRF/PCDBTRF.

JA   (global input) INTEGER
        The index in the global array A that points to the start of the matrix
        to be operated on (which may be either all of A or a submatrix of A).

DESCA   (global and local input) INTEGER array, dimension (DLEN_)
        The array descriptor for the distributed matrix A.
        If DESCA(DTYPE_)=501 then DLEN_$\geq$7,
        else if DESCA(DTYPE_)=1 then DLEN_$\geq$9.

B   (local input/local output) REAL/COMPLEX pointer into the local
        memory to an array of dimension (LLD_B, LOCc(NRHS))
        On entry, this array contains the local pieces of the right hand sides
        $B(IB:IB+N-1, 1:NRHS)$.

On exit, this array contains the local pieces of the distributed solution
matrix X.

IB   (global input) INTEGER
        The row index in the global array B that points to the first row of the
        matrix to be operated on (which may be either all of B or a submatrix
        of B).

DESCB   (global and local input) INTEGER array, dimension (DLEN_)
        The array descriptor for the distributed matrix B.
        If DESCB(DTYPE_)=502 then DLEN_$\geq$7,
        else if DESCB(DTYPE_)=1 then DLEN_$\geq$9.

AF   (local input) REAL/COMPLEX array, dimension (LAF)
        Auxiliary Fillin space. Fillin is created during the factorization routine
        PSDBTRF/PCDBTRF and this is stored in AF.

LAF   (local input) INTEGER
        The dimension of the array AF.
        $LAF \geq NB*(BWL+BWU)+6*\max(BWL,BWU)*\max(BWL,BWU)$.
        If LAF is not large enough, an error code will be returned and the
        minimum acceptable size will be returned in AF(1).

WORK   (local workspace/local output) REAL/COMPLEX array, dimension
        (LWORK)
        On exit, WORK(1) returns the minimal and optimal LWORK.

LWORK   (local or global input) INTEGER
        The dimension of the array WORK.
        LWORK is local input and must be at least
        $LWORK \geq \max(BWL,BWU)*\max(BWL,BWU)$.
        If LWORK = −1, then LWORK is global input and a workspace query is
        assumed; the routine only calculates the minimum and optimal size for
        all work arrays. Each of these values is returned in the first entry of the
        corresponding work array, and no error message is issued by PXERBLA.

INFO   (global output) INTEGER
        = 0: successful exit
        < 0: If the ith argument is an array and the j-entry had an illegal
              value, then INFO = −(i*100+j), if the ith argument is a scalar
              and had an illegal value, then INFO = −i.

PSDTSV/PCDTSV

```
      SUBROUTINE PSDTSV( N, NRHS, DL, D, DU, JA, DESCA, B, IB, DESCB,
     $                   WORK, LWORK, INFO )
      INTEGER            IB, INFO, JA, LWORK, N, NRHS
      INTEGER            DESCA( * ), DESCB( * )
      REAL               B( * ), D( * ), DL( * ), DU( * ), WORK( * )
```

```
SUBROUTINE PCDTSV( N, NRHS, DL, D, DU, JA, DESCA, B, IB, DESCB,
$                  WORK, LWORK, INFO )
INTEGER           IB, INFO, JA, LWORK, N, NRHS
INTEGER           DESCA( * ), DESCB( * )
COMPLEX           B( * ), D( * ), DL( * ), DU( * ), WORK( * )
```

Purpose

PSDTSV/PCDTSV solves a system of linear equations

$A(1:N, JA:JA+N-1) * X = B(IB:IB+N-1, 1:NRHS),$

where $A(1:N, JA:JA+N-1)$ is an n-by-n real/complex diagonally dominant-like tridiagonal distributed matrix, and X and B are n-by-nrhs distributed matrices.

Gaussian elimination without pivoting is used to factor the matrix into $P*L*U*P^T$, where P is a permutation matrix and L and U are banded lower and upper triangular matrices respectively. The permutations are performed for the sake of parallelism.

Arguments

N   (global input) INTEGER
         The number of rows and columns to be operated on, i.e. the order of
         the distributed submatrix $A(1:N, JA:JA+N-1)$. $N \geq 0$.

NRHS   (global input) INTEGER
         The number of right hand sides, i.e., the number of columns of the
         distributed submatrix B(IB:IB+N-1, 1:NRHS). NRHS ≥ 0.

DL   (local input/local output) REAL/COMPLEX pointer to the local array
         of dimension (DESCA(NB-))
         On entry, the local part of the global vector storing the subdiagonal
         elements of the matrix. Globally, DL(1) is not referenced, and DL must
         be aligned with D.
         On exit, this array contains information containing the factors of the
         matrix.

D   (local input/local output) REAL/COMPLEX pointer to the local array
         of dimension (DESCA(NB-))
         On entry, the local part of the global vector storing the diagonal ele-
         ments of the matrix.
         On exit, this array contains information containing the factors of the
         matrix.

DU   (local input/local output) REAL/COMPLEX pointer to the local array
         of dimension (DESCA(NB-))
         On entry, the local part of the global vector storing the super-diagonal
         elements of the matrix. DU(n) is not referenced, and DU must be
         aligned with D.
         On exit, this array contains information containing the factors of the
         matrix.

JA   (global input) INTEGER
         The index in the global array A that points to the start of the matrix
         to be operated on (which may be either all of A or a submatrix of A).

DESCA   (global and local input) INTEGER array, dimension (DLEN-)
         The array descriptor for the distributed matrix A.
         If DESCA(DTYPE-)=501 or 502 then DLEN_$\geq$7,
         else if DESCA(DTYPE-)=1 then DLEN-$\geq$9.

B   (local input/local output) REAL/COMPLEX pointer into the local
         memory to an array of dimension (LLD-B, LOCc(NRHS)).
         On entry, this array contains the the local pieces of the right hand sides
         B(IB:IB+N-1, 1:NRHS).
         On exit, this array contains the local pieces of the distributed solution
         matrix X.

IB   (global input) INTEGER
         The row index in the global array B that points to the first row of the
         matrix to be operated on (which may be either all of B or a submatrix
         of B).

DESCB   (global and local input) INTEGER array, dimension (DLEN-)
         The array descriptor for the distributed matrix B.
         If DESCB(DTYPE-)=502 then DLEN_$\geq$7,
         else if DESCB(DTYPE-)=1 then DLEN-$\geq$9.

WORK   (local workspace/local output) REAL/COMPLEX array, dimension
         (LWORK)
         On exit, WORK(1) returns the minimal and optimal LWORK.

LWORK   (local or global input) INTEGER
         The dimension of the array WORK.
         LWORK is local input and must be at least
         LWORK $\geq$ (12*NPCOL+3+NB)+max(10*NPCOL+4*NRHS, 8*NPCOL).
         If LWORK = −1, then LWORK is global input and a workspace query is
         assumed; the routine only calculates the minimum and optimal size for
         all work arrays. Each of these values is returned in the first entry of the
         corresponding work array, and no error message is issued by PXERBLA.

INFO   (global output) INTEGER
         = 0: successful exit
         < 0: If the ith argument is an array and the j-entry had an illegal
               value, then INFO = $-(i*100+j)$, if the ith argument is a scalar
               and had an illegal value, then INFO = −i.
         > 0: If INFO = K $\leq$ NPROCS, the submatrix stored on proces-
               sor INFO−NPROCS and factored locally was not diagonally
               dominant-like, and the factorization was not completed.
               If INFO = K > NPROCS, the submatrix stored on processor
               INFO−NPROCS representing interactions with other processors
               was not nonsingular, and the factorization was not completed.

PSDTTRF/PCDTTRF

```
SUBROUTINE PSDTTRF( N, DL, D, DU, JA, DESCA, AF, LAF, WORK, LWORK,
$                   INFO )
INTEGER            INFO, JA, LAF, LWORK, N
INTEGER            DESCA( * )
REAL               AF( * ), D( * ), DL( * ), DU( * ), WORK( * )

SUBROUTINE PCDTTRF( N, DL, D, DU, JA, DESCA, AF, LAF, WORK, LWORK,
$                   INFO )
INTEGER            INFO, JA, LAF, LWORK, N
INTEGER            DESCA( * )
COMPLEX            AF( * ), D( * ), DL( * ), DU( * ), WORK( * )
```

Purpose

PSDTTRF/PCDTTRF computes an LU factorization of an n-by-n real/complex diagonally dominant-like tridiagonal distributed matrix A without pivoting for stability.

The factorization has the form $A = P*L*U*P^T$, where P is a permutation matrix and L and U are banded lower and upper triangular matrices respectively.

The resulting factorization is *not* the same factorization as returned from LAPACK. Additional permutations are performed on the matrix for the sake of parallelism.

Arguments

N   (global input) INTEGER
        The number of rows and columns to be operated on, i.e. the order of the distributed submatrix A(1:N, JA:JA+N−1). N ≥ 0.

DL   (local input/local output) REAL/COMPLEX pointer to the local array of dimension (DESCA(NB_))
        On entry, the local part of the global vector storing the subdiagonal elements of the matrix. Globally, DL(1) is not referenced, and DL must be aligned with D.
        On exit, this array contains information containing the factors of the matrix.

D   (local input/local output) REAL/COMPLEX pointer to the local array of dimension (DESCA(NB_))
        On entry, the local part of the global vector storing the diagonal elements of the matrix.
        On exit, this array contains information containing the factors of the matrix.

DU   (local input/local output) REAL/COMPLEX pointer to the local array of dimension (DESCA(NB_))
        On entry, the local part of the global vector storing the super-diagonal elements of the matrix. DU(n) is not referenced, and DU must be

aligned with D.
        On exit, this array contains information containing the factors of the matrix.

JA   (global input) INTEGER
        The index in the global array A that points to the start of the matrix to be operated on (which may be either all of A or a submatrix of A).

DESCA   (global and local input) INTEGER array, dimension (DLEN_)
        The array descriptor for the distributed matrix A.
        If DESCA(DTYPE_)=501 or 502 then DLEN_≥7,
        else if DESCA(DTYPE_)=1 then DLEN_≥9.

AF   (local output) REAL/COMPLEX array, dimension (LAF)
        Auxiliary Fillin space. Fillin is created during the factorization routine PSDTTRF/PCDTTRF and this is stored in AF. If a linear system is to be solved using PSDTTRS/PCDTTRS after the factorization routine, AF must not be altered.

LAF   (local input) INTEGER
        The dimension of the array AF. LAF ≥ 2*(NB+2).
        If LAF is not large enough, an error code will be returned and the minimum acceptable size will be returned in AF(1).

WORK   (local workspace/local output) REAL/COMPLEX array, dimension (LWORK)
        On exit, WORK(1) returns the minimal and optimal LWORK.

LWORK   (local or global input) INTEGER
        The dimension of the array WORK.
        LWORK is local input and must be at least
        LWORK ≥ 8*NPCOL.
        If LWORK = −1, then LWORK is global input and a workspace query is assumed; the routine only calculates the minimum and optimal size for all work arrays. Each of these values is returned in the first entry of the corresponding work array, and no error message is issued by PXERBLA.

INFO   (global output) INTEGER
        = 0: successful exit
        < 0: If the ith argument is an array and the j-entry had an illegal value, then INFO = −(i*100+j), if the ith argument is a scalar and had an illegal value, then INFO = −i.
        > 0: If INFO = K ≤ NPROCS, the submatrix stored on processor INFO−NPROCS and factored locally was not diagonally dominant-like, and the factorization was not completed.
              If INFO = K > NPROCS, the submatrix stored on processor INFO−NPROCS representing interactions with other processors was not nonsingular, and the factorization was not completed.

PSDTTRS/PCDTTRS

```
SUBROUTINE PSDTTRS( TRANS, N, NRHS, DL, D, DU, JA, DESCA, B, IB,
$                   DESCB, AF, LAF, WORK, LWORK, INFO )
    CHARACTER        TRANS
    INTEGER          IB, INFO, JA, LAF, LWORK, N, NRHS
    INTEGER          DESCA( * ), DESCB( * )
    REAL             AF( * ), B( * ), D( * ), DL( * ), DU( * ),
$                    WORK( * )

SUBROUTINE PCDTTRS( TRANS, N, NRHS, DL, D, DU, JA, DESCA, B, IB,
$                   DESCB, AF, LAF, WORK, LWORK, INFO )
    CHARACTER        TRANS
    INTEGER          IB, INFO, JA, LAF, LWORK, N, NRHS
    INTEGER          DESCA( * ), DESCB( * )
    COMPLEX          AF( * ), B( * ), D( * ), DL( * ), DU( * ),
$                    WORK( * )
```

Purpose

PSDTTRS/PCDTTRS solves one of the systems of equations

$A(1:N, JA:JA+N-1) * X = B(IB:IB+N-1, 1:NRHS)$,
$A(1:N, JA:JA+N-1)^T * X = B(IB:IB+N-1, 1:NRHS)$, or
$A(1:N, JA:JA+N-1)^H * X = B(IB:IB+N-1, 1:NRHS)$,

with a diagonally dominant-like tridiagonal distributed matrix A using the LU factorization computed by PSDTTRF/PCDTTRF.

Arguments

TRANS   (global input) CHARACTER*1
        = 'N':   Solve with $A(1:N, JA:JA+N-1)$;
        = 'T':   Solve with $A(1:N, JA:JA+N-1)^T$;
        = 'C':   Solve with $A(1:N, JA:JA+N-1)^H$;

N   (global input) INTEGER
        The number of rows and columns to be operated on, i.e. the order of
        the distributed submatrix $A(1:N, JA:JA+N-1)$. $N \geq 0$.

DL   (local input) REAL/COMPLEX pointer to the local array of dimension
        (DESCA(NB_))
        On entry, details of the factorization. Globally, DL(1) is not referenced,
        and DL must be aligned with D.

D   (local input) REAL/COMPLEX pointer to the local array of dimension
        (DESCA(NB_))
        On entry, details of the factorization.

DU   (local input) REAL/COMPLEX pointer to the local array of dimension
        (DESCA(NB_))
        On entry, details of the factorization. DU(n) is not referenced, and DU

JA   (global input) INTEGER
        The index in the global array A that points to the start of the matrix
        to be operated on (which may be either all of A or a submatrix of A).

DESCA   (global and local input) INTEGER array, dimension (DLEN_)
        The array descriptor for the distributed matrix A.
        If DESCA(DTYPE_)=501 or 502 then DLEN_$\geq$7,
        else if DESCA(DTYPE_)=1 then DLEN_$\geq$9.

B   (local input/local output) REAL/COMPLEX pointer into the local
        memory to an array of dimension (LLD_B, LOCc(NRHS))
        On entry, this array contains the local pieces of the right hand sides
        B(IB:IB+N-1, 1:NRHS).
        On exit, this array contains the local pieces of the distributed solution
        matrix X.

IB   (global input) INTEGER
        The row index in the global array B that points to the first row of the
        matrix to be operated on (which may be either all of B or a submatrix
        of B).

DESCB   (global and local input) INTEGER array, dimension (DLEN_)
        The array descriptor for the distributed matrix B.
        If DESCB(DTYPE_)=502 then DLEN_$\geq$7,
        else if DESCB(DTYPE_)=1 then DLEN_$\geq$9.

AF   (local input) REAL/COMPLEX array, dimension (LAF)
        Auxiliary Fillin space. Fillin is created during the factorization routine
        PSDTTRF/PCDTTRF and this is stored in AF. If a linear system is to
        be solved using PSDTTRS/PCDTTRS after the factorization routine,
        AF *must not be altered*.

LAF   (local input) INTEGER
        The dimension of the array AF.
        LAF $\geq$ NB*(BWL+BWU)+6*(BWL+BWU)*(BWL+2*BWU).
        If LAF is not large enough, an error code will be returned and the
        minimum acceptable size will be returned in AF(1).

WORK   (local workspace/local output) REAL/COMPLEX array, dimension
        (LWORK)
        On exit, WORK(1) returns the minimal and optimal LWORK.

LWORK   (local or global input) INTEGER
        The dimension of the array WORK.
        LWORK is local input and must be at least
        LWORK $\geq$ 10*NPCOL+4*NRHS.
        If LWORK = -1, then LWORK is global input and a workspace query is
        assumed; the routine only calculates the minimum and optimal size for
        all work arrays. Each of these values is returned in the first entry of the
        corresponding work array, and no error message is issued by PXERBLA.

must be aligned with D.

INFO   (global output) INTEGER
= 0: successful exit
< 0: If the ith argument is an array and the j-entry had an illegal value, then INFO = $-(i*100+j)$, if the ith argument is a scalar and had an illegal value, then INFO = $-i$.

PSGBSV/PCGBSV

```
SUBROUTINE PSGBSV( N, BWL, BWU, NRHS, A, JA, DESCA, IPIV, B, IB,
$                  DESCB, WORK, LWORK, INFO )
INTEGER           BWL, BWU, INFO, IB, JA, LWORK, N, NRHS
INTEGER           DESCA( * ), DESCB( * ), IPIV( * )
REAL              A( * ), B( * ), WORK( * )

SUBROUTINE PCGBSV( N, BWL, BWU, NRHS, A, JA, DESCA, IPIV, B, IB,
$                  DESCB, WORK, LWORK, INFO )
INTEGER           BWL, BWU, INFO, IB, JA, LWORK, N, NRHS
INTEGER           DESCA( * ), DESCB( * ), IPIV( * )
COMPLEX           A( * ), B( * ), WORK( * )
```

Purpose

PSGBSV/PCGBSV solves a system of linear equations

$A(1:N, JA:JA+N-1) * X = B(IB:IB+N-1, 1:NRHS),$

where A(1:N, JA:JA+N-1) is an n-by-n real/complex general banded distributed matrix with bwl subdiagonals and bwu superdiagonals, and X and B are n-by-nrhs distributed matrices.

The LU decomposition with partial pivoting and row interchanges is used to factor A as A = P*L*U*Q, where P and Q are permutation matrices and L and U are banded lower and upper triangular matrices respectively. The matrix Q represents reordering of columns for the sake of parallelism, while P represents reordering of rows for numerical stability using classic partial pivoting.

Arguments

N   (global input) INTEGER
The number of rows and columns to be operated on, i.e. the order of the distributed submatrix A(1:N, JA:JA+N-1). N ≥ 0.

BWL   (global input) INTEGER
The number of subdiagonals within the band of A. 0 ≤ BWL ≤ N-1.

BWU   (global input) INTEGER
The number of superdiagonals within the band of A. 0 ≤ BWU ≤ N-1.

NRHS   (global input) INTEGER
The number of right hand sides, i.e, the number of columns of the distributed submatrix B(IB:IB+N-1, 1:NRHS). NRHS ≥ 0.

A   (local input/local output) REAL/COMPLEX pointer into the local memory to an array of dimension (LLD_A, LOCc(JA+N-1)).
LLD_A ≥ (2*BWL+2*BWU+1).
On entry, this array contains the local pieces of the global array A.
On exit, details of the factorization. Note, the resulting factorization is not the same factorization as returned from LAPACK. Additional permutations are performed on the matrix for the sake of parallelism.

JA   (global input) INTEGER
The index in the global array A that points to the start of the matrix to be operated on (which may be either all of A or a submatrix of A).

DESCA   (global and local input) INTEGER array, dimension (DLEN_).
The array descriptor for the distributed matrix A.
If DESCA(DTYPE_)=501 then DLEN_≥7,
else if DESCA(DTYPE_)=1 then DLEN_≥9.

IPIV   (local output) INTEGER array, dimension ≥ DESCA(NB).
Pivot indices for local factorizations. Users *should not* alter the contents between factorization and solve.

B   (local input/local output) REAL/COMPLEX pointer into the local memory to an array of dimension (LLD_B, LOCc(NRHS))
LLD_B ≥ NB.
On entry, this array contains the local pieces of the right hand sides B(IB:IB+N-1, 1:NRHS).
On exit, this array contains the local pieces of the distributed solution matrix X.

IB   (global input) INTEGER
The row index in the global array B that points to the first row of the matrix to be operated on (which may be either all of B or a submatrix of B).

DESCB   (global and local input) INTEGER array, dimension (DLEN_).
The array descriptor for the distributed matrix B.
If DESCB(DTYPE_)=502 then DLEN_≥7,
else if DESCB(DTYPE_)=1 then DLEN_≥9.

WORK   (local workspace/local output) REAL/COMPLEX array, dimension (LWORK)
On exit, WORK(1) returns the minimal and optimal LWORK.

LWORK   (local or global input) INTEGER
The dimension of the array WORK.
LWORK is local input and must be at least
LWORK ≥ (NB+BWU)*(BWL+BWU)+6*(BWL+BWU)*(BWL+2*BWU)
+max(NRHS*(NB+2*BWL+4*BWU), 1).
If LWORK = -1, then LWORK is global input and a workspace query is assumed; the routine only calculates the minimum and optimal size for

Arguments

N   (global input) INTEGER
The number of rows and columns to be operated on, i.e. the order of the distributed submatrix A(1:N, JA:JA+N−1). N ≥ 0.

BWL   (global input) INTEGER
The number of subdiagonals within the band of A. 0 ≤ BWL ≤ N−1.

BWU   (global input) INTEGER
The number of superdiagonals within the band of A. 0 ≤ BWU ≤ N−1.

A   (local input/local output) REAL/COMPLEX pointer into the local memory to an array of dimension (LLD_A, LOCc(JA+N−1))
LLD_A ≥ (2*BWL+2*BWU+1).
On entry, this array contains the local pieces of the N-by-N general banded distributed matrix A(1:N, JA:JA+N−1) to be factored.
On exit, details of the factorization. Note, the resulting factorization is *not* the same factorization as returned from LAPACK. Additional permutations are performed on the matrix for the sake of parallelism.

JA   (global input) INTEGER
The index in the global array A that points to the start of the matrix to be operated on (which may be either all of A or a submatrix of A).

DESCA (global and local input) INTEGER array, dimension (DLEN_)
The array descriptor for the distributed matrix A.
If DESCA(DTYPE_)=501 then DLEN_≥7,
else if DESCA(DTYPE_)=1 then DLEN_≥9.

IPIV   (local output) INTEGER array, dimension ≥ DESCA(NB).
Pivot indices for local factorizations. Users *should not* alter the contents between factorization and solve.

AF   (local output) REAL/COMPLEX array of dimension (LAF)
Auxiliary Fillin space. Fillin is created during the factorization routine PSDBTRF/PCDBTRF and this is stored in AF. If a linear system is to be solved using PSGBTRS/PCGBTRS after the factorization routine, AF *must not be altered*.

LAF   (local input) INTEGER
The dimension of the array AF.
LAF ≥ (NB+BWU)*(BWL+BWU)+6*(BWL+BWU)*(BWL+2*BWU).
If LAF is not large enough, an error code will be returned and the minimum acceptable size will be returned in AF(1).

WORK   (local workspace/local output) REAL/COMPLEX array of dimension (LWORK)
On exit, WORK(1) returns the minimal and optimal LWORK.

LWORK (local or global input) INTEGER
The dimension of the array WORK.
LWORK is local input and must be at least
LWORK ≥ 1.
If LWORK = −1, then LWORK is global input and a workspace query is

all work arrays. Each of these values is returned in the first entry of the corresponding work array, and no error message is issued by PXERBLA.

INFO   (global output) INTEGER
= 0:   successful exit
< 0:   If the ith argument is an array and the j-entry had an illegal value, then INFO = −(i*100+j), if the ith argument is a scalar and had an illegal value, then INFO = −i.
> 0:   If INFO = K ≤ NPROCS, the submatrix stored on processor INFO−NPROCS and factored locally was not nonsingular, and the factorization was not completed.
If INFO = K > NPROCS, the submatrix stored on processor INFO−NPROCS representing interactions with other processors was not nonsingular, and the factorization was not completed.

PSGBTRF/PCGBTRF

```
SUBROUTINE PSGBTRF( N, BWL, BWU, A, JA, DESCA, IPIV, AF, LAF,
$                    WORK, LWORK, INFO )
INTEGER       BWL, BWU, INFO, JA, LAF, LWORK, N
INTEGER       DESCA( * ), IPIV( * )
REAL          A( * ), AF( * ), WORK( * )

SUBROUTINE PCGBTRF( N, BWL, BWU, A, JA, DESCA, IPIV, AF, LAF,
$                    WORK, LWORK, INFO )
INTEGER       BWL, BWU, INFO, JA, LAF, LWORK, N
INTEGER       DESCA( * ), IPIV( * )
COMPLEX       A( * ), AF( * ), WORK( * )
```

Purpose

PSGBTRF/PCGBTRF computes an LU factorization of a real/complex n-by-n general banded distributed matrix A(1:N, JA:JA+N−1) using partial pivoting with row interchanges.

The resulting factorization is *not* the same factorization as returned from LAPACK. Additional permutations are performed on the matrix for the sake of parallelism.

The factorization has the form

A(1:N, JA:JA+N−1) = P * L * U * Q

where P and Q are permutation matrices and L and U are banded lower and upper triangular matrices respectively. The matrix Q represents reordering of columns for the sake of parallelism, while P represents reordering of rows for numerical stability using classic partial pivoting.

assumed; the routine only calculates the minimum and optimal size for all work arrays. Each of these values is returned in the first entry of the corresponding work array, and no error message is issued by PXERBLA.

INFO   (global output) INTEGER
= 0: successful exit
< 0: If the ith argument is an array and the j-entry had an illegal value, then INFO = $-(i*100+j)$, if the ith argument is a scalar and had an illegal value, then INFO = $-i$.
> 0: If INFO = K ≤ NPROCS, the submatrix stored on processor INFO−NPROCS and factored locally was not nonsingular, and the factorization was not completed.
If INFO = K > NPROCS, the submatrix stored on processor INFO−NPROCS representing interactions with other processors was not nonsingular, and the factorization was not completed.

PSGBTRS/PCGBTRS

```
SUBROUTINE PSGBTRS( TRANS, N, BWL, BWU, NRHS, A, JA, DESCA, IPIV,
$                    B, IB, DESCB, AF, LAF, WORK, LWORK, INFO )
    CHARACTER       TRANS
    INTEGER         BWU, BWL, IB, INFO, JA, LAF, LWORK, N, NRHS
    INTEGER         DESCA( * ), DESCB( * ), IPIV(*)
    REAL            A( * ), AF( * ), B( * ), WORK( * )

SUBROUTINE PCGBTRS( TRANS, N, BWL, BWU, NRHS, A, JA, DESCA, IPIV,
$                    B, IB, DESCB, AF, LAF, WORK, LWORK, INFO )
    CHARACTER       TRANS
    INTEGER         BWU, BWL, IB, INFO, JA, LAF, LWORK, N, NRHS
    INTEGER         DESCA( * ), DESCB( * ), IPIV(*)
    COMPLEX         A( * ), AF( * ), B( * ), WORK( * )
```

Purpose

PSGBTRS/PCGBTRS solves a system of linear equations

A(1:N, JA:JA+N−1) * X = B(IB:IB+N−1, 1:NRHS),
A(1:N, JA:JA+N−1)^T * X = B(IB:IB+N−1, 1:NRHS), or
A(1:N, JA:JA+N−1)^H * X = B(IB:IB+N−1, 1:NRHS),

with a general band matrix A using the LU factorization computed by PSGBTRF/PCGBTRF.

Arguments

TRANS   (global input) CHARACTER*1
= 'N': Solve with A(1:N, JA:JA+N−1);
= 'T': Solve with A(1:N, JA:JA+N−1)^T;
= 'C': Solve with A(1:N, JA:JA+N−1)^H;

N   (global input) INTEGER
The number of rows and columns to be operated on, i.e. the order of the distributed submatrix A(1:N, JA:JA+N−1). N ≥ 0.

BWL   (global input) INTEGER
The number of subdiagonals within the band of A. 0 ≤ BWL ≤ N−1.

BWU   (global input) INTEGER
The number of superdiagonals within the band of A. 0 ≤ BWU ≤ N−1.

A   (local input) REAL/COMPLEX pointer into the local memory to an array of dimension (LLD_A, LOCc(JA+N−1))
This array contains details of the LU factorization of the band matrix A.

JA   (global input) INTEGER
The index in the global array A that points to the start of the matrix to be operated on (which may be either all of A or a submatrix of A).

DESCA   (global and local input) INTEGER array, dimension (DLEN_)
The array descriptor for the distributed matrix A.
If DESCA(DTYPE_)=501 then DLEN_≥7,
else if DESCA(DTYPE_)=1 then DLEN_≥9.

IPIV   (local output) INTEGER array, dimension ≥ DESCA(NB).
Pivot indices for local factorizations. Users *should not* alter the contents between factorization and solve.

B   (local input/local output) REAL/COMPLEX pointer into the local memory to an array of dimension (LLD_B, LOCc(NRHS))
On entry, this array contains the local pieces of the right hand sides B(IB:IB+N−1, 1:NRHS).
On exit, this array contains the local pieces of the solution distributed matrix X.

IB   (global input) INTEGER
The row index in the global array B that points to the first row of the matrix to be operated on (which may be either all of B or a submatrix of B).

DESCB   (global and local input) INTEGER array, dimension (DLEN_)
The array descriptor for the distributed matrix B.
If DESCB(DTYPE_)=502 then DLEN_≥7,
else if DESCB(DTYPE_)=1 then DLEN_≥9.

AF   (local output) REAL/COMPLEX array, dimension (LAF)
Auxiliary Fillin space. Fillin is created during the factorization routine PSGBTRF/PCGBTRF and this is stored in AF. If a linear system is to be solved using PSGBTRS/PCGBTRS after the factorization routine, AF *must not be altered*.

LAF   (local input) INTEGER
The dimension of the array AF.

M   (global input) INTEGER
The number of rows to be operated on, i.e. the number of rows of the distributed submatrix sub(A). M ≥ 0.

N   (global input) INTEGER
The number of columns to be operated on, i.e. the number of columns of the distributed submatrix sub(A). N ≥ 0.

A   (local input/local output) REAL/COMPLEX pointer into the local memory to an array of dimension (LLD_A,LOCc(JA+N−1)).
On entry, this array contains the local pieces of the general distributed matrix sub(A).
On exit, if m ≥ n, the diagonal and the first superdiagonal of sub(A) are overwritten with the upper bidiagonal matrix B; the elements below the diagonal, with the array TAUQ, represent the unitary matrix Q as a product of elementary reflectors, and the elements above the first superdiagonal, with the array TAUP, represent the orthogonal matrix P as a product of elementary reflectors. If m < n, the diagonal and the first subdiagonal are overwritten with the lower bidiagonal matrix B; the elements below the first subdiagonal, with the array TAUQ, represent the unitary matrix Q as a product of elementary reflectors, and the elements above the diagonal, with the array TAUP, represent the orthogonal matrix P as a product of elementary reflectors.

IA   (global input) INTEGER
The row index in the global array A indicating the first row of sub(A).

JA   (global input) INTEGER
The column index in the global array A indicating the first column of sub(A).

DESCA   (global and local input) INTEGER array, dimension (DLEN_).
The array descriptor for the distributed matrix A.

D   (local output) REAL array, dimension
LOCc(JA+MIN(M,N)−1) if m ≥ n; LOCr(IA+MIN(M,N)−1) otherwise. The distributed diagonal elements of the bidiagonal matrix B: D(i) = A(i,i). D is tied to the distributed matrix A.

E   (local output) REAL array, dimension
LOCr(IA+MIN(M,N)−1) if M ≥ N; LOCc(JA+MIN(M,N)−2) otherwise. The distributed off-diagonal elements of the bidiagonal distributed matrix B:
if m ≥ n, E(i) = A(i,i+1) for i = 1,2,...,n−1;
if m < n, E(i) = A(i+1,i) for i = 1,2,...,m−1.
E is tied to the distributed matrix A.

TAUQ   (local output) REAL/COMPLEX array, dimension LOCc(JA+MIN(M,N)−1).
The scalar factors of the elementary reflectors which represent the orthogonal/unitary matrix Q. TAUQ is tied to the distributed matrix A.

LAF ≥ NB*(BWL+BWU)+6*(BWL+BWU)*(BWL+2*BWU).
If LAF is not large enough, an error code will be returned and the minimum acceptable size will be returned in AF(1).

WORK   (local workspace/local output) REAL/COMPLEX array, dimension (LWORK)
On exit, WORK(1) returns the minimal and optimal LWORK.

LWORK   (local or global input) INTEGER
The dimension of the array WORK.
LWORK is local input and must be at least
LWORK ≥ NRHS*(NB+2+BWL+4*BWU).
If LWORK = −1, then LWORK is global input and a workspace query is assumed; the routine only calculates the minimum and optimal size for all work arrays. Each of these values is returned in the first entry of the corresponding work array, and no error message is issued by PXERBLA.

INFO   (global output) INTEGER
= 0: successful exit
< 0: If the ith argument is an array and the j-entry had an illegal value, then INFO = −(i*100+j), if the ith argument is a scalar and had an illegal value, then INFO = −i.

PSGEBRD/PCGEBRD

```
SUBROUTINE PSGEBRD( M, N, A, IA, JA, DESCA, D, E, TAUQ, TAUP,
$                    WORK, LWORK, INFO )
    INTEGER          IA, INFO, JA, LWORK, M, N
    INTEGER          DESCA( * )
    REAL             A( * ), D( * ), E( * ), TAUP( * ), TAUQ( * ),
$                    WORK( * )

SUBROUTINE PCGEBRD( M, N, A, IA, JA, DESCA, D, E, TAUQ, TAUP,
$                    WORK, LWORK, INFO )
    INTEGER          IA, INFO, JA, LWORK, M, N
    INTEGER          DESCA( * )
    REAL             D( * ), E( * )
    COMPLEX          A( * ), TAUP( * ), TAUQ( * ), WORK( * )
```

Purpose

PSGEBRD/PCGEBRD reduces a real/complex general M-by-N distributed matrix sub(A) = A(IA:IA+M−1,JA:JA+N−1) to upper or lower bidiagonal form B by an orthogonal/unitary transformation: Q^H*sub(A)*P = B.

If m ≥ n, B is upper bidiagonal; if m < n, B is lower bidiagonal.

TAUP   (local output) REAL/COMPLEX array, dimension LOCr(IA+MIN(M,N)−1). The scalar factors of the elementary reflectors which represent the orthogonal/unitary matrix P. TAUP is tied to the distributed matrix A.

WORK   (local workspace/local output) REAL/COMPLEX array, dimension (LWORK). On exit, WORK(1) returns the minimal and optimal LWORK.

LWORK   (local or global input) INTEGER
The dimension of the array WORK.
LWORK is local input and must be at least
LWORK ≥ NB*(MpA0 + NqA0 + 1) + NqA0
where NB = MB_A = NB_A,

IROFFA = mod(IA−1, NB), ICOFFA = mod(JA−1, NB),
IAROW = INDXG2P(IA, NB, MYROW, RSRC_A, NPROW),
IACOL = INDXG2P(JA, NB, MYCOL, CSRC_A, NPCOL),
MpA0 = NUMROC(M+IROFFA, NB, MYROW, IAROW, NPROW),
NqA0 = NUMROC(N+ICOFFA, NB, MYCOL, IACOL, NPCOL).

INDXG2P and NUMROC are ScaLAPACK tool functions; MYROW, MYCOL, NPROW and NPCOL can be determined by calling the subroutine BLACS_GRIDINFO.

If LWORK = −1, then LWORK is global input and a workspace query is assumed; the routine only calculates the minimum and optimal size for all work arrays. Each of these values is returned in the first entry of the corresponding work array, and no error message is issued by PXERBLA.

INFO   (global output) INTEGER
= 0: successful exit
< 0: If the ith argument is an array and the j-entry had an illegal value, then INFO = −(i*100+j), if the ith argument is a scalar and had an illegal value, then INFO = −i.

PSGECON/PCGECON

```
SUBROUTINE PSGECON( NORM, N, A, IA, JA, DESCA, ANORM, RCOND, WORK,
$                    LWORK, IWORK, LIWORK, INFO )
     CHARACTER         NORM
     INTEGER           IA, INFO, JA, LIWORK, LWORK, N
     REAL              ANORM, RCOND
     INTEGER           DESCA( * ), IWORK( * )
     REAL              A( * ), WORK( * )
```

```
SUBROUTINE PCGECON( NORM, N, A, IA, JA, DESCA, ANORM, RCOND, WORK,
$                    LWORK, RWORK, LRWORK, INFO )
     CHARACTER         NORM
     INTEGER           IA, INFO, JA, LRWORK, LWORK, N
     REAL              ANORM, RCOND
     INTEGER           DESCA( * )
     REAL              RWORK( * )
     COMPLEX           A( * ), WORK( * )
```

Purpose

PSGECON/PCGECON estimates the reciprocal of the condition number of a general distributed real/complex matrix A(IA:IA+N−1,JA:JA+N−1), in either the 1-norm or the infinity-norm, using the LU factorization computed by PSGETRF/PCGETRF.

An estimate is obtained for $\|A(IA : IA + N − 1, JA : JA + N − 1)^{-1}\|$, and the reciprocal of the condition number is computed as

$$RCOND = \frac{1}{\|A(IA:IA+N−1,JA:JA+N−1)\| * \|A(IA:IA+N−1,JA:JA+N−1)^{-1}\|}.$$

Arguments

NORM   (global input) CHARACTER*1
Specifies whether the 1−norm condition number or the infinity-norm condition number is required:
= '1' or 'O': 1-norm;
= 'I': Infinity-norm.

N   (global input) INTEGER
The order of the distributed matrix A(IA:IA+N−1,JA:JA+N−1). N ≥ 0.

A   (local input) REAL/COMPLEX pointer into the local memory to an array of dimension (LLD_A, LOCc(JA+N−1)). On entry, this array contains the local pieces of the factors L and U from the factorization A(IA:IA+N−1,JA:JA+N−1) = P*L*U; the unit diagonal elements of L are not stored.

IA   (global input) INTEGER
The row index in the global array A indicating the first row of sub(A).

JA   (global input) INTEGER
The column index in the global array A indicating the first column of sub(A).

DESCA   (global and local input) INTEGER array, dimension (DLEN_). The array descriptor for the distributed matrix A.

ANORM   (global input) REAL
If NORM = '1' or 'O', the 1-norm of the original distributed matrix A(IA:IA+N−1,JA:JA+N−1).
If NORM = 'I', the infinity-norm of the original distributed matrix A(IA:IA+N−1,JA:JA+N−1).

RCOND (global output) REAL
The reciprocal of the condition number of the distributed matrix A(IA:IA+N−1,JA:JA+N−1), computed as RCOND =

$$\frac{1}{\left(\|A(IA:IA+N-1,JA:JA+N-1)\|\cdot\|A(IA:IA+N-1,JA:JA+N-1)^{-1}\|\right)}.$$

WORK (local workspace/local output) REAL/COMPLEX array, dimension (LWORK)
On exit, WORK(1) returns the minimal and optimal LWORK.

LWORK (local or global input) INTEGER
The dimension of the array WORK.
LWORK is local input and must be at least
PSGECON
LWORK $\geq$ 2*LOCr(N+mod(IA−1,MB_A)) + 2*LOCc(N+mod(JA−1,NB_A)) + max(2,max(NB_A*max(1,CEIL(NPROW−1,NPCOL)), LOCc(N+mod(JA−1,NB_A)) + NB_A*max(1, CEIL(NPCOL−1,NPROW)))).
PCGECON
LWORK $\geq$ 2*LOCr(N+mod(IA−1,MB_A)) + max(2,max(NB_A*CEIL(NPROW−1,NPCOL), LOCc(N+mod(JA−1,NB_A)) + NB_A*CEIL(NPCOL−1,NPROW))).

LOCr and LOCc values can be computed using the ScaLAPACK tool function NUMROC; NPROW and NPCOL can be determined by calling the subroutine BLACS_GRIDINFO.

If LWORK = −1, then LWORK is global input and a workspace query is assumed; the routine only calculates the minimum and optimal size for all work arrays. Each of these values is returned in the first entry of the corresponding work array, and no error message is issued by PXERBLA.

IWORK PSGECON only (local workspace/local output) INTEGER array, dimension (LIWORK)
On exit, IWORK(1) returns the minimal and optimal LIWORK.

LIWORK PSGECON only (local or global input) INTEGER
The dimension of the array IWORK.
LIWORK is local input and must be at least
LIWORK $\geq$ LOCr(N+mod(IA−1,MB_A)).
If LIWORK = −1, then LIWORK is global input and a workspace query is assumed; the routine only calculates the minimum and optimal size for all work arrays. Each of these values is returned in the first entry of the corresponding work array, and no error message is issued by PXERBLA.

RWORK PCGECON only (local workspace/local output) REAL array, dimension (LRWORK)
On exit, RWORK(1) returns the minimal and optimal LRWORK.

LRWORK PCGECON only (local or global input) INTEGER
The dimension of the array RWORK.
LRWORK is local input and must be at least
LRWORK $\geq$ 2*LOCc(N+mod(JA−1,NB_A)).
If LRWORK = −1, then LRWORK is global input and a workspace query is assumed; the routine only calculates the minimum and optimal

size for all work arrays. Each of these values is returned in the first entry of the corresponding work array, and no error message is issued by PXERBLA.

INFO (global output) INTEGER
= 0: successful exit
< 0: If the ith argument is an array and the j-entry had an illegal value, then INFO = −(i*100+j), if the ith argument is a scalar and had an illegal value, then INFO = −i.

PSGEEQU/PCGEEQU

```
SUBROUTINE PSGEEQU( M, N, A, IA, JA, DESCA, R, C, ROWCND, COLCND,
$                   AMAX, INFO )
INTEGER        IA, INFO, JA, M, N
REAL           AMAX, COLCND, ROWCND
INTEGER        DESCA( * )
REAL           A( * ), C( * ), R( * )

SUBROUTINE PCGEEQU( M, N, A, IA, JA, DESCA, R, C, ROWCND, COLCND,
$                   AMAX, INFO )
INTEGER        IA, INFO, JA, M, N
REAL           AMAX, COLCND, ROWCND
INTEGER        DESCA( * )
REAL           C( * ), R( * )
COMPLEX        A( * )
```

Purpose

PSGEEQU/PCGEEQU computes row and column scalings intended to equilibrate an M-by-N distributed matrix sub(A) = A(IA:IA+N−1,JA:JA+N−1) and reduce its condition number. R returns the row scale factors and C the column scale factors, chosen to try to make the largest entry in each row and column of the distributed matrix B with elements B(i,j) = R(i)*A(i,j)*C(j) have absolute value 1.

R(i) and C(j) are restricted to be between SMLNUM = smallest safe number and BIGNUM = largest safe number. Use of these scaling factors is not guaranteed to reduce the condition number of sub(A) but works well in practice.

Arguments

M (global input) INTEGER
The number of rows to be operated on i.e., the number of rows of the distributed submatrix sub(A). M $\geq$ 0.

N (global input) INTEGER
The number of columns to be operated on i.e., the number of columns of the distributed submatrix sub(A). N $\geq$ 0.

A   (local input) REAL/COMPLEX pointer into the local memory to an array of dimension (LLD_A, LOCc(JA+N-1)).
The local pieces of the m-by-n distributed matrix whose equilibration factors are to be computed.

IA   (global input) INTEGER
The row index in the global array A indicating the first row of sub(A).

JA   (global input) INTEGER
The column index in the global array A indicating the first column of sub(A).

DESCA (global and local input) INTEGER array, dimension (DLEN_).
The array descriptor for the distributed matrix A.

R   (local output) REAL array, dimension LOCr(M_A)
If INFO = 0 or INFO > IA+M-1, R(IA:IA+M-1) contains the row scale factors for sub(A). R is aligned with the distributed matrix A, and replicated across every process column. R is tied to the distributed matrix A.

C   (local output) REAL array, dimension LOCc(N_A)
If INFO = 0, C(JA:JA+N-1) contains the column scale factors for sub(A). C is aligned with the distributed matrix A, and replicated down every process row. C is tied to the distributed matrix A.

ROWCND (global output) REAL
If INFO = 0 or INFO > IA+M-1, ROWCND contains the ratio of the smallest R(i) to the largest R(i) (IA $\leq$ i $\leq$ IA+M-1). If ROWCND $\geq$ 0.1 and AMAX is neither too large nor too small, it is not worth scaling by R(IA:IA+M-1).

COLCND (global output) REAL
If INFO = 0, COLCND contains the ratio of the smallest C(j) to the largest C(j) (JA $\leq$ j $\leq$ JA+N-1). If COLCND $\geq$ 0.1, it is not worth scaling by C(JA:JA+N-1).

AMAX (global output) REAL
Absolute value of largest matrix element. If AMAX is very close to overflow or very close to underflow, the matrix should be scaled.

INFO (global output) INTEGER
= 0: successful exit
< 0: If the ith argument is an array and the j-entry had an illegal value, then INFO = $-(i*100+j)$, if the ith argument is a scalar and had an illegal value, then INFO = $-i$.
> 0: if INFO = i, and i is
$\leq$ M: the ith row of the distributed matrix sub(A) is exactly zero,
> M: the $(i-M)^{th}$ column of distributed matrix sub(A) is exactly zero.

PSGEHRD/PCGEHRD

```
SUBROUTINE PSGEHRD( N, ILO, IHI, A, IA, JA, DESCA, TAU, WORK,
$                   LWORK, INFO )

    INTEGER         IA, IHI, ILO, INFO, JA, LWORK, N
    INTEGER         DESCA( * )
    REAL            A( * ), TAU( * ), WORK( * )

SUBROUTINE PCGEHRD( N, ILO, IHI, A, IA, JA, DESCA, TAU, WORK,
$                   LWORK, INFO )

    INTEGER         IA, IHI, ILO, INFO, JA, LWORK, N
    INTEGER         DESCA( * )
    COMPLEX         A( * ), TAU( * ), WORK( * )
```

Purpose

PSGEHRD/PCGEHRD reduces a real/complex general distributed matrix sub(A) to upper Hessenberg form H by an orthogonal/unitary similarity transformation: $Q^H * sub(A) * Q = H$, where sub(A) = A(IA:IA+N-1,JA:JA+N-1).

Arguments

N   (global input) INTEGER
The number of rows and columns to be operated on, i.e., the order of the distributed submatrix sub(A). N $\geq$ 0.

ILO, IHI (global input) INTEGER
It is assumed that sub(A) is already upper triangular in rows IA:IA+ILO-2 and IA+IHI:IA+N-1 and columns JA:JA+ILO-2 and JA+IHI:JA+N-1.
If N > 0, 1 $\leq$ ILO $\leq$ IHI $\leq$ N; otherwise set ILO = 1, IHI = N.

A   (local input/local output) REAL/COMPLEX pointer into the local memory to an array of dimension (LLD_A,LOCc(JA+N-1)). On entry, this array contains the local pieces of the n-by-n general distributed matrix sub(A) to be reduced. On exit, the upper triangle and the first subdiagonal of sub(A) are overwritten with the upper Hessenberg matrix H, and the elements below the first subdiagonal, with the array TAU, represent the orthogonal/unitary matrix Q as a product of elementary reflectors.

IA   (global input) INTEGER
The row index in the global array A indicating the first row of sub(A).

JA   (global input) INTEGER
The column index in the global array A indicating the first column of sub(A).

DESCA (global and local input) INTEGER array, dimension (DLEN_).
The array descriptor for the distributed matrix A.

TAU (local output) REAL/COMPLEX array, dimension LOCc(JA+N-2).
The scalar factors of the elementary reflectors. Elements JA:JA+ILO-2

and JA+IHI:JA+N−2 of TAU are set to zero. TAU is tied to the distributed matrix A.

WORK   (local workspace/local output) REAL/COMPLEX array, dimension (LWORK)
On exit, WORK(1) returns the minimal and optimal LWORK.

LWORK   (local or global input) INTEGER
The dimension of the array WORK.
LWORK is local input and must be at least

LWORK $\geq$ NB*NB + NB*max(IHIP+1, IHLP+INLQ)

where NB = MB_A = NB_A, IROFFA = mod(IA−1, NB),
ICOFFA = mod(JA−1, NB), IOFF = mod(IA+ILO−2, NB),
IAROW = INDXG2P(IA, NB, MYROW, RSRC_A, NPROW),
IHIP = NUMROC(IHI+IROFFA, NB, MYROW, IAROW, NPROW),
ILROW = INDXG2P(IA+ILO−1, NB, MYROW, RSRC_A, NPROW),
IHLP = NUMROC(IHI−ILO+IOFF+1, NB, MYROW, ILROW, NPROW),
ILCOL = INDXG2P(JA+ILO−1, NB, MYCOL, CSRC_A, NPCOL),
INLQ = NUMROC(N−ILO+IOFF+1, NB, MYCOL, ILCOL, NPCOL).
INDXG2P and NUMROC are ScaLAPACK tool functions; MYROW, MYCOL, NPROW and NPCOL can be determined by calling the subroutine BLACS_GRIDINFO.

If LWORK = −1, then LWORK is global input and a workspace query is assumed; the routine only calculates the minimum and optimal size for all work arrays. Each of these values is returned in the first entry of the corresponding work array, and no error message is issued by PXERBLA.

INFO   (global output) INTEGER
= 0: successful exit
< 0: If the ith argument is an array and the j-entry had an illegal value, then INFO = −(i*100+j), if the ith argument is a scalar and had an illegal value, then INFO = −i.

PSGELQF/PCGELQF

```
SUBROUTINE PSGELQF( M, N, A, IA, JA, DESCA, TAU, WORK, LWORK,
$                    INFO )
INTEGER            IA, INFO, JA, LWORK, M, N
INTEGER            DESCA( * )
REAL               A( * ), TAU( * ), WORK( * )

SUBROUTINE PCGELQF( M, N, A, IA, JA, DESCA, TAU, WORK, LWORK,
$                    INFO )
INTEGER            IA, INFO, JA, LWORK, M, N
INTEGER            DESCA( * )
COMPLEX            A( * ), TAU( * ), WORK( * )
```

Purpose

PSGELQF/PCGELQF computes a LQ factorization of a real/complex distributed m-by-n matrix sub(A) = A(IA:IA+M−1,JA:JA+N−1) = L*Q.

Arguments

M   (global input) INTEGER
The number of rows to be operated on, i.e., the number of rows of the distributed submatrix sub(A). $M \geq 0$.

N   (global input) INTEGER
The number of columns to be operated on, i.e., the number of columns of the distributed submatrix sub(A). $N \geq 0$.

A   (local input/local output) REAL/COMPLEX pointer into the local memory to an array of dimension (LLD_A, LOCc(JA+N−1)). On entry, the local pieces of the m-by-n distributed matrix sub(A) which is to be factored.
On exit, the elements on and below the diagonal of sub(A) contain the m-by-min(m,n) lower trapezoidal matrix L (L is lower triangular if m ≤ n); the elements above the diagonal, with the array TAU, represent the orthogonal/unitary matrix Q as a product of elementary reflectors.

IA   (global input) INTEGER
The row index in the global array A indicating the first row of sub(A).

JA   (global input) INTEGER
The column index in the global array A indicating the first column of sub(A).

DESCA   (global and local input) INTEGER array, dimension (DLEN_).
The array descriptor for the distributed matrix A.

TAU   (local output) REAL/COMPLEX array, dimension LOCr(IA+MIN(M,N)−1).
This array contains the scalar factors of the elementary reflectors. TAU is tied to the distributed matrix A.

WORK   (local workspace/local output) REAL/COMPLEX array, dimension (LWORK)
On exit, WORK(1) returns the minimal and optimal LWORK.

LWORK   (local or global input) INTEGER
The dimension of the array WORK.
LWORK is local input and must be at least

LWORK $\geq$ MB_A * (Mp0 + Nq0 + MB_A), where

IROFF = mod(IA−1, MB_A), ICOFF = mod(JA−1, NB_A),
IAROW = INDXG2P(IA, MB_A, MYROW, RSRC_A, NPROW),
IACOL = INDXG2P(JA, NB_A, MYCOL, CSRC_A, NPCOL),
Mp0 = NUMROC(M+IROFF, MB_A, MYROW, IAROW, NPROW),
Nq0 = NUMROC(N+ICOFF, NB_A, MYCOL, IACOL, NPCOL).
NUMROC and INDXG2P are ScaLAPACK tool functions; MYROW, MYCOL, NPROW and NPCOL can be determined by calling the subroutine BLACS_GRIDINFO.

If LWORK = −1, then LWORK is global input and a workspace query is assumed; the routine only calculates the minimum and optimal size for all work arrays. Each of these values is returned in the first entry of the corresponding work array, and no error message is issued by PXERBLA.

INFO (global output) INTEGER
= 0: successful exit
< 0: If the ith argument is an array and the j-entry had an illegal value, then INFO = −(i*100+j), if the ith argument is a scalar and had an illegal value, then INFO = −i.

PSGELS/PCGELS

```
SUBROUTINE PSGELS( TRANS, M, N, NRHS, A, IA, JA, DESCA, B, IB, JB,
$                   DESCB, WORK, LWORK, INFO )
     CHARACTER   TRANS
     INTEGER     IA, IB, INFO, JA, JB, LWORK, M, N, NRHS
     INTEGER     DESCA( * ), DESCB( * )
     REAL        A( * ), B( * ), WORK( * )

SUBROUTINE PCGELS( TRANS, M, N, NRHS, A, IA, JA, DESCA, B, IB, JB,
$                   DESCB, WORK, LWORK, INFO )
     CHARACTER   TRANS
     INTEGER     IA, IB, INFO, JA, JB, LWORK, M, N, NRHS
     INTEGER     DESCA( * ), DESCB( * )
     COMPLEX     A( * ), B( * ), WORK( * )
```

Purpose

PSGELS/PCGELS solves overdetermined or underdetermined real/complex linear systems involving an M-by-N matrix sub(A) = A(IA:IA+M−1,JA:JA+N−1), or its transpose/conjugate-transpose, using a QR or LQ factorization of sub(A). It is assumed that sub(A) has full rank.

The following options are provided:

1. If TRANS = 'N' and m ≥ n: find the least squares solution of an overdetermined system, i.e., solve the least squares problem

$$\text{minimize } \| \text{sub}(B) - \text{sub}(A)*x \|_2 .$$

2. If TRANS = 'N' and m < n: find the minimum norm solution of an underdetermined system sub(A)*x = sub(B).

3. If TRANS = 'T'/'C' and m ≥ n: find the minimum norm solution of an undetermined system sub(A)^H*X = sub(B).

4. If TRANS = 'T'/'C' and m < n: find the least squares solution of an overdetermined system, i.e., solve the least squares problem

$$\text{minimize } \| \text{sub}(B) - \text{sub}(A)^H*x \|_2 .$$

where sub(B) denotes B(IB:IB+M−1, JB:JB+NRHS−1) when TRANS = 'N' and B(IB:IB+N−1, JB:JB+NRHS−1) otherwise. Several right hand side vectors b and solution vectors x can be handled in a single call; When TRANS = 'N', the solution vectors are stored as the columns of the n-by-nrhs right hand side matrix sub(B) and the m-by-nrhs right hand side matrix sub(B) otherwise.

Arguments

TRANS (global input) CHARACTER
= 'N': the linear system involves sub(A);
= 'T': the linear system involves sub(A)^T (PSGELS);
= 'C': the linear system involves sub(A)^H (PCGELS).

M (global input) INTEGER
The number of rows to be operated on, i.e., the number of rows of the distributed submatrix sub(A). M ≥ 0.

N (global input) INTEGER
The number of columns to be operated on, i.e., the number of columns of the distributed submatrix sub(A). N ≥ 0.

NRHS (global input) INTEGER
The number of right hand sides, i.e., the number of columns of the distributed submatrices sub(B) and X. NRHS ≥ 0.

A (local input/local output) REAL/COMPLEX pointer into the local memory to an array of local dimension (LLD_A, LOCc(JA+N−1)).
On entry, the m-by-n matrix A.
On exit,
if m ≥ n, sub(A) is overwritten by details of its QR factorization as returned by PSGEQRF/PCGEQRF;
if m < n, sub(A) is overwritten by details of its LQ factorization as returned by PSGELQF/PCGELQF.

IA (global input) INTEGER
The row index in the global array A indicating the first row of sub(A).

JA (global input) INTEGER
The column index in the global array A indicating the first column of sub(A).

DESCA (global and local input) INTEGER array, dimension (DLEN_).
The array descriptor for the distributed matrix A.

B (input/output) REAL/COMPLEX pointer into the local memory to an array of local dimension (LLD_B, LOCc(JB+NRHS−1)).
On entry, this array contains the local pieces of the distributed matrix B of right hand side vectors, stored columnwise; sub(B) is m-by-nrhs if TRANS='N', and n-by-nrhs otherwise.
On exit, sub(B) is overwritten by the solution vectors, stored columnwise: if TRANS = 'N' and m ≥ n, rows 1 to n of sub(B) contain the least squares solution vectors; the residual sum of squares for the

for all work arrays. Each of these values is returned in the first entry of the corresponding work array, and no error message is issued by PXERBLA.

INFO   (global output) INTEGER
        = 0: successful exit
        < 0: If the ith argument is an array and the j-entry had an illegal value, then INFO $= -(i*100+j)$, if the ith argument is a scalar and had an illegal value, then INFO $= -i$.

PSGEQLF/PCGEQLF

```
SUBROUTINE PSGEQLF( M, N, A, IA, JA, DESCA, TAU, WORK, LWORK,
$                    INFO )
    INTEGER         IA, INFO, JA, LWORK, M, N
    INTEGER         DESCA( * )
    REAL            A( * ), TAU( * ), WORK( * )

SUBROUTINE PCGEQLF( M, N, A, IA, JA, DESCA, TAU, WORK, LWORK,
$                    INFO )
    INTEGER         IA, INFO, JA, LWORK, M, N
    INTEGER         DESCA( * )
    COMPLEX         A( * ), TAU( * ), WORK( * )
```

Purpose

PSGEQLF/PCGEQLF computes a QL factorization of a real/complex distributed m-by-n matrix sub(A) = A(IA:IA+M−1,JA:JA+N−1) = Q*L.

Arguments

M   (global input) INTEGER
        The number of rows to be operated on, i.e, the number of rows of the distributed submatrix sub(A). M ≥ 0.

N   (global input) INTEGER
        The number of columns to be operated on, i.e., the number of columns of the distributed submatrix sub(A). N ≥ 0.

A   (local input/local output) REAL/COMPLEX pointer into the local memory to an array of dimension (LLD_A, LOCc(JA+N−1)).
        On entry, the local pieces of the m-by-n distributed matrix sub(A) which is to be factored.
        On exit,
        if m ≥ n, the lower triangle of the distributed submatrix A(IA+M−N:IA+M−1, JA:JA+N−1) contains the n-by-n lower triangular matrix L;
        if m ≥ n, the elements on and below the $(n-m)^{th}$ superdiagonal contain the m by n lower trapezoidal matrix L;

---

solution in each column is given by the sum of squares of elements n+1 to m in that column; if TRANS = 'N' and m < n, rows 1 to of sub(B) contain the minimum norm solution vectors; if TRANS = 'T' and m ≥ n, rows 1 to m of sub(B) contain the minimum norm solution vectors; if TRANS = 'T' and m < n, rows 1 to m of sub(B) contain the least squares solution vectors; the residual sum of squares for the solution in each column is given by the sum of squares of elements m+1 to n in that column.

IB   (global input) INTEGER
        The row index in the global array B indicating the first row of sub(B).

JB   (global input) INTEGER
        The column index in the global array B indicating the first column of sub(B).

DESCB   (global and local input) INTEGER array, dimension (DLEN_).
        The array descriptor for the distributed matrix B.

WORK   (local workspace/local output) REAL/COMPLEX array, dimension (LWORK)
        On exit, WORK(1) returns the minimal and optimal LWORK.

LWORK   (local or global input) INTEGER
        The dimension of the array WORK.
        LWORK is local input and must be at least
        LWORK ≥ LTAU + max(LWF, LWS) where
        If M ≥ N, then
        LTAU = NUMROC(JA+MIN(M,N)−1, NB_A, MYCOL, CSRC_A, NPCOL),
        LWF = NB_A * (MpA0 + NqA0 + NB_A)
        LWS = max((NB_A*(NB_A−1))/2, (NRHSqB0 + MpB0)*NB_A) + NB_A * NB_A
        Else
        LTAU = NUMROC(IA+MIN(M,N)−1, MB_A, MYROW, RSRC_A, NPROW),
        LWF = MB_A * (MpA0 + NqA0 + MB_A)
        LWS = max((MB_A*(MB_A−1))/2, (NpB0 + max(NqA0 +
              NUMROC(NUMROC(N+IROFFB, MB_A, 0, 0, NPROW),
              MB_A, 0, 0, LCMP), NRHSqB0))*MB_A) + MB_A * MB_A
        End if
        where LCMP = LCM / NPROW with LCM = ILCM(NPROW, NPCOL),
        IROFFA = mod(IA−1, MB_A), ICOFFA = mod(JA−1, NB_A),
        IAROW = INDXG2P(IA, MB_A, MYROW, RSRC_A, NPROW),
        IACOL = INDXG2P(JA, NB_A, MYCOL, CSRC_A, NPCOL),
        MpA0 = NUMROC(M+IROFFA, MB_A, MYROW, IAROW, NPROW),
        NqA0 = NUMROC(N+ICOFFA, NB_A, MYCOL, IACOL, NPCOL),
        IROFFB = mod(IB−1, MB_B), ICOFFB = mod(JB−1, NB_B),
        IBROW = INDXG2P(IB, MB_B, MYROW, RSRC_B, NPROW),
        IBCOL = INDXG2P(JB, NB_B, MYCOL, CSRC_B, NPCOL),
        MpB0 = NUMROC(M+IROFFB, MB_B, MYROW, IBROW, NPROW),
        NpB0 = NUMROC(N+IROFFB, MB_B, MYROW, IBROW, NPROW),
        NRHSqB0 = NUMROC(NRHS+ICOFFB, NB_B, MYCOL, ICOL, NPCOL).
        ILCM, INDXG2P and NUMROC are ScaLAPACK tool functions; MYROW, MYCOL, NPROW and NPCOL can be determined by calling the subroutine BLACS_GRIDINFO.
        If LWORK = −1, then LWORK is global input and a workspace query is assumed; the routine only calculates the minimum and optimal size

PSGEQPF/PCGEQPF

```
SUBROUTINE PSGEQPF( M, N, A, IA, JA, DESCA, IPIV, TAU, WORK,
$                   LWORK, INFO )
    INTEGER           IA, JA, INFO, LWORK, M, N
    INTEGER           DESCA( * ), IPIV( * )
    REAL              A( * ), TAU( * ), WORK( * )

SUBROUTINE PCGEQPF( M, N, A, IA, JA, DESCA, IPIV, TAU, WORK,
$                   LWORK, RWORK, LRWORK, INFO )
    INTEGER           IA, JA, INFO, LRWORK, LWORK, M, N
    INTEGER           DESCA( * ), IPIV( * )
    REAL              RWORK( * )
    COMPLEX           A( * ), TAU( * ), WORK( * )
```

Purpose

PSGEQPF/PCGEQPF computes a QR factorization with column pivoting of a real/complex m-by-n distributed matrix sub(A) = A(IA:IA+M-1,JA:JA+N-1): sub(A)*P = Q*R.

Arguments

M   (global input) INTEGER
        The number of rows to be operated on, i.e., the number of rows of the distributed submatrix sub(A). M $\geq$ 0.

N   (global input) INTEGER
        The number of columns to be operated on, i.e., the number of columns of the distributed submatrix sub(A). N $\geq$ 0.

A   (local input/local output) REAL/COMPLEX pointer into the local memory to an array of dimension (LLD_A, LOCc(JA+N-1)).
        On entry, the local pieces of the m-by-n distributed matrix sub(A) which is to be factored.
        On exit, the elements on and above the diagonal of sub(A) contain the min(m,n) by n upper trapezoidal matrix R (R is upper triangular if m $\geq$ n); the elements below the diagonal, with the array TAU, represent the orthogonal/unitary matrix Q as a product of elementary reflectors.

IA   (global input) INTEGER
        The row index in the global array A indicating the first row of sub(A).

JA   (global input) INTEGER
        The column index in the global array A indicating the first column of sub(A).

DESCA   (global and local input) INTEGER array, dimension (DLEN_).
        The array descriptor for the distributed matrix A.

IPIV   (local output) INTEGER array, dimension LOCc(JA+N-1).
        On exit, if IPIV(I) = K, the local i^{th} column of sub(A)*P was the global K^{th} column of sub(A). IPIV is tied to the distributed matrix A.

the remaining elements, with the array TAU, represent the orthogonal/unitary matrix Q as a product of elementary reflectors.

IA   (global input) INTEGER
        The row index in the global array A indicating the first row of sub(A).

JA   (global input) INTEGER
        The column index in the global array A indicating the first column of sub(A).

DESCA   (global and local input) INTEGER array, dimension (DLEN_).
        The array descriptor for the distributed matrix A.

TAU   (local output) REAL/COMPLEX array, dimension LOCc(JA+N-1).
        This array contains the scalar factors of the elementary reflectors. TAU is tied to the distributed matrix A.

WORK   (local workspace/local output) REAL/COMPLEX array, dimension (LWORK)
        On exit, WORK(1) returns the minimal and optimal LWORK.

LWORK   (local or global input) INTEGER
        The dimension of the array WORK.
        LWORK is local input and must be at least
        LWORK $\geq$ NB_A * (Mp0 + Nq0 + NB_A), where
        IROFF = mod(IA-1, MB_A), ICOFF = mod(JA-1, NB_A),
        IAROW = INDXG2P(IA, MB_A, MYROW, RSRC_A, NPROW),
        IACOL = INDXG2P(JA, NB_A, MYCOL, CSRC_A, NPCOL),
        Mp0 = NUMROC(M+IROFF, MB_A, MYROW, IAROW, NPROW),
        Nq0 = NUMROC(N+ICOFF, NB_A, MYCOL, IACOL, NPCOL).
        NUMROC and INDXG2P are ScaLAPACK tool functions; MYROW, MYCOL, NPROW and NPCOL can be determined by calling the subroutine BLACS_GRIDINFO.
        If LWORK = -1, then LWORK is global input and a workspace query is assumed; the routine only calculates the minimum and optimal size for all work arrays. Each of these values is returned in the first entry of the corresponding work array, and no error message is issued by PXERBLA.

INFO   (global output) INTEGER
        = 0: successful exit
        < 0: If the ith argument is an array and the j-entry had an illegal value, then INFO = -(i*100+j), if the ith argument is a scalar and had an illegal value, then INFO = -i.

TAU (local output) REAL/COMPLEX array, dimension LOCc(JA+MIN(M,N)-1).
This array contains the scalar factors TAU of the elementary reflectors. TAU is tied to the distributed matrix A.

WORK (local workspace/local output) REAL/COMPLEX array, dimension (LWORK)
On exit, WORK(1) returns the minimal and optimal LWORK.

LWORK (local or global input) INTEGER
The dimension of the array WORK.
LWORK is local input and must be at least
PSGEQPF
LWORK ≥ max(3,Mp0 + Nq0) + LOCc(JA+N-1)+Nq0.
PCGEQPF
LWORK ≥ max(3,Mp0 + Nq0).

IROFF = mod(IA-1, MB_A), ICOFF = mod(JA-1, NB_A),
IAROW = INDXG2P(IA, MB_A, MYROW, RSRC_A, NPROW),
IACOL = INDXG2P(JA, NB_A, MYCOL, CSRC_A, NPCOL),
Mp0 = NUMROC(M+IROFF, MB_A, MYROW, IAROW, NPROW),
Nq0 = NUMROC(N+ICOFF, NB_A, MYCOL, IACOL, NPCOL),

LOCc(JA+N-1) = NUMROC(JA+N-1, NB_A, MYCOL, CSRC_A, NPCOL).

NUMROC and INDXG2P are ScaLAPACK tool functions; MYROW, MYCOL, NPROW and NPCOL can be determined by calling the subroutine BLACS_GRIDINFO.

If LWORK = -1, then LWORK is global input and a workspace query is assumed; the routine only calculates the minimum and optimal size for all work arrays. Each of these values is returned in the first entry of the corresponding work array, and no error message is issued by PXERBLA.

RWORK PCGEQPF only (local workspace/local output) REAL array, dimension (LRWORK)
On exit, RWORK(1) returns the minimal and optimal LRWORK.

LRWORK PCGEQPF only (local or global input) INTEGER
The dimension of the array RWORK.
LRWORK is local input and must be at least
LRWORK ≥ LOCc(JA+N-1)+Nq0.

IROFF = mod(IA-1, MB_A), ICOFF = mod(JA-1, NB_A),
IAROW = INDXG2P(IA, MB_A, MYROW, RSRC_A, NPROW),
IACOL = INDXG2P(JA, NB_A, MYCOL, CSRC_A, NPCOL),
Mp0 = NUMROC(M+IROFF, MB_A, MYROW, IAROW, NPROW),
Nq0 = NUMROC(N+ICOFF, NB_A, MYCOL, IACOL, NPCOL),

LOCc(JA+N-1) = NUMROC(JA+N-1, NB_A, MYCOL, CSRC_A, NPCOL).

NUMROC and INDXG2P are ScaLAPACK tool functions; MYROW, MYCOL, NPROW and NPCOL can be determined by calling the subroutine BLACS_GRIDINFO.

If LRWORK = -1, then LRWORK is global input and a workspace query is assumed; the routine only calculates the minimum and optimal size for all work arrays. Each of these values is returned in the first

entry of the corresponding work array, and no error message is issued by PXERBLA.

INFO (global output) INTEGER
= 0: successful exit
< 0: If the ith argument is an array and the j-entry had an illegal value, then INFO = -(i*100+j); if the ith argument is a scalar and had an illegal value, then INFO = -i.

PSGEQRF/PCGEQRF

```
SUBROUTINE PSGEQRF( M, N, A, IA, JA, DESCA, TAU, WORK, LWORK,
$                    INFO )
INTEGER     IA, INFO, JA, LWORK, M, N
INTEGER     DESCA( * )
REAL        A( * ), TAU( * ), WORK( * )

SUBROUTINE PCGEQRF( M, N, A, IA, JA, DESCA, TAU, WORK, LWORK,
$                    INFO )
INTEGER     IA, INFO, JA, LWORK, M, N
INTEGER     DESCA( * )
COMPLEX     A( * ), TAU( * ), WORK( * )
```

Purpose

PSGEQRF/PCGEQRF computes a QR factorization of a real/complex distributed m-by-n matrix sub(A) = A(IA:IA+M-1,JA:JA+N-1) = Q*R.

Arguments

M (global input) INTEGER
The number of rows to be operated on, i.e., the number of rows of the distributed submatrix sub(A). M ≥ 0.

N (global input) INTEGER
The number of columns to be operated on, i.e., the number of columns of the distributed submatrix sub(A). N ≥ 0.

A (local input/local output) REAL/COMPLEX pointer into the local memory to an array of dimension (LLD_A, LOCc(JA+N-1)). On entry, the local pieces of the m-by-n distributed matrix sub(A) which is to be factored.
On exit,
the elements on and above the diagonal of sub(A) contain the min(m,n) by n upper trapezoidal matrix R (R is upper triangular if m ≥ n); the elements below the diagonal, with the array TAU, represent the orthogonal/unitary matrix Q as a product of elementary reflectors.

IA (global input) INTEGER
The row index in the global array A indicating the first row of sub(A).

PSGERFS/PCGERFS

```
SUBROUTINE PSGERFS( TRANS, N, NRHS, A, IA, JA, DESCA, AF, IAF,
$                   JAF, DESCAF, IPIV, B, IB, JB, DESCB, X, IX,
$                   JX, DESCX, FERR, BERR, WORK, LWORK, IWORK,
$                   LIWORK, INFO )

CHARACTER          TRANS
INTEGER            IA, IAF, IB, IX, INFO, JA, JAF, JB, JX,
$                  LIWORK, LWORK, N, NRHS
INTEGER            DESCA( * ), DESCAF( * ), DESCB( * ),
$                  DESCX( * ),IPIV( * ), IWORK( * )
REAL               A( * ), AF( * ), B( * ), BERR( * ), FERR( * ),
$                  WORK( * ), X( * )

SUBROUTINE PCGERFS( TRANS, N, NRHS, A, IA, JA, DESCA, AF, IAF,
$                   JAF, DESCAF, IPIV, B, IB, JB, DESCB, X, IX,
$                   JX, DESCX, FERR, BERR, WORK, LWORK, RWORK,
$                   LRWORK, INFO )

CHARACTER          TRANS
INTEGER            IA, IAF, IB, IX, INFO, JA, JAF, JB, JX,
$                  LRWORK, LWORK, N, NRHS
INTEGER            DESCA( * ), DESCAF( * ), DESCB( * ),
$                  DESCX( * ), IPIV( * )
REAL               BERR( * ), FERR( * ), RWORK( * )
COMPLEX            A( * ), AF( * ), B( * ), WORK( * ), X( * )
```

Purpose

PSGERFS/PCGERFS improves the computed solution to a system of linear equations and provides error bounds and backward error estimates for the solution.

Arguments

TRANS   (global input) CHARACTER*1
        Specifies the form of the system of equations:
        = 'N':   sub(A)*sub(X) = sub(B) (No transpose)
        = 'T':   sub(A)^T*sub(X) = sub(B) (Transpose)
        = 'C':   sub(A)^H*sub(X) = sub(B) (Conjugate transpose)

N   (global input) INTEGER
        The order of the matrix sub(A). N ≥ 0.

NRHS   (global input) INTEGER
        The number of right hand sides, i.e., the number of columns of the
        matrices sub(B) and sub(X). NRHS ≥ 0.

A   (local input) REAL/COMPLEX pointer into the local memory to an
        array of local dimension (LLD_A,LOCc(JA+N−1)).
        This array contains the local pieces of the distributed matrix sub(A).

IA   (global input) INTEGER
        The row index in the global array A indicating the first row of sub(A).

JA   (global input) INTEGER
        The column index in the global array A indicating the first column of
        sub(A).

DESCA   (global and local input) INTEGER array, dimension (DLEN_).
        The array descriptor for the distributed matrix A.

TAU   (local output) REAL/COMPLEX array, dimension
        LOCc(JA+MIN(M,N)−1).
        This array contains the scalar factors TAU of the elementary reflectors.
        TAU is tied to the distributed matrix A.

WORK   (local workspace/local output) REAL/COMPLEX array, dimension
        (LWORK)
        On exit, WORK(1) returns the minimal and optimal LWORK.

LWORK   (local or global input) INTEGER
        The dimension of the array WORK.
        LWORK is local input and must be at least
        LWORK ≥ NB_A * (Mp0 + Nq0 + NB_A), where
        IROFF = mod(IA−1, MB_A), ICOFF = mod(JA−1, NB_A),
        IAROW = INDXG2P(IA, MB_A, MYROW, RSRC_A, NPROW),
        IACOL = INDXG2P(JA, NB_A, MYCOL, CSRC_A, NPCOL),
        Mp0 = NUMROC(M+IROFF, MB_A, MYROW, IAROW, NPROW),
        Nq0 = NUMROC(N+ICOFF, NB_A, MYCOL, IACOL, NPCOL),
        and NUMROC, INDXG2P are ScaLAPACK tool functions; MYROW,
        MYCOL, NPROW and NPCOL can be determined by calling the sub-
        routine BLACS_GRIDINFO.
        If LWORK = −1, then LWORK is global input and a workspace query
        is assumed; the routine only calculates the minimum and optimal size
        for all work arrays. Each of these values is returned in the first en-
        try of the corresponding work array, and no error message is issued by
        PXERBLA.

INFO   (global output) INTEGER
        = 0:   successful exit
        < 0:   If the ith argument is an array and the j-entry had an illegal
               value, then INFO = −(i*100+j), if the ith argument is a scalar
               and had an illegal value, then INFO = −i.

DESCX (global and local input) INTEGER array, dimension (DLEN_).
The array descriptor for the distributed matrix X.

FERR (local output) REAL array, dimension LOCc(JB+NRHS−1).
The estimated forward error bound for each solution vector of sub(X). If XTRUE is the true solution corresponding to sub(X), FERR is an estimated upper bound for the magnitude of the largest element in (sub(X) − XTRUE) divided by the magnitude of the largest element in sub(X). The estimate is as reliable as the estimate for RCOND, and is almost always a slight overestimate of the true error. This array is tied to the distributed matrix X.

BERR (local output) REAL array, dimension LOCc(JB+NRHS−1).
The componentwise relative backward error of each solution vector (i.e., the smallest relative change in any entry of sub(A) or sub(B) that makes sub(X) an exact solution). This array is tied to the distributed matrix X.

WORK (local workspace/local output) REAL/COMPLEX array, dimension (LWORK)
On exit, WORK(1) returns the minimal and optimal LWORK.

LWORK (local or global input) INTEGER
The dimension of the array WORK.
LWORK is local input and must be at least
PSGERFS
LWORK $\geq$ 3*LOCr(N + mod(IA−1,MB_A)).
PCGERFS
LWORK $\geq$ 2*LOCr(N + mod(IA−1,MB_A)).
If LWORK = −1, then LWORK is global input and a workspace query is assumed; the routine only calculates the minimum and optimal size for all work arrays. Each of these values is returned in the first entry of the corresponding work array, and no error message is issued by PXERBLA.

IWORK *PSGERFS only* (local workspace/local output) INTEGER array, dimension (LIWORK)
On exit, IWORK(1) returns the minimal and optimal LIWORK.

LIWORK (local or global input) INTEGER
The dimension of the array IWORK.
LIWORK is local input and must be at least
LIWORK $\geq$ LOCr(N + mod(IB−1,MB_B)).
If LIWORK = −1, then LIWORK is global input and a workspace query is assumed; the routine only calculates the minimum and optimal size for all work arrays. Each of these values is returned in the first entry of the corresponding work array, and no error message is issued by PXERBLA.

RWORK *PCGERFS only* (local workspace/local output) REAL array, dimension (LRWORK)
On exit, RWORK(1) returns the minimal and optimal LRWORK.

LRWORK *PCGERFS only* (local or global input) INTEGER
The dimension of the array RWORK.

JA (global input) INTEGER
The column index in the global array A indicating the first column of sub(A).

DESCA (global and local input) INTEGER array, dimension (DLEN_).
The array descriptor for the distributed matrix A.

AF (local input) REAL/COMPLEX pointer into the local memory to an array of local dimension (LLD_AF,LOCc(JA+N−1)).
This array contains the local pieces of the distributed factors of the matrix sub(A) = P*L*U as computed by PSGETRF/PCGETRF.

IAF (global input) INTEGER
The row index in the global array AF indicating the first row of sub(AF).

JAF (global input) INTEGER
The column index in the global array AF indicating the first column of sub(AF).

DESCAF (global and local input) INTEGER array, dimension (DLEN_).
The array descriptor for the distributed matrix AF.

IPIV (local input) INTEGER array, dimension LOCr(M_AF)+MB_AF.
This array contains the pivoting information as computed by PS-GETRF/PCGETRF. IPIV(i) = the global row local row i was swapped with. This array is tied to the distributed matrix A.

B (local input) REAL/COMPLEX pointer into the local memory to an array of local dimension (LLD_B,LOCc(JB+NRHS−1)).
This array contains the local pieces of the distributed matrix of right hand sides sub(B).

IB (global input) INTEGER
The row index in the global array B indicating the first row of sub(B).

JB (global input) INTEGER
The column index in the global array B indicating the first column of sub(B).

DESCB (global and local input) INTEGER array, dimension (DLEN_).
The array descriptor for the distributed matrix B.

X (local input and output) REAL/COMPLEX pointer into the local memory to an array of local dimension (LLD_X,LOCc(JX+NRHS−1)).
On entry, this array contains the local pieces of the distributed matrix solution sub(X).
On exit, the improved solution vectors.

IX (global input) INTEGER
The row index in the global array X indicating the first row of sub(X).

JX (global input) INTEGER
The column index in the global array X indicating the first column of sub(X).

LRWORK is local input and must be at least

LRWORK $\geq$ LOCr(N + mod(IB$-$1,MB_B)).

If LRWORK = -1, then LRWORK is global input and a workspace query is assumed; the routine only calculates the minimum and optimal size for all work arrays. Each of these values is returned in the first entry of the corresponding work array, and no error message is issued by PXERBLA.

INFO   (global output) INTEGER
- = 0:   successful exit
- < 0:   If the ith argument is an array and the j-entry had an illegal value, then INFO = $-(i*100+j)$, if the ith argument is a scalar and had an illegal value, then INFO = $-i$.

PSGERQF/PCGERQF

```
SUBROUTINE PSGERQF( M, N, A, IA, JA, DESCA, TAU, WORK, LWORK,
$                    INFO )
    INTEGER    IA, INFO, JA, LWORK, M, N
    INTEGER    DESCA( * )
    REAL       A( * ), TAU( * ), WORK( * )

SUBROUTINE PCGERQF( M, N, A, IA, JA, DESCA, TAU, WORK, LWORK,
$                    INFO )
    INTEGER    IA, INFO, JA, LWORK, M, N
    INTEGER    DESCA( * )
    COMPLEX    A( * ), TAU( * ), WORK( * )
```

Purpose

PSGERQF/PCGERQF computes a RQ factorization of a real/complex distributed m-by-n matrix sub(A) = A(IA:IA+M-1,JA:JA+N-1) = R*Q.

Arguments

M   (global input) INTEGER.
The number of rows to be operated on, i.e., the number of rows of the distributed submatrix sub(A). M $\geq$ 0.

N   (global input) INTEGER.
The number of columns to be operated on, i.e., the number of columns of the distributed submatrix sub(A). N $\geq$ 0.

A   (local input/local output) REAL/COMPLEX pointer into the local memory to an array of dimension (LLD_A, LOCc(JA+N-1)).
On entry, the local pieces of the m-by-n distributed matrix sub(A) which is to be factored.
On exit,
if m $\leq$ n, the upper triangle of A(IA:IA+M-1, JA+N$-$M:JA+N-1)
contains the m by m upper triangular matrix R;
if m $\geq$ n, the elements on and above the $(m-n)^{th}$ subdiagonal contain the m by n upper trapezoidal matrix R;
the remaining elements, with the array TAU, represent the orthogonal/unitary matrix Q as a product of elementary reflectors.

IA   (global input) INTEGER
The row index in the global array A indicating the first row of sub(A).

JA   (global input) INTEGER
The column index in the global array A indicating the first column of sub(A).

DESCA   (global and local input) INTEGER array, dimension (DLEN_).
The array descriptor for the distributed matrix A.

TAU   (local output) REAL/COMPLEX array, dimension LOCr(IA+M-1).
This array contains the scalar factors of the elementary reflectors.
TAU is tied to the distributed matrix A.

WORK   (local workspace/local output) REAL/COMPLEX array, dimension (LWORK)
On exit, WORK(1) returns the minimal and optimal LWORK.

LWORK   (local or global input) INTEGER
The dimension of the array WORK.
LWORK is local input and must be at least
LWORK $\geq$ MB_A * (Mp0 + Nq0 + MB_A), where
IROFF = mod(IA-1, MB_A), ICOFF = mod(JA-1, NB_A),
IAROW = INDXG2P(IA, MB_A, MYROW, RSRC_A, NPROW),
IACOL = INDXG2P(JA, NB_A, MYCOL, CSRC_A, NPCOL),
Mp0 = NUMROC(M+IROFF, MB_A, MYROW, IAROW, NPROW),
Nq0 = NUMROC(N+ICOFF, NB_A, MYCOL, IACOL, NPCOL),
and NUMROC, INDXG2P are ScaLAPACK tool functions; MYROW, MYCOL, NPROW and NPCOL can be determined by calling the subroutine BLACS_GRIDINFO.
If LWORK = -1, then LWORK is global input and a workspace query is assumed; the routine only calculates the minimum and optimal size for all work arrays. Each of these values is returned in the first entry of the corresponding work array, and no error message is issued by PXERBLA.

INFO   (global output) INTEGER
- = 0:   successful exit
- < 0:   If the ith argument is an array and the j-entry had an illegal value, then INFO = $-(i*100+j)$, if the ith argument is a scalar and had an illegal value, then INFO = $-i$.

PSGESV/PCGESV

```
SUBROUTINE PSGESV( N, NRHS, A, IA, JA, DESCA, IPIV, B, IB, JB,
$                  DESCB, INFO )
    INTEGER        IA, IB, INFO, JA, JB, N, NRHS
    INTEGER        DESCA( * ), DESCB( * ), IPIV( * )
    REAL           A( * ), B( * )

SUBROUTINE PCGESV( N, NRHS, A, IA, JA, DESCA, IPIV, B, IB, JB,
$                  DESCB, INFO )
    INTEGER        IA, IB, INFO, JA, JB, N, NRHS
    INTEGER        DESCA( * ), DESCB( * ), IPIV( * )
    COMPLEX        A( * ), B( * )
```

Purpose

PSGESV/PCGESV computes the solution to a real/complex system of linear equations $sub(A) * X = sub(B)$, where $sub(A) = A(IA:IA+N-1,JA:JA+N-1)$ is an n-by-n distributed matrix and X and $sub(B) = B(IB:IB+N-1,JB:JB+NRHS-1)$ are n-by-nrhs distributed matrices.

The LU decomposition with partial pivoting and row interchanges is used to factor $sub(A)$ as $sub(A) = P*L*U$, where P is a permutation matrix, L is unit lower triangular, and U is upper triangular. L and U are stored in $sub(A)$. The factored form of $sub(A)$ is then used to solve the system of equations $sub(A)*X = sub(B)$.

Arguments

N (global input) INTEGER
The number of rows and columns to be operated on, i.e., the order of the distributed submatrix $sub(A)$. $N \geq 0$.

NRHS (global input) INTEGER
The number of right hand sides, i.e., the number of columns of the distributed submatrix $sub(A)$. $NRHS \geq 0$.

A (local input/local output) REAL/COMPLEX pointer into the local memory to an array of dimension (LLD_A,LOCc(JA+N-1)).
On entry, the local pieces of the n-by-n distributed matrix $sub(A)$ to be factored.
On exit, this array contains the local pieces of the factors L and U from the factorization $sub(A) = P*L*U$; the unit diagonal elements of L are not stored.

IA (global input) INTEGER
The row index in the global array A indicating the first row of $sub(A)$.

JA (global input) INTEGER
The column index in the global array A indicating the first column of $sub(A)$.

DESCA (global and local input) INTEGER array, dimension (DLEN_).
The array descriptor for the distributed matrix A.

IPIV (local output) INTEGER array, dimension (LOCr(M_A)+MB_A)
This array contains the pivoting information. IPIV(i) = The global row local row i was swapped with. This array is tied to the distributed matrix A.

B (local input/local output) REAL/COMPLEX pointer into the local memory to an array of dimension (LLD_B,LOCc(JB+NRHS-1)).
On entry, the right hand side distributed matrix $sub(B)$.
On exit, if INFO = 0, $sub(B)$ is overwritten by the solution distributed matrix X.

IB (global input) INTEGER
The row index in the global array B indicating the first row of $sub(B)$.

JB (global input) INTEGER
The column index in the global array B indicating the first column of $sub(B)$.

DESCB (global and local input) INTEGER array, dimension (DLEN_).
The array descriptor for the distributed matrix B.

INFO (global output) INTEGER
= 0: successful exit
< 0: If the ith argument is an array and the j-entry had an illegal value, then INFO = -(i*100+j), if the ith argument is a scalar and had an illegal value, then INFO = -i.
> 0: If INFO = K, $U(IA+K-1,JA+K-1)$ is exactly zero. The factorization has been completed, but the factor U is exactly singular, so the solution could not be computed.

PSGESVD

```
SUBROUTINE PSGESVD( JOBU, JOBVT, M, N, A, IA, JA, DESCA, S, U,
$                   IU, JU, DESCU, VT, IVT, JVT, DESCVT,
$                   WORK, LWORK, INFO )
    CHARACTER      JOBU, JOBVT
    INTEGER        IA, INFO, IU, IVT, JA, JU, JVT, LWORK, M, N
    INTEGER        DESCA( * ), DESCU( * ), DESCVT( * )
    REAL           A( * ), S( * ), U( * ), VT( * ), WORK( * )
```

Purpose

PSGESVD computes the singular value decomposition (SVD) of a real m-by-n distributed matrix A, optionally computing the left and/or right singular vectors. The SVD is written as

$$A = U*\Sigma*V^H$$

where Σ is an m-by-n matrix which is zero except for its min(m,n) diagonal elements, U is an m-by-m orthogonal matrix, and V is an n-by-n orthogonal matrix. The diagonal elements of Σ are the singular values of A and the columns of U and

V are the corresponding right and left singular vectors, respectively. The singular values are returned in array S in decreasing order and only the first $\min(m,n)$ columns of U and rows of $VT = V^T$ are computed.

Note that the routine returns V^T, not V.

MP = number of local rows in A and U
NQ = number of local columns in A and VT
SIZE = $\min(M, N)$
SIZEQ = number of local columns in U
SIZEP = number of local rows in VT

Arguments

JOBU   (global input) CHARACTER*1
Specifies options for computing all or part of the matrix U:
= 'V':   the first $\min(m,n)$ columns of U (the left singular vectors) are returned in the array U;
= 'N':   no columns of U (no left singular vectors) are computed.

JOBVT   (global input) CHARACTER*1
Specifies options for computing all or part of the matrix V^T:
= 'V':   the first $\min(m,n)$ rows of V^T (the right singular vectors) are returned in the array VT;
= 'N':   no rows of V^T (no right singular vectors) are computed.

M   (global input) INTEGER
The number of rows of the input matrix A. $M \geq 0$.

N   (global input) INTEGER
The number of columns of the input matrix A. $N \geq 0$.

A   (local input/local output) REAL pointer into the local memory to an array of dimension (LLD_A,LOCc(JA+N−1)).
On entry, the m-by-n distributed matrix A.
On exit, the contents of A are destroyed.

IA   (global input) INTEGER
The row index in the global array A indicating the first row of sub(A).

JA   (global input) INTEGER
The column index in the global array A indicating the first column of sub(A).

DESCA   (global and local input) INTEGER array, dimension (DLEN_).
The array descriptor for the distributed matrix A.

S   (global output) REAL array, dimension $\min(M,N)$
The singular values of A, sorted so that $S(i) \geq S(i+1)$.

U   (local output) REAL pointer into the local memory to an array of dimension (LLD_U,LOCc(JU+min(M,N)−1)).

If JOBU = 'V', U contains the first $\min(m,n)$ columns of U.
If JOBU = 'N', U is not referenced.

IU   (global input) INTEGER
The row index in the global array U indicating the first row of sub(U).

JU   (global input) INTEGER
The column index in the global array U indicating the first column of sub(U).

DESCU   (global and local input) INTEGER array, dimension (DLEN_).
The array descriptor for the distributed matrix U.

VT   (local output) REAL pointer into the local memory to an array of dimension (LLD_VT,LOCc(JVT+N−1)).
Note that LLD_VT = LOCr(IVT+min(M,N)−1)).
If JOBVT = 'V', VT contains the first $\min(m,n)$ rows of V^T.
If JOBVT = 'N', VT is not referenced.

IVT   (global input) INTEGER
The row index in the global array VT indicating the first row of sub(VT).

JVT   (global input) INTEGER
The column index in the global array VT indicating the first column of sub(VT).

DESCVT (global and local input) INTEGER array, dimension (DLEN_).
The array descriptor for the distributed matrix VT.

WORK   (local workspace/local output) REAL array, dimension (LWORK)
On exit, WORK(1) returns the minimal and optimal LWORK.

LWORK   (local or global input) INTEGER
The dimension of the array WORK.
LWORK is local input and must be at least
LWORK $\geq 2 + 6 \ast$SIZEB $+ \max($WATOBD, WBDTOSVD$)$,
where SIZEB $= \max(M,N)$, and WATOBD and WBDTOSVD refer, respectively, to the workspace required to bidiagonalize the matrix A and to go from the bidiagonal matrix to the singular value decomposition U∗S∗VT.
For WATOBD, the following holds:
WATOBD $= \max(\max($WPSLANGE,WPSGEBRD$)$,
$\qquad \max($WPSLARED2D,WPSLARED1D$))$,
where WPSLANGE, WPSLARED1D, WPSLARED2D, WPSGEBRD are the workspaces required respectively for the subprograms PSLANGE, PSLARED1D, PSLARED2D, PSGEBRD. Using the standard notation
MP = NUMROC(M, MB, MYROW, DESCA(RSRC_), NPROW),
NQ = NUMROC(N, NB, MYCOL, DESCA(CSRC_), NPCOL),
the workspaces required for the above subprograms are
WPSLANGE = MP,
WPSLARED1D = NQ0,
WPSLARED2D = MP0,

PSGESVX/PCGESVX

```
SUBROUTINE PSGESVX( FACT, TRANS, N, NRHS, A, IA, JA, DESCA, AF,
$                   IAF, JAF, DESCAF, IPIV, EQUED, R, C, B, IB,
$                   JB, DESCB, X, IX, JX, DESCX, RCOND, FERR,
$                   BERR, WORK, LWORK, IWORK, LIWORK, INFO )
      CHARACTER      EQUED, FACT, TRANS
      INTEGER        IA, IAF, IB, INFO, IX, JA, JAF, JB, JX, LIWORK,
$                    LWORK, N, NRHS
      REAL           RCOND
      INTEGER        DESCA( * ), DESCAF( * ), DESCB( * ),
$                    DESCX( * ), IPIV( * ), IWORK( * )
      REAL           A( * ), AF( * ), B( * ), BERR( * ), C( * ),
$                    FERR( * ), R( * ), WORK( * ), X( * )

SUBROUTINE PCGESVX( FACT, TRANS, N, NRHS, A, IA, JA, DESCA, AF,
$                   IAF, JAF, DESCAF, IPIV, EQUED, R, C, B, IB,
$                   JB, DESCB, X, IX, JX, DESCX, RCOND, FERR,
$                   BERR, WORK, LWORK, RWORK, LRWORK, INFO )
      CHARACTER      EQUED, FACT, TRANS
      INTEGER        IA, IAF, IB, INFO, IX, JA, JAF, JB, JX, LRWORK,
$                    LWORK, N, NRHS
      REAL           RCOND
      INTEGER        DESCA( * ), DESCAF( * ), DESCB( * ),
$                    DESCX( * ), IPIV( * )
      REAL           BERR( * ), C( * ), FERR( * ), R( * ),
$                    RWORK( * )
      COMPLEX        A( * ), AF( * ), B( * ), WORK( * ), X( * )
```

Purpose

PSGESVX/PCGESVX uses the LU factorization to compute the solution to a real/complex system of linear equations
$A(IA:IA+N-1,JA:JA+N-1) * X = B(IB:IB+N-1,JB:JB+NRHS-1),$
where $A(IA:IA+N-1,JA:JA+N-1)$ is an n-by-n matrix and X and
$B(IB:IB+N-1,JB:JB+NRHS-1)$ are n-by-nrhs matrices.

Error bounds on the solution and a condition estimate are also provided.

In the following description, A denotes $A(IA:IA+N-1,JA:JA+N-1)$, B denotes $B(IB:IB+N-1,JB:JB+NRHS-1)$ and X denotes $X(IX:IX+N-1,JX:JX+NRHS-1)$.

Description

The following steps are performed:

1. If FACT = 'E', real scaling factors are computed to equilibrate the system:

   TRANS = 'N': $diag(R)*A*diag(C)*diag(C)^{-1}*X = diag(R)*B$
   TRANS = 'T': $(diag(R)*A*diag(C))^{T}*diag(R)^{-1}*X = diag(C)*B$
   TRANS = 'C': $(diag(R)*A*diag(C))^{H}*diag(R)^{-1}*X = diag(C)*B$

$WPSGEBRD = NB*(MP + NQ + 1) + NQ,$
where NQ0 and MP0 refer, respectively, to the values obtained at MY-COL = 0 and MYROW = 0. In general, the upper limit for the workspace is given by a workspace required on processor (0,0):
$WATOBD \leq NB*(MP0 + NQ0 + 1) + NQ0.$
In case of a homogeneous process grid this upper limit can be used as an estimate of the minimum workspace for every processor.
For WBDTOSVD, the following holds:
$WBDTOSVD = SIZE*(WANTU*NRU + WANTVT*NCVT) +$
$\qquad max(WBDSQR, max(WANTU*WPSORMBRQLN,$
$\qquad\qquad WANTVT*WPSORMBRPRT)),$
where
$WANTU(WANTVT) = 1,$ if left(right) singular vectors are wanted
$\qquad\qquad\qquad 0,$ otherwise

and WDBDSQR, WPSORMBRQLN and WPSORMBRPRT refer respectively to the workspace required for the subprograms DBDSQR, PSORMBR(QLN), and PSORMBR(PRT), where QLN and PRT are the values of the arguments VECT, SIDE, and TRANS in the call to PSORMBR. NRU is equal to the local number of rows of the matrix U when distributed 1-dimensional "column" of processes. Analogously, NCVT is equal to the local number of columns of the matrix VT when distributed across 1-dimensional "row" of processes. Calling the LA-PACK procedure DBDSQR requires
$WDBDSQR = max(1, 2*SIZE + (2*SIZE - 4)*max(WANTU, WANTVT))$
on every processor. Finally,
$WPSORMBRQLN = max((NB*(NB-1))/2, (SIZEQ+MP)*NB)+NB*NB,$
$WPSORMBRPRT = max((MB*(MB-1))/2, (SIZEP+NQ)*MB)+MB*MB,$
If LWORK = −1, then LWORK is global input and a workspace query is assumed; the routine only calculates the minimum and optimal size for all work arrays. Each of these values is returned in the first entry of the corresponding work array, and no error message is issued by PXERBLA.

INFO   (global output) INTEGER
= 0:   successful exit
< 0:   If the ith argument is an array and the j-entry had an illegal value, then INFO = $-(i*100+j)$, if the ith argument is a scalar and had an illegal value, then INFO = −i.
> 0:   if SBDSQR did not converge:
        If INFO = MIN(M,N) + 1, then PSGESVD has detected heterogenity by finding that eigenvalues were not identical across the process grid. In this case, the accuracy of the results from PSGESVD cannot be guaranteed.

Whether or not the system will be equilibrated depends on the scaling of the matrix A, but if equilibration is used, A is overwritten by diag(R)*A*diag(C) and B by diag(R)*B (if TRANS='N') or diag(C)*B (if TRANS = 'T' or 'C').

2. If FACT = 'N' or 'E', the LU decomposition is used to factor the matrix A (after equilibration if FACT = 'E') as A = P*L*U, where P is a permutation matrix, L is a unit lower triangular matrix, and U is upper triangular.

3. The factored form of A is used to estimate the condition number of the matrix A. If the reciprocal of the condition number is less than relative machine precision, steps 4–6 are skipped.

4. The system of equations is solved for X using the factored form of A.

5. Iterative refinement is applied to improve the computed solution matrix and calculate error bounds and backward error estimates for it.

6. If equilibration was used, the matrix X is premultiplied by diag(C) (if TRANS = 'N') or diag(R) (if TRANS = 'T' or 'C') so that it solves the original system before equilibration.

Arguments

FACT   (global input) CHARACTER*1
Specifies whether or not the factored form of the matrix A(IA:IA+N−1,JA:JA+N−1) is supplied on entry, and if not, whether the matrix A(IA:IA+N−1,JA:JA+N−1) should be equilibrated before it is factored.

= 'F':   On entry, AF(IAF:IAF+N−1,JAF:JAF+N−1) and IPIV contain the factored form of A(IA:IA+N−1,JA:JA+N−1). If EQUED is not 'N', the matrix A(IA:IA+N−1,JA:JA+N−1) has been equilibrated with scaling factors given by R and C. A(IA:IA+N−1,JA:JA+N−1), AF(IAF:IAF+N−1,JAF:JAF+N−1), and IPIV are not modified.

= 'N':   The matrix A(IA:IA+N−1,JA:JA+N−1) will be copied to AF(IAF:IAF+N−1,JAF:JAF+N−1) and factored.

= 'E':   The matrix A(IA:IA+N−1,JA:JA+N−1) will be equilibrated if necessary, then copied to AF(IAF:IAF+N−1,JAF:JAF+N−1) and factored.

TRANS   (global input) CHARACTER*1
Specifies the form of the system of equations:

= 'N':   A(IA:IA+N−1,JA:JA+N−1) * X(IX:IX+N−1,JX:JX+NRHS−1) = B(IB:IB+N−1,JB:JB+NRHS−1) (No transpose)

= 'T':   A(IA:IA+N−1,JA:JA+N−1)^T * X(IX:IX+N−1,JX:JX+NRHS−1) = B(IB:IB+N−1,JB:JB+NRHS−1) (Transpose)

= 'C':   A(IA:IA+N−1,JA:JA+N−1)^H * X(IX:IX+N−1,JX:JX+NRHS−1) = B(IB:IB+N−1,JB:JB+NRHS−1) (Conjugate transpose)

N   (global input) INTEGER
The number of rows and columns to be operated on, i.e. the order of the distributed submatrix A(IA:IA+N−1,JA:JA+N−1). N ≥ 0.

NRHS   (global input) INTEGER
The number of right-hand sides, i.e., the number of columns of the distributed submatrices B(IB:IB+N−1,JB:JB+NRHS−1) and X(IX:IX+N−1,JX:JX+NRHS−1). NRHS ≥ 0.

A   (local input/local output) REAL/COMPLEX pointer into the local memory to an array of local dimension (LLD_A,LOCc(JA+N−1)).
On entry, the n-by-n matrix A(IA:IA+N−1,JA:JA+N−1). If FACT = 'F' and EQUED is not 'N', then A(IA:IA+N−1,JA:JA+N−1) must have been equilibrated by the scaling factors in R and/or C. A(IA:IA+N−1,JA:JA+N−1) is not modified if FACT = 'F' or 'N', or if FACT = 'E' and EQUED = 'N' on exit.
On exit, if EQUED ≠ 'N', A(IA:IA+N−1,JA:JA+N−1) is scaled as follows:

EQUED = 'R': A(IA:IA+N−1,JA:JA+N−1) := diag(R) * A(IA:IA+N−1,JA:JA+N−1)

EQUED = 'C': A(IA:IA+N−1,JA:JA+N−1) := A(IA:IA+N−1,JA:JA+N−1) * diag(C)

EQUED = 'B': A(IA:IA+N−1,JA:JA+N−1) := diag(R) * A(IA:IA+N−1,JA:JA+N−1) * diag(C).

IA   (global input) INTEGER
The row index in the global array A indicating the first row of sub(A).

JA   (global input) INTEGER
The column index in the global array A indicating the first column of sub(A).

DESCA   (global and local input) INTEGER array, dimension (DLEN_).
The array descriptor for the distributed matrix A.

AF   (local input or local output) REAL/COMPLEX pointer into the local memory to an array of local dimension (LLD_AF,LOCc(JA+N−1)).
If FACT = 'F', then AF(IAF:IAF+N−1,JAF:JAF+N−1) is an input argument and on entry contains the factors L and U from the factorization A(IA:IA+N−1,JA:JA+N−1) = P*L*U as computed by PS-GETRF/PCGETRF. If EQUED ≠ 'N', then AF is the factored form of the equilibrated matrix A(IA:IA+N−1,JA:JA+N−1).
If FACT = 'N', then AF(IAF:IAF+N−1,JAF:JAF+N−1) is an output argument and on exit returns the factors L and U from the factorization A(IA:IA+N−1,JA:JA+N−1) = P*L*U of the original matrix A(IA:IA+N−1,JA:JA+N−1).
If FACT = 'E', then AF(IAF:IAF+N−1,JAF:JAF+N−1) is an output argument and on exit returns the factors L and U from the factorization A(IA:IA+N−1,JA:JA+N−1) = P*L*U of the equili-

C (local input or local output) REAL array, dimension LOCc(N_A). The column scale factors for A(IA:IA+N−1,JA:JA+N−1) (see the description of EQUED). If EQUED = 'C' or 'B', A(IA:IA+N−1,JA:JA+N−1) is multiplied on the right by diag(C); if EQUED = 'N' or 'R', C is not accessed. C is an input variable if FACT = 'F'; otherwise, C is an output variable. If FACT = 'F' and EQUED = 'C' or 'B', each element of C must be positive. C is replicated in every process row, and is aligned with the distributed matrix A.

B (local input/local output) REAL/COMPLEX pointer into the local memory to an array of local dimension (LLD_B,LOCc(JB+NRHS−1)). On entry, the n-by-nrhs right-hand side matrix B(IB:IB+N−1,JB:JB+NRHS−1). On exit,
if EQUED = 'N', B(IB:IB+N−1,JB:JB+NRHS−1) is not modified;
if TRANS = 'N' and EQUED = 'R' or 'B', B is overwritten by diag(R)*B(IB:IB+N−1,JB:JB+NRHS−1);
if TRANS = 'T' or 'C' and EQUED = 'C' or 'B', B(IB:IB+N−1,JB:JB+NRHS−1) is over-written by diag(C)*B(IB:IB+N−1,JB:JB+NRHS−1).

IB (global input) INTEGER
The row index in the global array B indicating the first row of sub(B).

JB (global input) INTEGER
The column index in the global array B indicating the first column of sub(B).

DESCB (global and local input) INTEGER array, dimension (DLEN_).
The array descriptor for the distributed matrix B.

X (local input/local output) REAL/COMPLEX pointer into the local memory to an array of local dimension (LLD_X, LOCc(JX+NRHS−1)).
If INFO = 0, the n-by-nrhs solution matrix X(IX:IX+N−1,JX:JX+NRHS−1) to the original system of equations. Note that A(IA:IA+N−1,JA:JA+N−1) and B(IB:IB+N−1,JB:JB+NRHS−1) are modified on exit if EQUED ≠ 'N', and the solution to the equilibrated system is $\text{diag}(C)^{-1}$ * X(IX:IX+N−1,JX:JX+NRHS−1) if TRANS = 'N' and EQUED = 'C' or 'B', or $\text{diag}(R)^{-1}$ * X(IX:IX+N−1,JX:JX+NRHS−1) if TRANS = 'T' or 'C' and EQUED = 'R' or 'B'.

IX (global input) INTEGER
The row index in the global array X indicating the first row of sub(X).

JX (global input) INTEGER
The column index in the global array X indicating the first column of sub(X).

DESCX (global and local input) INTEGER array, dimension (DLEN_).
The array descriptor for the distributed matrix X.

RCOND (global output) REAL
The estimate of the reciprocal condition number of the matrix

brated matrix A(IA:IA+N−1,JA:JA+N−1) (see the description of A(IA:IA+N−1,JA:JA+N−1) for the form of the equilibrated matrix).

IAF (global input) INTEGER
The row index in the global array AF indicating the first row of sub(AF).

JAF (global input) INTEGER
The column index in the global array AF indicating the first column of sub(AF).

DESCAF (global and local input) INTEGER array, dimension (DLEN_).
The array descriptor for the distributed matrix AF.

IPIV (local input or local output) INTEGER array, dimension LOCr(M_A)+MB_A.
If FACT = 'F', then IPIV is an input argument and on entry contains the pivot indices from the factorization A(IA:IA+N−1,JA:JA+N−1) = P*L*U as computed by PSGETRF/PCGETRF; IPIV(i) = the global row local row i was swapped with. This array must be aligned with A(IA:IA+N−1, *).
If FACT = 'N', then IPIV is an output argument and on exit contains the pivot indices from the factorization A(IA:IA+N−1,JA:JA+N−1) = P*L*U of the original matrix A(IA:IA+N−1,JA:JA+N−1).
If FACT = 'E', then IPIV is an output argument and on exit contains the pivot indices from the factorization A(IA:IA+N−1,JA:JA+N−1) = P*L*U of the equilibrated matrix A(IA:IA+N−1,JA:JA+N−1).

EQUED (global input or global output) CHARACTER*1
Specifies the form of equilibration that was done.
= 'N': No equilibration (always true if FACT = 'N').
= 'R': Row equilibration, i.e., A has been premultiplied by diag(R).
= 'C': Column equilibration, i.e., A has been postmultiplied by diag(C).
= 'B': Both row and column equilibration, i.e., A has been replaced by diag(R)*A(IA:IA+N−1,JA:JA+N−1)*diag(C).
EQUED is an input argument if FACT = 'F'; otherwise, it is an output argument.

R (local input or local output) REAL array, dimension LOCr(M_A).
The row scale factors for A(IA:IA+N−1,JA:JA+N−1). If EQUED = 'R' or 'B', A(IA:IA+N−1,JA:JA+N−1) is multiplied on the left by diag(R); if EQUED='N' or 'C', R is not accessed. R is an input variable if FACT = 'F'; otherwise, R is an output variable. If FACT = 'F' and EQUED = 'R' or 'B', each element of R must be positive. R is replicated in every process column, and is aligned with the distributed matrix A.

A(IA:IA+N−1,JA:JA+N−1) after equilibration (if done). If RCOND is less than the relative machine precision (in particular, if RCOND = 0), the matrix is singular to working precision. This condition is indicated by a return code of INFO > 0.

FERR (local output) REAL array, dimension LOCc(N_B).
The estimated forward error bounds for each solution vector X(j) (the jth column of the solution matrix X(IX:IX+N−1,JX:JX+NRHS−1). If XTRUE is the true solution, FERR(j) bounds the magnitude of the largest entry in (X(j) − XTRUE) divided by the magnitude of the largest entry in X(j). The estimate is as reliable as the estimate for RCOND, and is almost always a slight overestimate of the true error. FERR is replicated in every process row, and is aligned with the matrices B and X.

BERR (local output) REAL array, dimension LOCc(N_B).
The componentwise relative backward error of each solution vector X(j) (i.e., the smallest relative change in any entry of A(IA:IA+N−1,JA:JA+N−1) or B(IB:IB+N−1,JB:JB+NRHS−1) that makes X(j) an exact solution). BERR is replicated in every process row, and is aligned with the matrices B and X.

WORK (local workspace/local output) REAL/COMPLEX array, dimension (LWORK)
On exit, if INFO = 0, WORK(1) returns the optimal LWORK, otherwise WORK(1) returns the minimal LWORK.

LWORK (local or global input) INTEGER
The leading dimension of the array WORK.
LWORK is local input and must be at least
PSGESVX
LWORK = max(PSGECON(LWORK), PSGERFS(LWORK)) + LOCr(N_A).
PCGESVX
LWORK = max(PCGECON(LWORK), PCGERFS(LWORK)) + LOCr(N_A).
If LWORK = −1, then LWORK is global input and a workspace query is assumed; the routine only calculates the minimum and optimal size for all work arrays. Each of these values is returned in the first entry of the corresponding work array, and no error message is issued by PXERBLA.

IWORK PSGESVX only (local workspace/local output) INTEGER array, dimension (LIWORK)
On exit, if INFO = 0, IWORK(1) returns the optimal LIWORK, otherwise IWORK(1) returns the minimal LIWORK.

LIWORK PSGESVX only (local input) INTEGER
The dimension of the array IWORK.
LIWORK is local input and must be at least
LIWORK = LOCr(N_A).
If LIWORK = −1, then LIWORK is global input and a workspace query is assumed; the routine only calculates the minimum and optimal size for all work arrays. Each of these values is returned in the first entry of the corresponding work array, and no error message is issued by PXERBLA.

RWORK PCGESVX only (local workspace/local output) REAL array, dimension (LRWORK)
On exit, if INFO = 0, RWORK(1) returns the optimal LRWORK, otherwise RWORK(1) returns the minimal LRWORK.

LRWORK (local or global input) INTEGER
The dimension of the array RWORK.
LRWORK is local input and must be at least
LRWORK = 2*LOCc(N_A).
If LRWORK = −1, then LRWORK is global input and a workspace query is assumed; the routine only calculates the minimum and optimal size for all work arrays. Each of these values is returned in the first entry of the corresponding work array, and no error message is issued by PXERBLA.

INFO (global output) INTEGER
= 0: successful exit
< 0: If the ith argument is an array and the j-entry had an illegal value, then INFO = −(i*100+j), if the ith argument is a scalar and had an illegal value, then INFO = −i.
> 0: if INFO = i, and i is
     ≤ N: U(i,i) is exactly zero. The factorization has been completed, but the factor U is exactly singular, so the solution and error bounds could not be computed.
     = N+1: U is nonsingular, but RCOND is less than relative machine precision. The factorization has been completed, but the matrix is singular to working precision, and the solution and error bounds have not been computed.

PSGETRF/PCGETRF

```
SUBROUTINE PSGETRF( M, N, A, IA, JA, DESCA, IPIV, INFO )
INTEGER          IA, INFO, JA, M, N
INTEGER          DESCA( * ), IPIV( * )
REAL             A( * )

SUBROUTINE PCGETRF( M, N, A, IA, JA, DESCA, IPIV, INFO )
INTEGER          IA, INFO, JA, M, N
INTEGER          DESCA( * ), IPIV( * )
COMPLEX          A( * )
```

Purpose

PSGETRF/PCGETRF computes an LU factorization of a general M-by-N distributed matrix sub(A) = (IA:IA+M−1,JA:JA+N−1) using partial pivoting with row interchanges.

The factorization has the form sub(A) = P*L*U, where P is a permutation matrix, L is lower triangular with unit diagonal elements (lower trapezoidal if m > n), and

U is upper triangular (upper trapezoidal if m < n). L and U are stored in sub(A).

This is the right-looking Parallel Level 3 BLAS version of the algorithm.

Arguments

M (global input) INTEGER
The number of rows to be operated on, i.e., the number of rows of the distributed submatrix sub(A). $M \geq 0$.

N (global input) INTEGER
The number of columns to be operated on, i.e., the number of columns of the distributed submatrix sub(A). $N \geq 0$.

A (local input/local output) REAL/COMPLEX pointer into the local memory to an array of dimension (LLD_A, LOCc(JA+N−1)).
On entry, this array contains the local pieces of the m-by-n distributed matrix sub(A) to be factored.
On exit, this array contains the local pieces of the factors L and U from the factorization sub(A) = P*L*U; the unit diagonal elements of L are not stored.

IA (global input) INTEGER
The row index in the global array A indicating the first row of sub(A).

JA (global input) INTEGER
The column index in the global array A indicating the first column of sub(A).

DESCA (global and local input) INTEGER array, dimension (DLEN_).
The array descriptor for the distributed matrix A.

IPIV (local output) INTEGER array, dimension (LOCr(M_A)+MB_A)
This array contains the pivoting information. IPIV(i) = the global row local row i was swapped with. This array is tied to the distributed matrix A.

INFO (global output) INTEGER
= 0: successful exit
< 0: If the ith argument is an array and the j-entry had an illegal value, then INFO = −(i*100+j), if the ith argument is a scalar and had an illegal value, then INFO = −i.
> 0: if INFO = i, U(i,i) is exactly zero. The factorization has been completed, but the factor U is exactly singular, and division by zero will occur if it is used to solve a system of equations.

PSGETRI/PCGETRI

```
SUBROUTINE PSGETRI( N, A, IA, JA, DESCA, IPIV, WORK, LWORK,
$                   IWORK, LIWORK, INFO )
     INTEGER          IA, INFO, JA, LIWORK, LWORK, N
     INTEGER          DESCA( * ), IPIV( * ), IWORK( * )
     REAL             A( * ), WORK( * )

SUBROUTINE PCGETRI( N, A, IA, JA, DESCA, IPIV, WORK, LWORK,
$                   IWORK, LIWORK, INFO )
     INTEGER          IA, INFO, JA, LIWORK, LWORK, N
     INTEGER          DESCA( * ), IPIV( * ), IWORK( * )
     COMPLEX          A( * ), WORK( * )
```

Purpose

PSGETRI/PCGETRI computes the inverse of a distributed matrix using the LU factorization computed by PSGETRF/PCGETRF. This method inverts U and then computes the inverse of sub(A) = A(IA:IA+N−1,JA:JA+N−1) denoted InvA by solving the system InvA*L = U^{-1} for InvA.

Arguments

N (global input) INTEGER
The number of rows and columns to be operated on, i.e., the order of the distributed submatrix sub(A). $N \geq 0$.

A (local input/local output) REAL/COMPLEX pointer into the local memory to an array of dimension (LLD_A,LOCc(JA+N−1)).
On entry, the local pieces of the L and U obtained by the factorization sub(A) = P*L*U computed by PSGETRF/PCGETRF.
On exit, if INFO = 0, sub(A) contains the inverse of the original distributed matrix sub(A).

IA (global input) INTEGER
The row index in the global array A indicating the first row of sub(A).

JA (global input) INTEGER
The column index in the global array A indicating the first column of sub(A).

DESCA (global and local input) INTEGER array, dimension (DLEN_).
The array descriptor for the distributed matrix A.

IPIV (local input) INTEGER array, dimension LOCr(M_A)+MB_A
Keeps track of the pivoting information. IPIV(i) is the global row index the local row i was swapped with. This array is tied to the distributed matrix A.

WORK (local workspace/local output) REAL/COMPLEX array, dimension (LWORK)
On exit, WORK(1) returns the minimal and optimal LWORK.

LWORK (local or global input) INTEGER
The dimension of the array WORK.

```
SUBROUTINE PCGETRS( TRANS, N, NRHS, A, IA, JA, DESCA, IPIV, B,
$                    IB, JB, DESCB, INFO )
    CHARACTER         TRANS
    INTEGER           IA, IB, INFO, JA, JB, N, NRHS
    INTEGER           DESCA( * ), DESCB( * ), IPIV( * )
    COMPLEX           A( * ), B( * )
```

Purpose

PSGETRS/PCGETRS solves a system of distributed linear equations op(sub(A))
* X = sub(B) with a general n-by-n distributed matrix sub(A) using the LU
factorization computed by PSGETRF/PCGETRF.

sub(A) denotes A(IA:IA+N−1,JA:JA+N−1), op(A) = A, A^T or A^H and sub(B)
denotes B(IB:IB+N−1,JB:JB+NRHS−1).

Arguments

TRANS   (global input) CHARACTER*1
     Specifies the form of the system of equations:
     = 'N':   sub(A)*X = sub(B) (No transpose)
     = 'T':   sub(A)^T*X = sub(B) (Transpose)
     = 'C':   sub(A)^H*X = sub(B) (Conjugate transpose)

N   (global input) INTEGER
     The number of rows and columns to be operated on, i.e., the order of
     the distributed submatrix sub(A). N ≥ 0.

NRHS   (global input) INTEGER
     The number of right hand sides, i.e., the number of columns of the
     distributed submatrix sub(B). NRHS ≥ 0.

A   (local input) REAL/COMPLEX pointer into the local memory to an
     array of dimension (LLD_A, LOCc(JA+N−1)).
     On entry, this array contains the local pieces of the factors L and U
     from the factorization sub(A) = P*L*U; the unit diagonal elements of
     L are not stored.

IA   (global input) INTEGER
     The row index in the global array A indicating the first row of sub(A).

JA   (global input) INTEGER
     The column index in the global array A indicating the first column of
     sub(A).

DESCA   (global and local input) INTEGER array, dimension (DLEN_).
     The array descriptor for the distributed matrix A.

IPIV   (local input) INTEGER array, dimension (LOCr(M_A)+MB_A)
     This array contains the pivoting information. IPIV(i) = The global
     row local row i was swapped with. This array is tied to the distributed
     matrix A.

B   (local input/local output) REAL/COMPLEX pointer into the local
     memory to an array of dimension (LLD_B,LOCc(JB+NRHS−1)).

     LWORK is local input and must be at least
     LWORK = LOCr(N+mod(IA−1,MB_A))*NB_A.
     WORK is used to keep a copy of at most an entire column block of
     sub(A).
     If LWORK = −1, then LWORK is global input and a workspace query is
     assumed; the routine only calculates the minimum and optimal size for
     all work arrays. Each of these values is returned in the first entry of the
     corresponding work array, and no error message is issued by PXERBLA.

IWORK   (local workspace/local output) INTEGER array, dimension (LIWORK)
     On exit, IWORK(1) returns the minimal and optimal LIWORK.

LIWORK   (local or global input) INTEGER
     The dimension of the array IWORK used as workspace for physically
     transposing the pivots.
     LIWORK is local input and must be at least
     If NPROW == NPCOL then
     LIWORK = LOCc(N_A + mod(JA−1, NB_A)) + NB_A,
     Else
     LIWORK = LOCc(N_A + mod(JA−1, NB_A)) +
        max(CEIL(CEIL(LOCr(M_A)/MB_A)/(LCM/NPROW)), NB_A)
     End if
     where LCM is the least common multiple of process rows and columns
     (NPROW and NPCOL).
     If LIWORK = −1, then LIWORK is global input and a workspace
     query is assumed; the routine only calculates the minimum and optimal
     size for all work arrays. Each of these values is returned in the first
     entry of the corresponding work array, and no error message is issued
     by PXERBLA.

INFO   (global output) INTEGER
     = 0:   successful exit
     < 0:   If the ith argument is an array and the j-entry had an illegal
        value, then INFO = −(i*100+j), if the ith argument is a scalar
        and had an illegal value, then INFO = −i.
     > 0:   If INFO = K, U(IA+K−1,IA+K−1) is exactly zero; the matrix
        is singular and its inverse could not be computed.

PSGETRS/PCGETRS

```
SUBROUTINE PSGETRS( TRANS, N, NRHS, A, IA, JA, DESCA, IPIV, B,
$                    IB, JB, DESCB, INFO )
    CHARACTER         TRANS
    INTEGER           IA, IB, INFO, JA, JB, N, NRHS
    INTEGER           DESCA( * ), DESCB( * ), IPIV( * )
    REAL              A( * ), B( * )
```

On entry, the right hand sides sub(B).
On exit, sub(B) is overwritten by the solution distributed matrix X.

IB   (global input) INTEGER
The row index in the global array B indicating the first row of sub(B).

JB   (global input) INTEGER
The column index in the global array B indicating the first column of sub(B).

DESCB (global and local input) INTEGER array, dimension (DLEN_).
The array descriptor for the distributed matrix B.

INFO (global output) INTEGER
= 0: successful exit
< 0: If the ith argument is an array and the j-entry had an illegal value, then INFO = $-(i*100+j)$, if the ith argument is a scalar and had an illegal value, then INFO = $-i$.

PSGGQRF/PCGGQRF

```
SUBROUTINE PSGGQRF( N, M, P, A, IA, JA, DESCA, TAUA, B, IB, JB,
$                   DESCB, TAUB, WORK, LWORK, INFO )
     INTEGER   IA, IB, INFO, JA, JB, LWORK, M, N, P
     INTEGER   DESCA( * ), DESCB( * )
     REAL      A( * ), B( * ), TAUA( * ), TAUB( * ), WORK( * )

SUBROUTINE PCGGQRF( N, M, P, A, IA, JA, DESCA, TAUA, B, IB, JB,
$                   DESCB, TAUB, WORK, LWORK, INFO )
     INTEGER   IA, IB, INFO, JA, JB, LWORK, M, N, P
     INTEGER   DESCA( * ), DESCB( * )
     COMPLEX   A( * ), B( * ), TAUA( * ), TAUB( * ), WORK( * )
```

Purpose

PSGGQRF/PCGGQRF computes a generalized QR factorization of an n-by-m matrix sub(A) = A(IA:IA+N−1,JA:JA+M−1) and an n-by-p matrix sub(B) = B(IB:IB+N−1,JB:JB+P−1):

$$\text{sub}(A) = Q*R, \qquad \text{sub}(B) = Q*T*Z,$$

where Q is an n-by-n orthogonal/unitary matrix, Z is a p-by-p orthogonal/unitary matrix, and R and T assume one of the forms:

$$\text{if } n \ge m, \qquad R = \begin{array}{c} m \\ n-m \end{array}\begin{pmatrix} R_{11} \\ 0 \end{pmatrix} m$$

or

$$\text{if } n < m, \qquad R = n\begin{pmatrix} \overset{n}{R_{11}} & \overset{m-n}{R_{12}} \end{pmatrix}$$

where R_{11} is upper triangular, and

$$\text{if } n \le p, \qquad T = n\begin{pmatrix} \overset{p-n}{0} & \overset{n}{T_{12}} \end{pmatrix}$$

or

$$\text{if } n > p, \qquad T = \begin{array}{c} n-p \\ p \end{array}\begin{pmatrix} T_{11} \\ T_{21} \end{pmatrix} p$$

where T_{12} or T_{21} is a p-by-p upper triangular matrix.

In particular, if sub(B) is square and nonsingular, the GQR factorization of sub(A) and sub(B) implicitly gives the QR factorization of sub(B)$^{-1}$*sub(A):

$$\text{sub}(B)^{-1}*\text{sub}(A) = Z^H*(T^{-1}*R).$$

Arguments

N   (global input) INTEGER
The number of rows to be operated on i.e., the number of rows of the distributed submatrices sub(A) and sub(B). N $\geq$ 0.

M   (global input) INTEGER
The number of columns to be operated on i.e., the number of columns of the distributed submatrix sub(A). M $\geq$ 0.

P   (global input) INTEGER
The number of columns to be operated on i.e., the number of columns of the distributed submatrix sub(B). P $\geq$ 0.

A   (local input/local output) REAL/COMPLEX pointer into the local memory to an array of dimension (LLD_A, LOCc(JA+M−1)).
On entry, the local pieces of the n-by-m distributed matrix sub(A) which is to be factored.
On exit, the elements on and above the diagonal of sub(A) contain the min(n,m) by m upper trapezoidal matrix R (R is upper triangular if n $\geq$ m); the elements below the diagonal, with the array TAUA, represent the orthogonal/unitary matrix Q as a product of min(n,m) elementary reflectors.

IA   (global input) INTEGER
The row index in the global array A indicating the first row of sub(A).

JA   (global input) INTEGER
The column index in the global array A indicating the first column of sub(A).

DESCA (global and local input) INTEGER array, dimension (DLEN_).
The array descriptor for the distributed matrix A.

TAUA (local output) REAL/COMPLEX array, dimension LOCc(JA+MIN(N,M)−1).
This array contains the scalar factors TAUA of the elementary reflectors

which represent the orthogonal/unitary matrix Q. TAUA is tied to the distributed matrix A.

B   (local input/local output) REAL/COMPLEX pointer into the local memory to an array of dimension (LLD_B, LOCc(JB+P−1)).
On entry, the local pieces of the n-by-p distributed matrix sub(B) which is to be factored.
On exit,
if n $\leq$ p, the upper triangle of B(IB:IB+N−1,JB+P−N:JB+P−1) contains the n by n upper triangular matrix T;
if n $>$ p, the elements on and above the $(n-p)^{th}$ subdiagonal contain the n-by-p upper trapezoidal matrix T; the remaining elements, with the array TAUB, represent the orthogonal/unitary matrix Z as a product of elementary reflectors.

IB   (global input) INTEGER
The row index in the global array B indicating the first row of sub(B).

JB   (global input) INTEGER
The column index in the global array B indicating the first column of sub(B).

DESCB   (global and local input) INTEGER array, dimension (DLEN_).
The array descriptor for the distributed matrix B.

TAUB   (local output) REAL/COMPLEX array, dimension LOCr(IB+N−1).
This array contains the scalar factors of the elementary reflectors which represent the orthogonal/unitary matrix Z. TAUB is tied to the distributed matrix B.

WORK   (local workspace/local output) REAL/COMPLEX array, dimension (LWORK)
On exit, WORK(1) returns the minimal and optimal LWORK.

LWORK   (local or global input) INTEGER
The dimension for the array WORK.
LWORK is local input and must be at least
LWORK $\geq$ max(NB_A * (NpA0 + MqA0 + NB_A),
    max((NB_A*(NB_A−1))/2, (PqB0 + NpB0)*NB_A) + NB_A * NB_A,
    MB_B * (NpB0 + PqB0 + MB_B)), where

IROFFA = mod(IA−1, MB_A), ICOFFA = mod(JA−1, NB_A),
IAROW = INDXG2P(IA, MB_A, MYROW, RSRC_A, NPROW),
IACOL = INDXG2P(JA, NB_A, MYCOL, CSRC_A, NPCOL),
NpA0 = NUMROC(N+IROFFA, MB_A, MYROW, IAROW, NPROW),
MqA0 = NUMROC(M+ICOFFA, NB_A, MYCOL, IACOL, NPCOL),
IROFFB = mod(IB−1, MB_B), ICOFFB = mod(JB−1, NB_B),
IBROW = INDXG2P(IB, MB_B, MYROW, RSRC_B, NPROW),
IBCOL = INDXG2P(JB, NB_B, MYCOL, CSRC_B, NPCOL),
NpB0 = NUMROC(N+IROFFB, MB_B, MYROW, IBROW, NPROW),
PqB0 = NUMROC(P+ICOFFB, NB_B, MYCOL, IBCOL, NPCOL),
and NUMROC, INDXG2P are ScaLAPACK tool functions; MYROW, MYCOL, NPROW and NPCOL can be determined by calling the subroutine BLACS_GRIDINFO.
If LWORK = −1, then LWORK is global input and a workspace query

is assumed; the routine only calculates the minimum and optimal size for all work arrays. Each of these values is returned in the first entry of the corresponding work array, and no error message is issued by PXERBLA.

INFO   (global output) INTEGER
= 0:   successful exit
< 0:   If the ith argument is an array and the j-entry had an illegal value, then INFO = −(i*100+j), if the ith argument is a scalar and had an illegal value, then INFO = −i.

PSGGRQF/PCGGRQF

```
SUBROUTINE PSGGRQF( M, P, N, A, IA, JA, DESCA, TAUA, B, IB, JB,
$                   DESCB, TAUB, WORK, LWORK, INFO )
INTEGER            IA, IB, INFO, JA, JB, LWORK, M, N, P
INTEGER            DESCA( * ), DESCB( * )
REAL               A( * ), B( * ), TAUA( * ), TAUB( * ), WORK( * )

SUBROUTINE PCGGRQF( M, P, N, A, IA, JA, DESCA, TAUA, B, IB, JB,
$                   DESCB, TAUB, WORK, LWORK, INFO )
INTEGER            IA, IB, INFO, JA, JB, LWORK, M, N, P
INTEGER            DESCA( * ), DESCB( * )
COMPLEX            A( * ), B( * ), TAUA( * ), TAUB( * ), WORK( * )
```

Purpose

PSGGRQF/PCGGRQF computes a generalized RQ factorization of an m-by-n matrix sub(A) = A(IA:IA+M−1,JA:JA+N−1) and a p-by-n matrix sub(B) = B(IB:IB+P−1,JB:JB+N−1):

$$\text{sub}(A) = R*Q, \quad \text{sub}(B) = Z*T*Q,$$

where Q is an n-by-n orthogonal/unitary matrix, and Z is a p-by-p orthogonal/unitary matrix, and R and T assume one of the forms:

$$\text{if } m \leq n, \quad R = \begin{array}{c} \\ m \end{array}\!\begin{array}{cc} n-m & m \\ \left(0 \right. & \left. R_{12} \right) \end{array}$$

or

$$\text{if } m > n, \quad R = \begin{array}{c} m-n \\ n \end{array}\!\begin{array}{c} n \\ \left(\begin{matrix} R_{11} \\ R_{21} \end{matrix} \right) \end{array}$$

where R_{12} or R_{21} is upper triangular, and

$$\text{if } p \geq n, \quad T = \begin{array}{c} n \\ p-n \end{array}\!\begin{array}{c} n \\ \left(\begin{matrix} T_{11} \\ 0 \end{matrix} \right) \end{array}$$

or

$$\text{if } p < n, \qquad T = \begin{matrix} & p & n-p \\ p & (T_{11} & T_{12}) \end{matrix}$$

where T_{11} is upper triangular.

In particular, if sub(B) is square and nonsingular, the GRQ factorization of sub(A) and sub(B) implicitly gives the RQ factorization of sub(A)*sub(B)$^{-1}$:

$$\text{sub}(A)*\text{sub}(B)^{-1} = (R*T^{-1})*Z^H$$

Arguments

M
(global input) INTEGER
The number of rows to be operated on i.e., the number of rows of the distributed submatrix sub(A). $M \geq 0$.

P
(global input) INTEGER
The number of rows to be operated on i.e., the number of rows of the distributed submatrix sub(B). $P \geq 0$.

N
(global input) INTEGER
The number of columns to be operated on i.e., the number of columns of the distributed submatrices sub(A) and sub(B). $N \geq 0$.

A
(local input/local output) REAL/COMPLEX pointer into the local memory to an array of dimension (LLD_A, LOCc(JA+N−1)).
On entry, the local pieces of the m-by-n distributed matrix sub(A) which is to be factored.
On exit,
if $m \leq n$, the upper triangle of A(IA:IA+M−1, JA+N−M:JA+N−1) contains the m-by-m upper triangular matrix R; if $m \geq n$, the elements on and above the $(m-n)^{th}$ subdiagonal contain the m-by-n upper trapezoidal matrix R; the remaining elements, with the array TAUA, represent the orthogonal/unitary matrix Q as a product of elementary reflectors.

IA
(global input) INTEGER
The row index in the global array A indicating the first row of sub(A).

JA
(global input) INTEGER
The column index in the global array A indicating the first column of sub(A).

DESCA
(global and local input) INTEGER array, dimension (DLEN_).
The array descriptor for the distributed matrix A.

TAUA
(local output) REAL/COMPLEX array, dimension LOCr(IA+M−1).
This array contains the scalar factors of the elementary reflectors which represent the orthogonal unitary matrix Q. TAUA is tied to the distributed matrix A.

B
(local input/local output) REAL/COMPLEX pointer into the local memory to an array of dimension (LLD_B, LOCc(JB+N−1)).
On entry, the local pieces of the p-by-n distributed matrix sub(B) which is to be factored.
On exit,
the elements on and above the diagonal of sub(B) contain the min(p,n)-by-n upper trapezoidal matrix T (T is upper triangular if $p \geq n$); the elements below the diagonal, with the array TAUB, represent the orthogonal/unitary matrix Z as a product of elementary reflectors.

IB
(global input) INTEGER
The row index in the global array B indicating the first row of sub(B).

JB
(global input) INTEGER
The column index in the global array B indicating the first column of sub(B).

DESCB
(global and local input) INTEGER array, dimension (DLEN_).
The array descriptor for the distributed matrix B.

TAUB
(local output) REAL/COMPLEX array, dimension LOCc(JB+MIN(P,N)−1).
This array contains the scalar factors TAUB of the elementary reflectors which represent the orthogonal/unitary matrix Z. TAUB is tied to the distributed matrix B.

WORK
(local workspace/local output) REAL/COMPLEX array, dimension (LWORK)
On exit, WORK(1) returns the minimal and optimal LWORK.

LWORK
(local or global input) INTEGER.
The dimension of the array WORK.
LWORK $\geq$ max(MB_A * (MpA0 + NqA0 + MB_A),
max((MB_A*(MB_A−1))/2, (PpB0 + NqB0)*MB_A) +
MB_A * MB_A, NB_B * (PpB0 + NqB0 + NB_B)), where

IROFFA = mod(IA−1, MB_A), ICOFFA = mod(JA−1, NB_A),
IAROW = INDXG2P(IA, MB_A, MYROW, RSRC_A, NPROW),
IACOL = INDXG2P(JA, NB_A, MYCOL, CSRC_A, NPCOL),
MpA0 = NUMROC(M+IROFFA, MB_A, MYROW, IAROW, NPROW),
NqA0 = NUMROC(N+ICOFFA, NB_A, MYCOL, IACOL, NPCOL),

IROFFB = mod(IB−1, MB_B), ICOFFB = mod(JB−1, NB_B),
IBROW = INDXG2P(IB, MB_B, MYROW, RSRC_B, NPROW),
IBCOL = INDXG2P(JB, NB_B, MYCOL, CSRC_B, NPCOL),
PpB0 = NUMROC(P+IROFFB, MB_B, MYROW, IBROW, NPROW),
NqB0 = NUMROC(N+ICOFFB, NB_B, MYCOL, IBCOL, NPCOL),

and NUMROC, INDXG2P are ScaLAPACK tool functions; MYROW, MYCOL, NPROW and NPCOL can be determined by calling the subroutine BLACS_GRIDINFO.
If LWORK = −1, then LWORK is global input and a workspace query is assumed; the routine only calculates the minimum and optimal size for all work arrays. Each of these values is returned in the first entry of the corresponding work array, and no error message is issued by PXERBLA.

INFO
(global output) INTEGER
= 0: successful exit

< 0:   If the ith argument is an array and the j-entry had an illegal value, then INFO = $-(i*100+j)$, if the ith argument is a scalar and had an illegal value, then INFO = $-i$.

PSLAHQR

```
SUBROUTINE PSLAHQR( WANTT, WANTZ, N, ILO, IHI, A, DESCA, WR, WI,
$                   ILOZ, IHIZ, Z, DESCZ, WORK, LWORK, IWORK,
$                   ILWORK, INFO )
    LOGICAL          WANTT, WANTZ
    INTEGER          IHI, IHIZ, ILO, ILOZ, ILWORK, INFO, LWORK, N
    INTEGER          DESCA( * ), DESCZ( * ), IWORK( * )
    REAL             A( * ), WI( * ), WORK( * ), WR( * ), Z( * )
```

Purpose

PSLAHQR is an auxiliary routine used to find the Schur decomposition and or eigenvalues of a matrix already in Hessenberg form from cols ILO to IHI.

Arguments

WANTT   (global input) LOGICAL
        = .TRUE. : the full Schur form T is required;
        = .FALSE.: only eigenvalues are required.

WANTZ   (global input) LOGICAL
        = .TRUE. : the matrix of Schur vectors Z is required;
        = .FALSE.: Schur vectors are not required.

N   (global input) INTEGER
        The order of the Hessenberg matrix A (and Z if WANTZ). N ≥ 0.

ILO, IHI (global input) INTEGER
        It is assumed that A is already upper quasi-triangular in rows and columns IHI+1:N, and that A(ILO,ILO−1) = 0 (unless ILO = 1). PSLAHQR works primarily with the Hessenberg submatrix in rows and columns ILO to IHI, but applies transformations to all of H if WANTT is .TRUE.. 1 ≤ ILO ≤ max(1,IHI); IHI ≤ N.

A   (global input/output) REAL array, dimension (DESCA(LLD_),*)
        On entry, the upper Hessenberg matrix A.
        On exit, if WANTT is .TRUE., A is upper quasi-triangular in rows and columns ILO:IHI, with any 2-by-2 or larger diagonal blocks not yet in standard form. If WANTT is .FALSE., the contents of A are unspecified on exit.

DESCA   (global and local input) INTEGER array, dimension (DLEN_).
        The array descriptor for the distributed matrix A.

WR   (global replicated output) REAL array, dimension (N)

WI   (global replicated output) REAL array, dimension (N)
        The real and imaginary parts, respectively, of the computed eigenvalues ILO to IHI are stored in the corresponding elements of WR and WI. If two eigenvalues are computed as a complex conjugate pair, they are stored in consecutive elements of WR and WI, say the ith and $(i+1)^{st}$, with WI(i) > 0 and WI(i+1) < 0. If WANTT is .TRUE., the eigenvalues are stored in the same order as on the diagonal of the Schur form returned in A. A may be returned with larger diagonal blocks until the next release.

ILOZ, IHIZ (global input) INTEGER
        Specify the rows of Z to which transformations must be applied if WANTZ is .TRUE..
        1 ≤ ILOZ ≤ ILO; IHI ≤ IHIZ ≤ N.

Z   (global input/output) REAL array.
        If WANTZ is .TRUE., on entry Z must contain the current matrix Z of transformations accumulated by PDHSEQR, and on exit Z has been updated; transformations are applied only to the submatrix Z(ILOZ:IHIZ,ILO:IHI). If WANTZ is .FALSE., Z is not referenced.

DESCZ   (global and local input) INTEGER array, dimension (DLEN_).
        The array descriptor for the distributed matrix Z.

WORK   (local workspace/local output) REAL array, dimension (LWORK)
        On exit, WORK(1) returns the minimal and optimal LWORK.

LWORK   (local or global input) INTEGER
        The dimension of the array WORK.
        LWORK is local input and must be at least
        LWORK ≥ 3*N + max(2*max(DESCZ(LLD_),DESCA(LLD_)) + 2*LOCc(N), 7*Ceil(N/HBL)/LCM(NPROW,NPCOL))).

        If LWORK = −1, then LWORK is global input and a workspace query is assumed; the routine only calculates the minimum and optimal size for all work arrays. Each of these values is returned in the first entry of the corresponding work array, and no error message is issued by PXERBLA.

IWORK   (global and local input) INTEGER array, dimension (LWORK)
        This will hold some of the IBLK integer arrays. This is held as a place holder for a future release. Currently unreferenced.

ILWORK (local input) INTEGER
        This holds the some of the IBLK integer arrays. This is held as a place holder for the next release. Currently unreferenced.

INFO   (global output) INTEGER
        = 0: successful exit
        < 0: If the ith argument is an array and the j-entry had an illegal value, then INFO = $-(i*100+j)$, if the ith argument is a scalar and had an illegal value, then INFO = $-i$.
        > 0: PSLAHQR failed to compute all the eigenvalues ILO to IHI in a total of 30*(IHI−ILO+1) iterations; if INFO = i, elements i+1:ihi of WR and WI contain those eigenvalues which have been successfully computed.

PSORGLQ/PCUNGLQ

```
SUBROUTINE PSORGLQ( M, N, K, A, IA, JA, DESCA, TAU, WORK, LWORK,
$                   INFO )
INTEGER           IA, INFO, JA, K, LWORK, M, N
INTEGER           DESCA( * )
REAL              A( * ), TAU( * ), WORK( * )

SUBROUTINE PCUNGLQ( M, N, K, A, IA, JA, DESCA, TAU, WORK, LWORK,
$                   INFO )
INTEGER           IA, INFO, JA, K, LWORK, M, N
INTEGER           DESCA( * )
COMPLEX           A( * ), TAU( * ), WORK( * )
```

Purpose

PSORGLQ/PCUNGLQ generates an m-by-n real/complex distributed matrix Q denoting A(IA:IA+M−1,JA:JA+N−1) with orthonormal rows, which is defined as the first m rows of a product of k elementary reflectors of order n

$Q = H_k \cdots H_2 H_1 \ (PSORGLQ)$

$Q = H_k^H \cdots H_2^H H_1^H \ (PCUNGLQ)$

as returned by PSGELQF/PCGELQF.

Arguments

M   (global input) INTEGER
     The number of rows to be operated on i.e., the number of rows of the distributed submatrix Q. M ≥ 0.

N   (global input) INTEGER
     The number of columns to be operated on i.e., the number of columns of the distributed submatrix Q. N ≥ M.

K   (global input) INTEGER
     The number of elementary reflectors whose product defines the matrix Q. M ≥ K ≥ 0.

A   (local input/local output) REAL/COMPLEX pointer into the local memory to an array of dimension (LLD_A,LOCc(JA+N−1)). On entry, the ith row must contain the vector which defines the elementary reflector H_i, IA ≤ i ≤ IA+K−1, as returned by PSGELQF/PCGELQF in the K rows of its distributed matrix argument A(IA:IA+K−1,JA:*). On exit, this array contains the local pieces of the m-by-n distributed matrix Q.

IA   (global input) INTEGER
     The row index in the global array A indicating the first row of sub(A).

JA   (global input) INTEGER
     The column index in the global array A indicating the first column of sub(A).

DESCA (global and local input) INTEGER array, dimension (DLEN_).
      The array descriptor for the distributed matrix A.

TAU (local input) REAL/COMPLEX array, dimension LOCr(IA+K−1). This array contains the scalar factors TAU(i) of the elementary reflectors H_i as returned by PSGELQF/PCGELQF. TAU is tied to the distributed matrix A.

WORK (local workspace/local output) REAL/COMPLEX array, dimension (LWORK)
     On exit, WORK(1) returns the minimal and optimal LWORK.

LWORK (local or global input) INTEGER
      The dimension of the array WORK.
      LWORK is local input and must be at least
      LWORK ≥ MB_A * (MpA0 + NqA0 + MB_A), where
      IROFFA = mod(IA−1, MB_A), ICOFFA = mod(JA−1, NB_A),
      IAROW = INDXG2P(IA, MB_A, MYROW, RSRC_A, NPROW),
      IACOL = INDXG2P(JA, NB_A, MYCOL, CSRC_A, NPCOL),
      MpA0 = NUMROC(M+IROFFA, MB_A, MYROW, IAROW, NPROW),
      NqA0 = NUMROC(N+ICOFFA, NB_A, MYCOL, IACOL, NPCOL),
      INDXG2P and NUMROC are ScaLAPACK tool functions; MYROW, MYCOL, NPROW and NPCOL can be determined by calling the subroutine BLACS_GRIDINFO.
      If LWORK = −1, then LWORK is global input and a workspace query is assumed; the routine only calculates the minimum and optimal size for all work arrays. Each of these values is returned in the first entry of the corresponding work array, and no error message is issued by PXERBLA.

INFO (global output) INTEGER
     = 0: successful exit
     < 0: If the ith argument is an array and the j-entry had an illegal value, then INFO = −(i*100+j), if the ith argument is a scalar and had an illegal value, then INFO = −i.

PSORGQL/PCUNGQL

```
SUBROUTINE PSORGQL( M, N, K, A, IA, JA, DESCA, TAU, WORK, LWORK,
$                   INFO )
INTEGER           IA, INFO, JA, K, LWORK, M, N
INTEGER           DESCA( * )
REAL              A( * ), TAU( * ), WORK( * )
```

```
SUBROUTINE PCUNGQL( M, N, K, A, IA, JA, DESCA, TAU, WORK, LWORK,
$                   INFO )
    INTEGER         IA, INFO, JA, K, LWORK, M, N
    INTEGER         DESCA( * )
    COMPLEX         A( * ), TAU( * ), WORK( * )
```

Purpose

PSORGQL/PCUNGQL generates an m-by-n real/complex matrix Q denoting A(IA:IA+M−1,JA:JA+N−1) with orthonormal columns, which is defined as the last n columns of a product of k elementary reflectors H_i of order m

$$Q = H_k \cdots H_2 H_1$$

as returned by PSGEQLF/PCGEQLF.

Arguments

M   (global input) INTEGER
        The number of rows to be operated on, i.e., the number of rows of the distributed submatrix Q. M ≥ 0.

N   (global input) INTEGER
        The number of columns to be operated on, i.e., the number of columns of the distributed submatrix Q. M ≥ N ≥ 0.

K   (global input) INTEGER
        The number of elementary reflectors whose product defines the matrix Q. N ≥ K ≥ 0.

A   (local input/local output) REAL/COMPLEX pointer into the local memory to an array of dimension (LLD_A,LOCc(JA+N−1)).
        On entry, the jth column must contain the vector which defines the elementary reflector H_j, JA+N−K ≤ j ≤ JA+N−1, as returned by PSGEQLF/PCGEQLF in the K columns of its distributed matrix argument A(IA:*,JA+N−K:JA+N−1).
        On exit, this array contains the local pieces of the m-by-n distributed matrix Q.

IA   (global input) INTEGER
        The row index in the global array A indicating the first row of sub(A).

JA   (global input) INTEGER
        The column index in the global array A indicating the first column of sub(A).

DESCA   (global and local input) INTEGER array, dimension (DLEN_).
        The array descriptor for the distributed matrix A.

TAU   (local input) REAL/COMPLEX array, dimension LOCc(JA+N−1).
        This array contains the scalar factors TAU(j) of the elementary reflectors H_j as returned by PSGEQLF/PCGEQLF. TAU is tied to the distributed matrix A.

WORK   (local workspace/local output) REAL/COMPLEX array, dimension (LWORK)
        On exit, WORK(1) returns the minimal and optimal LWORK.

LWORK   (local or global input) INTEGER
        The dimension of the array WORK.
        LWORK is local input and must be at least

        LWORK ≥ NB_A * (NqA0 + MpA0 + NB_A), where

        IROFFA = mod(IA−1, MB_A), ICOFFA = mod(JA−1, NB_A),
        IAROW = INDXG2P(IA, MB_A, MYROW, RSRC_A, NPROW),
        IACOL = INDXG2P(JA, NB_A, MYCOL, CSRC_A, NPCOL),
        MpA0 = NUMROC(M+IROFFA, MB_A, MYROW, IAROW, NPROW),
        NqA0 = NUMROC(N+ICOFFA, NB_A, MYCOL, IACOL, NPCOL),

        INDXG2P and NUMROC are ScaLAPACK tool functions; MYROW, MYCOL, NPROW and NPCOL can be determined by calling the subroutine BLACS_GRIDINFO.

        If LWORK = −1, then LWORK is global input and a workspace query is assumed; the routine only calculates the minimum and optimal size for all work arrays. Each of these values is returned in the first entry of the corresponding work array, and no error message is issued by PXERBLA.

INFO   (global output) INTEGER
        = 0:   successful exit
        < 0:   If the ith argument is an array and the j-entry had an illegal value, then INFO = −(i*100+j), if the ith argument is a scalar and had an illegal value, then INFO = −i.

PSORGQR/PCUNGQR

```
SUBROUTINE PSORGQR( M, N, K, A, IA, JA, DESCA, TAU, WORK, LWORK,
$                   INFO )
    INTEGER         IA, INFO, JA, K, LWORK, M, N
    INTEGER         DESCA( * )
    REAL            A( * ), TAU( * ), WORK( * )

SUBROUTINE PCUNGQR( M, N, K, A, IA, JA, DESCA, TAU, WORK, LWORK,
$                   INFO )
    INTEGER         IA, INFO, JA, K, LWORK, M, N
    INTEGER         DESCA( * )
    COMPLEX         A( * ), TAU( * ), WORK( * )
```

Purpose

PSORGQR/PCUNGQR generates an m-by-n real/complex matrix Q denoting A(IA:IA+M−1,JA:JA+N−1) with orthonormal columns, which is defined as the first n columns of a product of k elementary reflectors H_i of order m

$$Q = H_1 H_2 \cdots H_k$$

Arguments

M   (global input) INTEGER
The number of rows to be operated on, i.e., the number of rows of the distributed submatrix Q. M ≥ 0.

N   (global input) INTEGER
The number of columns to be operated on, i.e., the number of columns of the distributed submatrix Q. M ≥ N ≥ 0.

K   (global input) INTEGER
The number of elementary reflectors whose product defines the matrix Q. N ≥ K ≥ 0.

A   (local input/local output) REAL/COMPLEX pointer into the local memory to an array of dimension (LLD_A,LOCc(JA+N−1)).
On entry, the jth column must contain the vector which defines the elementary reflector H_j, JA ≤ j ≤ JA+K−1, as returned by PSGEQRF/PCGEQRF in the K columns of its distributed matrix argument A(IA:*,JA:JA+K−1).
On exit, this array contains the local pieces of the m-by-n distributed matrix Q.

IA   (global input) INTEGER
The row index in the global array A indicating the first row of sub(A).

JA   (global input) INTEGER
The column index in the global array A indicating the first column of sub(A).

DESCA   (global and local input) INTEGER array, dimension (DLEN_).
The array descriptor for the distributed matrix A.

TAU   (local input) REAL/COMPLEX array, dimension LOCc(JA+K−1).
This array contains the scalar factors TAU(j) of the elementary reflectors H_j, as returned by PSGEQRF/PCGEQRF. TAU is tied to the distributed matrix A.

WORK   (local workspace/local output) REAL/COMPLEX array, dimension (LWORK)
On exit, WORK(1) returns the minimal and optimal LWORK.

LWORK   (local or global input) INTEGER
The dimension of the array WORK.
LWORK is local input and must be at least
LWORK ≥ NB_A * (NqA0 + MpA0 + NB_A), where
IROFFA = mod(IA−1, MB_A), ICOFFA = mod(JA−1, NB_A),
IAROW = INDXG2P(IA, MB_A, MYROW, RSRC_A, NPROW),
IACOL = INDXG2P(JA, NB_A, MYCOL, CSRC_A, NPCOL),
MpA0 = NUMROC(M+IROFFA, MB_A, MYROW, IAROW, NPROW),
NqA0 = NUMROC(N+ICOFFA, NB_A, MYCOL, IACOL, NPCOL),
INDXG2P and NUMROC are ScaLAPACK tool functions; MYROW,

MYCOL, NPROW and NPCOL can be determined by calling the subroutine BLACS_GRIDINFO.
If LWORK = −1, then LWORK is global input and a workspace query is assumed; the routine only calculates the minimum and optimal size for all work arrays. Each of these values is returned in the first entry of the corresponding work array, and no error message is issued by PXERBLA.

INFO   (global output) INTEGER
= 0:   successful exit
< 0:   If the ith argument is an array and the j-entry had an illegal value, then INFO = −(i*100+j), if the ith argument is a scalar and had an illegal value, then INFO = −i.

PSORGRQ/PCUNGRQ

```
SUBROUTINE PSORGRQ( M, N, K, A, IA, JA, DESCA, TAU, WORK, LWORK,
$                    INFO )
INTEGER           IA, INFO, JA, K, LWORK, M, N
INTEGER           DESCA( * )
REAL              A( * ), TAU( * ), WORK( * )

SUBROUTINE PCUNGRQ( M, N, K, A, IA, JA, DESCA, TAU, WORK, LWORK,
$                    INFO )
INTEGER           IA, INFO, JA, K, LWORK, M, N
INTEGER           DESCA( * )
COMPLEX           A( * ), TAU( * ), WORK( * )
```

Purpose

PSORGRQ/PCUNGRQ generates an m-by-n real/complex matrix Q denoting A(IA:IA+M−1,JA:JA+N−1) with orthonormal rows, which is defined as the last m rows of a product of k elementary reflectors H_i of order n

$$Q = H_1 H_2 \cdots H_k \ (PSORGRQ)$$

$$Q = H_1^H H_2^H \cdots H_k^H \ (PCUNGRQ)$$

as returned by PSGERQF/PCGERQF.

Arguments

M   (global input) INTEGER
The number of rows to be operated on, i.e., the number of rows of the distributed submatrix Q. M ≥ 0.

N   (global input) INTEGER
The number of columns to be operated on, i.e., the number of columns of the distributed submatrix Q. N ≥ M.

< 0: If the ith argument is an array and the j-entry had an illegal value, then INFO = $-(i*100+j)$, if the ith argument is a scalar and had an illegal value, then INFO = $-i$.

PSORMBR/PCUNMBR

```
SUBROUTINE PSORMBR( VECT, SIDE, TRANS, M, N, K, A, IA, JA, DESCA,
$                    TAU, C, IC, JC, DESCC, WORK, LWORK, INFO )
    CHARACTER         SIDE, TRANS, VECT
    INTEGER           IA, IC, INFO, JA, JC, K, LWORK, M, N
    INTEGER           DESCA( * ), DESCC( * )
    REAL              A( * ), C( * ), TAU( * ), WORK( * )

SUBROUTINE PCUNMBR( VECT, SIDE, TRANS, M, N, K, A, IA, JA, DESCA,
$                    TAU, C, IC, JC, DESCC, WORK, LWORK, INFO )
    CHARACTER         SIDE, TRANS, VECT
    INTEGER           IA, IC, INFO, JA, JC, K, LWORK, M, N
    INTEGER           DESCA( * ), DESCC( * )
    COMPLEX           A( * ), C( * ), TAU( * ), WORK( * )
```

Purpose

If VECT = 'Q', PSORMBR/PCUNMBR overwrites the general real/complex distributed m-by-n matrix sub(C) = C(IC:IC+M−1,JC:JC+N−1) with

	SIDE = 'L'	SIDE = 'R'
TRANS = 'N':	Q*sub(C)	sub(C)*Q
TRANS = 'T':	Q^T*sub(C)	sub(C)*Q^T	(PSORMBR only)
TRANS = 'C':	Q^H*sub(C)	sub(C)*Q^H	(PCUNMBR only)

If VECT = 'P', PSORMBR/PCUNMBR overwrites the general real/complex distributed m-by-n matrix sub(C) with

	SIDE = 'L'	SIDE = 'R'
TRANS = 'N':	P*sub(C)	sub(C)*P
TRANS = 'T':	P^T*sub(C)	sub(C)*P^T	(PSORMBR only)
TRANS = 'C':	P^H*sub(C)	sub(C)*P^H	(PCUNMBR only)

Here Q and P^H are the orthogonal/unitary matrices determined by PSGE-BRD/PCGEBRD when reducing a real/complex distributed matrix A(IA:*,JA:*) to bidiagonal form: A(IA:*,JA:*) = $Q*B*P^H$. Q and P^H are defined as products of elementary reflectors H_i and G_i, respectively.

Let nq = m if SIDE = 'L' and nq = n if SIDE = 'R'. Thus nq is the order of the orthogonal/unitary matrix Q or P^H that is applied.

If VECT = 'Q', A(IA:*,JA:*) is assumed to have been an nq-by-k matrix:

if nq $\geq$ k, $Q = H_1 H_2 \cdots H_k$;

K   (global input) INTEGER
        The number of elementary reflectors whose product defines the matrix Q. M $\geq$ K $\geq$ 0.

A   (local input/local output) REAL/COMPLEX pointer into the local memory to an array of dimension (LLD_A,LOCc(JA+N−1)).
        On entry, the ith row must contain the vector which defines the elementary reflector H_i, IA+M−K $\leq$ i $\leq$ IA+M−1, as returned by PSGERQF/PCGERQF in the K rows of its distributed matrix argument A(IA+M−K:IA+M−1,JA:*).
        On exit, this array contains the local pieces of the m-by-n distributed matrix Q.

IA   (global input) INTEGER
        The row index in the global array A indicating the first row of sub(A).

JA   (global input) INTEGER
        The column index in the global array A indicating the first column of sub(A).

DESCA   (global and local input) INTEGER array, dimension (DLEN_).
        The array descriptor for the distributed matrix A.

TAU   (local input) REAL/COMPLEX array, dimension LOCr(IA+M-1).
        This array contains the scalar factors TAU(i) of the elementary reflectors H_i as returned by PSGERQF/PCGERQF. TAU is tied to the distributed matrix A.

WORK   (local workspace/local output) REAL/COMPLEX array, dimension (LWORK)
        On exit, WORK(1) returns the minimal and optimal LWORK.

LWORK   (local or global input) INTEGER
        The dimension of the array WORK.
        LWORK is local input and must be at least
        LWORK $\geq$ MB_A * (MpA0 + NqA0 + MB_A), where
        IROFFA = mod(IA−1, MB_A), ICOFFA = mod(JA−1, NB_A),
        IAROW = INDXG2P(IA, MB_A, MYROW, RSRC_A, NPROW),
        IACOL = INDXG2P(JA, NB_A, MYCOL, CSRC_A, NPCOL),
        MpA0 = NUMROC(M+IROFFA, MB_A, MYROW, IAROW, NPROW),
        NqA0 = NUMROC(N+ICOFFA, NB_A, MYCOL, IACOL, NPCOL),
        INDXG2P and NUMROC are ScaLAPACK tool functions; MYROW, MYCOL, NPROW and NPCOL can be determined by calling the subroutine BLACS_GRIDINFO.
        If LWORK = −1, then LWORK is global input and a workspace query is assumed; the routine only calculates the minimum and optimal size for all work arrays. Each of these values is returned in the first entry of the corresponding work array, and no error message is issued by PXERBLA.

INFO   (global output) INTEGER
        = 0: successful exit

if nq < k, $Q = H_1 H_2 \cdots H_{nq-1}$.

If VECT = 'P', A(IA:*,JA:*) is assumed to have been a k-by-nq matrix:

if k < nq, $P = G_1 G_2 \cdots G_k$;

if k ≥ nq, $P = G_1 G_2 \cdots G_{nq-1}$.

Arguments

VECT (global input) CHARACTER*1
= 'Q': apply Q or Q^H;
= 'P': apply P or P^H.

SIDE (global input) CHARACTER*1
= 'L': apply Q, Q^H, P or P^H from the Left;
= 'R': apply Q, Q^H, P or P^H from the Right.

TRANS (global input) CHARACTER*1
= 'N': No transpose, apply Q or P;
= 'T': Transpose, apply Q^T or P^T (*PSORMBR only*);
= 'C': Conjugate transpose, apply Q^H or P^H (*PCUNMBR only*).

M (global input) INTEGER
The number of rows to be operated on, i.e., the number of rows of the distributed submatrix sub(C). M ≥ 0.

N (global input) INTEGER
The number of columns to be operated on, i.e., the number of columns of the distributed submatrix sub(C). N ≥ 0.

K (global input) INTEGER
If VECT = 'Q', the number of columns in the original distributed matrix reduced by PSGEBRD/PCGEBRD.
If VECT = 'P', the number of rows in the original distributed matrix reduced by PSGEBRD/PCGEBRD.
K ≥ 0.

A (local input) REAL/COMPLEX pointer into the local memory to an array of dimension (LLD_A,LOCc(JA+MIN(NQ,K)-1)) if VECT='Q', and (LLD_A,LOCc(JA+NQ-1)) if VECT = 'P'.
NQ = M if SIDE = 'L', and NQ = N otherwise. The vectors which define the elementary reflectors H_i and G_i, whose products determine the matrices Q and P, as returned by PSGEBRD/PCGEBRD.
If VECT = 'Q', LLD_A ≥ max(1,LOCr(IA+NQ-1));
if VECT = 'P', LLD_A ≥ max(1,LOCr(IA+MIN(NQ,K)-1)).

IA (global input) INTEGER
The row index in the global array A indicating the first row of sub(A).

JA (global input) INTEGER
The column index in the global array A indicating the first column of sub(A).

DESCA (global and local input) INTEGER array, dimension (DLEN_).
The array descriptor for the distributed matrix A.

TAU (local input) REAL/COMPLEX array, dimension LOCc(JA+MIN(NQ,K)-1) if VECT = 'Q', LOCr(IA+MIN(NQ,K)-1) if VECT = 'P'.
TAU(i) must contain the scalar factor of the elementary reflector H_i or G_i, which determines Q or P, as returned by PSGEBRD/PCGEBRD in its array argument TAUQ or TAUP. TAU is tied to the distributed matrix A.

C (local input/local output) REAL/COMPLEX pointer into the local memory to an array of dimension (LLD_C,LOCc(JC+N-1)).
On entry, the local pieces of the distributed matrix sub(C).
On exit, if VECT='Q', sub(C) is overwritten by Q*sub(C) or Q^H*sub(C) or sub(C)*Q^H or sub(C)*Q, if VECT='P', sub(C) is overwritten by P*sub(C) or P^H*sub(C) or sub(C)*P or sub(C)*P^H.

IC (global input) INTEGER
The row index in the global array C indicating the first row of sub(C).

JC (global input) INTEGER
The column index in the global array C indicating the first column of sub(C).

DESCC (global and local input) INTEGER array, dimension (DLEN_).
The array descriptor for the distributed matrix C.

WORK (local workspace/local output) REAL/COMPLEX array, dimension (LWORK)
On exit, WORK(1) returns the minimal and optimal LWORK.

LWORK (local or global input) INTEGER.
The dimension of the array WORK.
LWORK is local input and must be at least

```
IF SIDE = 'L',
  NQ = M;
  if( (VECT = 'Q' and NQ ≥ K) or (VECT ≠ 'Q' and NQ > K) ),
    IAA=IA; JAA=JA; MI=M; NI=N; ICC=IC+1; JCC=JC;
  else
    IAA=IA+1; JAA=JA; MI=M-1; NI=N; ICC=IC+1; JCC=JC;
  end if
ELSE IF SIDE = 'R',
  NQ = N;
  if( (VECT = 'Q' and NQ ≥ K) or (VECT ≠ 'Q' and NQ > K) ),
    IAA=IA; JAA=JA; MI=M; NI=N; ICC=IC; JCC=JC;
  else
    IAA=IA; JAA=JA+1; MI=M; NI=N-1; ICC=IC; JCC=JC+1;
  end if
END IF

IF VECT = 'Q',
  If SIDE = 'L',
    LWORK ≥ max( (NB_A*(NB_A-1))/2, (NqC0 + MpC0)*NB_A ) +
            NB_A * NB_A

  else if SIDE = 'R',
    LWORK ≥ max( (NB_A*(NB_A-1))/2, ( NqC0 + max( NpA0 +
            NUMROC( NUMROC( NI+ICOFFC, NB_A, 0, 0, NPCOL ),
            NB_A, 0, 0, LCMQ ), MpC0 ) )*NB_A ) + NB_A * NB_A

  end if
```

SUBROUTINE PCUNMHR(SIDE, TRANS, M, N, ILO, IHI, A, IA, JA, DESCA,
$ TAU, C, IC, JC, DESCC, WORK, LWORK, INFO)
          CHARACTER   SIDE, TRANS
          INTEGER   IA, IC, IHI, ILO, INFO, JA, JC, LWORK, M, N
          INTEGER   DESCA(*), DESCC(*)
          COMPLEX   A(*), C(*), TAU(*), WORK(*)

Purpose

PSORMHR/PCUNMHR overwrites the general real/complex m-by-n distributed matrix sub(C) = C(IC:IC+M−1,JC:JC+N−1) with

	SIDE = 'L'	SIDE = 'R'	
TRANS = 'N':	$Q*$sub(C)	sub$(C)*Q$	
TRANS = 'T':	Q^T*sub(C)	sub$(C)*Q^T$	*(PSORMHR only)*
TRANS = 'C':	Q^H*sub(C)	sub$(C)*Q^H$	*(PCUNMHR only)*

where Q is a real/complex orthogonal/unitary distributed matrix of order nq, with nq = m if SIDE = 'L' and nq = n if SIDE = 'R'. Q is defined as the product of ihi-ilo elementary reflectors, as returned by PSGEHRD/PCGEHRD:

$$Q = H_{ilo} H_{ilo+1} \cdots H_{ihi-1}.$$

Arguments

SIDE   (global input) CHARACTER*1
        = 'L':   apply Q or Q^H from the Left;
        = 'R':   apply Q or Q^H from the Right.

TRANS   (global input) CHARACTER*1
        = 'N':   No transpose, apply Q
        = 'T':   Transpose, apply Q^T *(PSORMHR only)*
        = 'C':   Conjugate transpose, apply Q^H *(PCUNMHR only)*

M   (global input) INTEGER
        The number of rows to be operated on, i.e., the number of rows of the distributed submatrix sub(C). M ≥ 0.

N   (global input) INTEGER
        The number of columns to be operated on, i.e., the number of columns of the distributed submatrix sub(C). N ≥ 0.

ILO, IHI (global input) INTEGER
        ILO and IHI must have the same values as in the previous call of PSGEHRD/PCGEHRD. Q is equal to the unit matrix except in the distributed submatrix Q(ia+ilo:ia+ihi−1,ia+ilo:ia+ihi−1).
        If SIDE = 'L', then 1 ≤ ILO ≤ IHI ≤ M, if M > 0, and ILO = 1 and IHI = 0, if M = 0;
        if SIDE = 'R', then 1 ≤ ILO ≤ IHI ≤ N, if N > 0, and ILO = 1 and IHI = 0, if N = 0.

A   (local input) REAL/COMPLEX pointer into the local memory to an array of dimension (LLD_A,LOCc(JA+M−1)) if SIDE='L', and (LLD_A,LOCc(JA+N−1)) if SIDE = 'R'.

ELSE IF VECT ≠ 'Q',
   if SIDE = 'L',
      LWORK ≥ max((MB_A*(MB_A−1))/2, (MpC0 + max(MqA0 +
             NUMROC(NUMROC(MI+IROFFC, MB_A, 0, 0, NPROW),
             MB_A, 0, 0, LCMP), NqC0))*MB_A) + MB_A * MB_A
   else if SIDE = 'R',
      LWORK ≥ max((MB_A*(MB_A−1))/2, (MpC0 + NqC0)*MB_A) +
             MB_A * MB_A
   end if
END IF

where LCMP = LCM / NPROW, LCMQ = LCM / NPCOL,
with LCM = ILCM(NPROW, NPCOL),

IROFFA = mod(IAA−1, MB_A), ICOFFA = mod(JAA−1, NB_A),
IAROW = INDXG2P(IAA, MB_A, MYROW, RSRC_A, NPROW),
IACOL = INDXG2P(JAA, NB_A, MYCOL, CSRC_A, NPCOL),
MqA0 = NUMROC(MI+ICOFFA, NB_A, MYCOL, IACOL, NPCOL),
NpA0 = NUMROC(NI+IROFFA, MB_A, MYROW, IAROW, NPROW),
IROFFC = mod(ICC−1, MB_C), ICOFFC = mod(JCC−1, NB_C),
ICROW = INDXG2P(ICC, MB_C, MYROW, RSRC_C, NPROW),
ICCOL = INDXG2P(JCC, NB_C, MYCOL, CSRC_C, NPCOL),
MpC0 = NUMROC(MI+IROFFC, MB_C, MYROW, ICROW, NPROW),
NqC0 = NUMROC(NI+ICOFFC, NB_C, MYCOL, ICCOL, NPCOL),
INDXG2P and NUMROC are ScaLAPACK tool functions; MYROW, MYCOL, NPROW and NPCOL can be determined by calling the subroutine BLACS_GRIDINFO.

If LWORK = −1, then LWORK is global input and a workspace query is assumed; the routine only calculates the minimum and optimal size for all work arrays. Each of these values is returned in the first entry of the corresponding work array, and no error message is issued by PXERBLA.

INFO   (global output) INTEGER
        = 0:   successful exit
        < 0:   If the ith argument is an array and the j-entry had an illegal value, then INFO = −(i*100+j); if the ith argument is a scalar and had an illegal value, then INFO = −i.

PSORMHR/PCUNMHR

SUBROUTINE PSORMHR(SIDE, TRANS, M, N, ILO, IHI, A, IA, JA, DESCA,
$ TAU, C, IC, JC, DESCC, WORK, LWORK, INFO)
          CHARACTER   SIDE, TRANS
          INTEGER   IA, IC, IHI, ILO, INFO, JA, JC, LWORK, M, N
          INTEGER   DESCA(*), DESCC(*)
          REAL   A(*), C(*), TAU(*), WORK(*)

The vectors which define the elementary reflectors, as returned by PSGEHRD/PCGEHRD.

IA   (global input) INTEGER
The row index in the global array A indicating the first row of sub(A).

JA   (global input) INTEGER
The column index in the global array A indicating the first column of sub(A).

DESCA   (global and local input) INTEGER array, dimension (DLEN_).
The array descriptor for the distributed matrix A.

TAU   (local input) REAL/COMPLEX array, dimension LOCc(JA+M−2) if SIDE = 'L', and LOCc(JA+N−2) if SIDE = 'R'.
This array contains the scalar factors TAU(j) of the elementary reflectors H_j as returned by PSGEHRD/PCGEHRD. TAU is tied to the distributed matrix A.

C   (local input/local output) REAL/COMPLEX pointer into the local memory to an array of dimension (LLD_C,LOCc(JC+N−1)).
On entry, the local pieces of the distributed matrix sub(C).
On exit, sub(C) is overwritten by Q*sub(C) or Q^H*sub(C) or sub(C)*Q^H or sub(C)*Q.

IC   (global input) INTEGER
The row index in the global array C indicating the first row of sub(C).

JC   (global input) INTEGER
The column index in the global array C indicating the first column of sub(C).

DESCC   (global and local input) INTEGER array, dimension (DLEN_).
The array descriptor for the distributed matrix C.

WORK   (local workspace/local output) REAL/COMPLEX array, dimension (LWORK)
On exit, WORK(1) returns the minimal and optimal LWORK.

LWORK   (local or global input) INTEGER.
The dimension of the array WORK.
LWORK is local input and must be at least
IAA = IA + ILO; JAA = JA+ILO−1;
IF SIDE = 'L',
  MI = IHI−ILO; NI = N; ICC = IC + ILO; JCC = JC;
  LWORK ≥ max((NB_A*(NB_A−1))/2, (NqC0 + MpC0)*NB_A) + NB_A * NB_A
ELSE IF SIDE = 'R',
  MI = M; NI = IHI−ILO; ICC = IC; JCC = JC + ILO;
  LWORK ≥ max((NB_A*(NB_A−1))/2, (NqC0 + max(NpA0 + NUMROC(NUMROC(NI+ICOFFC, NB_A, 0, 0, NPCOL), NB_A, 0, 0, LCMQ), MpC0)*NB_A) + NB_A * NB_A
END IF
where LCMQ = LCM / NPCOL with LCM = ILCM(NPROW, NPCOL),
IROFFA = mod(IAA−1, MB_A), ICOFFA = mod(JAA−1, NB_A),
IAROW = INDXG2P(IAA, MB_A, MYROW, RSRC_A, NPROW),
NpA0 = NUMROC(NI+IROFFA, MB_A, MYROW, IAROW, NPROW),
IROFFC = mod(ICC−1, MB_C), ICOFFC = mod(JCC−1, NB_C),

ICROW = INDXG2P(ICC, MB_C, MYROW, RSRC_C, NPROW),
ICCOL = INDXG2P(JCC, NB_C, MYCOL, CSRC_C, NPCOL),
MpC0 = NUMROC(MI+IROFFC, MB_C, MYROW, ICROW, NPROW),
NqC0 = NUMROC(NI+ICOFFC, NB_C, MYCOL, ICCOL, NPCOL),
ILCM, INDXG2P and NUMROC are ScaLAPACK tool functions; MYROW, MYCOL, NPROW and NPCOL can be determined by calling the subroutine BLACS_GRIDINFO.
If LWORK = −1, then LWORK is global input and a workspace query is assumed; the routine only calculates the minimum and optimal size for all work arrays. Each of these values is returned in the first entry of the corresponding work array, and no error message is issued by PXERBLA.

INFO   (global output) INTEGER
    = 0: successful exit
    < 0: If the ith argument is an array and the j-entry had an illegal value, then INFO = −(i*100+j), if the ith argument is a scalar and had an illegal value, then INFO = −i.

PSORMLQ/PCUNMLQ

```
SUBROUTINE PSORMLQ( SIDE, TRANS, M, N, K, A, IA, JA, DESCA, TAU,
$                    C, IC, JC, DESCC, WORK, LWORK, INFO )
CHARACTER           SIDE, TRANS
INTEGER             IA, IC, INFO, JA, JC, K, LWORK, M, N
                    DESCA( * ), DESCC( * )
REAL                A( * ), C( * ), TAU( * ), WORK( * )

SUBROUTINE PCUNMLQ( SIDE, TRANS, M, N, K, A, IA, JA, DESCA, TAU,
$                    C, IC, JC, DESCC, WORK, LWORK, INFO )
CHARACTER           SIDE, TRANS
INTEGER             IA, IC, INFO, JA, JC, K, LWORK, M, N
                    DESCA( * ), DESCC( * )
COMPLEX             A( * ), C( * ), TAU( * ), WORK( * )
```

Purpose

PSORMLQ/PCUNMLQ overwrites the general real/complex m-by-n distributed matrix sub(C) = C(IC:IC+M−1,JC:JC+N−1) with

	SIDE = 'L'	SIDE = 'R'
TRANS = 'N':	Q*sub(C)	sub(C)*Q
TRANS = 'T':	Q^T*sub(C)	sub(C)*Q^T	(PSORMLQ only)
TRANS = 'C':	Q^H*sub(C)	sub(C)*Q^H	(PCUNMLQ only)

where Q is a real/complex orthogonal/unitary distributed matrix defined as the product of k elementary reflectors H_i

$$Q = H_k \cdots H_2 H_1 \quad (PSORMLQ)$$

$Q = H_k^H \cdots H_2^H H_1^H$ (PCUNMLQ)

as returned by PSGELQF/PCGELQF. Q is of order m if SIDE = 'L' and of order n if SIDE = 'R'.

Arguments

SIDE   (global input) CHARACTER*1
= 'L': apply Q or Q^H from the Left;
= 'R': apply Q or Q^H from the Right.

TRANS   (global input) CHARACTER*1
= 'N': No transpose, apply Q
= 'T': Transpose, apply Q^T (*PSORMLQ only*)
= 'C': Conjugate transpose, apply Q^H (*PCUNMLQ only*)

M   (global input) INTEGER
The number of rows to be operated on, i.e., the number of rows of the distributed submatrix sub(C). M ≥ 0.

N   (global input) INTEGER
The number of columns to be operated on, i.e, the number of columns of the distributed submatrix sub(C). N ≥ 0.

K   (input) INTEGER
The number of elementary reflectors whose product defines the matrix Q.
If SIDE = 'L', M ≥ K ≥ 0;
if SIDE = 'R', N ≥ K ≥ 0.

A   (local input) REAL/COMPLEX pointer into the local memory to an array of dimension (LLD_A,LOCc(JA+M−1)) if SIDE='L', and (LLD_A,LOCc(JA+N−1)) if SIDE='R', where LLD_A ≥ max(1,LOCr(IA+K−1)); On entry, the ith row must contain the vector which defines the elementary reflector H_i, IA ≤ i ≤ IA+K−1, as returned by PSGELQF/PCGELQF in the K rows of its distributed matrix argument A(IA:IA+K−1,JA:*). A(IA:IA+K−1,JA:*) is modified by the routine but restored on exit.

IA   (global input) INTEGER
The row index in the global array A indicating the first row of sub(A).

JA   (global input) INTEGER
The column index in the global array A indicating the first column of sub(A).

DESCA   (global and local input) INTEGER array, dimension (DLEN_).
The array descriptor for the distributed matrix A.

TAU   (local input) REAL/COMPLEX array, dimension LOCc(IA+K−1).
This array contains the scalar factors TAU(i) of the elementary reflectors H_i as returned by PSGELQF/PCGELQF. TAU is tied to the distributed matrix A.

C   (local input/local output) REAL/COMPLEX pointer into the local memory to an array of dimension (LLD_C,LOCc(JC+N−1)).
On entry, the local pieces of the distributed matrix sub(C).
On exit, sub(C) is overwritten by Q*sub(C) or Q^H*sub(C) or sub(C)*Q^H or sub(C)*Q.

IC   (global input) INTEGER
The row index in the global array C indicating the first row of sub(C).

JC   (global input) INTEGER
The column index in the global array C indicating the first column of sub(C).

DESCC   (global and local input) INTEGER array, dimension (DLEN_).
The array descriptor for the distributed matrix C.

WORK   (local workspace/local output) REAL/COMPLEX array, dimension (LWORK)
On exit, WORK(1) returns the minimal and optimal LWORK.

LWORK   (local or global input) INTEGER
The dimension of the array WORK.
LWORK is local input and must be at least
IF SIDE = 'L',
LWORK ≥ max((MB_A*(MB_A−1))/2, (MpC0 + max(MqA0 + NUMROC(NUMROC(M+IROFFC, MB_A, 0, 0, NPROW), MB_A, 0, 0, LCMP), NqC0))*MB_A) + MB_A * MB_A
ELSE IF SIDE = 'R',
LWORK ≥ max((MB_A*(MB_A−1))/2, (MpC0 + NqC0)*MB_A) + MB_A * MB_A
END IF

where LCMP = LCM / NPROW with LCM = ILCM(NPROW, NPCOL),
IROFFA = mod(IA−1, MB_A), ICOFFA = mod(JA−1, NB_A),
IACOL = INDXG2P(JA, NB_A, MYCOL, CSRC_A, NPCOL),
MqA0 = NUMROC(M+ICOFFA, NB_A, MYCOL, IACOL, NPCOL),
IROFFC = mod(IC−1, MB_C), ICOFFC = mod(JC−1, NB_C),
ICROW = INDXG2P(IC, MB_C, MYROW, RSRC_C, NPROW),
ICCOL = INDXG2P(JC, NB_C, MYCOL, CSRC_C, NPCOL),
MpC0 = NUMROC(M+IROFFC, MB_C, MYROW, ICROW, NPROW),
NqC0 = NUMROC(N+ICOFFC, NB_C, MYCOL, ICCOL, NPCOL),
ILCM, INDXG2P and NUMROC are ScaLAPACK tool functions; MYROW, MYCOL, NPROW and NPCOL can be determined by calling the subroutine BLACS_GRIDINFO.

If LWORK = −1, then LWORK is global input and a workspace query is assumed; the routine only calculates the minimum and optimal size for all work arrays. Each of these values is returned in the first entry of the corresponding work array, and no error message is issued by PXERBLA.

INFO   (global output) INTEGER
= 0: successful exit
< 0: If the ith argument is an array and the j-entry had an illegal value, then INFO = −(i*100+j), if the ith argument is a scalar and had an illegal value, then INFO = −i.

PSORMQL/PCUNMQL

```
SUBROUTINE PSORMQL( SIDE, TRANS, M, N, K, A, IA, JA, DESCA, TAU,
$                   C, IC, JC, DESCC, WORK, LWORK, INFO )
    CHARACTER      SIDE, TRANS
    INTEGER        IA, IC, JC, INFO, JA, JC, K, LWORK, M, N
    REAL           DESCA( * ), DESCC( * )
                   A( * ), C( * ), TAU( * ), WORK( * )

SUBROUTINE PCUNMQL( SIDE, TRANS, M, N, K, A, IA, JA, DESCA, TAU,
$                   C, IC, JC, DESCC, WORK, LWORK, INFO )
    CHARACTER      SIDE, TRANS
    INTEGER        IA, IC, INFO, JA, JC, K, LWORK, M, N
    COMPLEX        DESCA( * ), DESCC( * )
                   A( * ), C( * ), TAU( * ), WORK( * )
```

Purpose

PSORMQL/PCUNMQL overwrites the general real/complex m-by-n distributed matrix sub(C) = C(IC:IC+M−1,JC:JC+N−1) with

	SIDE = 'L'	SIDE = 'R'
TRANS = 'N':	Q*sub(C)	sub(C)*Q
TRANS = 'T':	Q^T*sub(C)	sub(C)*Q^T	(PSORMQL only)
TRANS = 'C':	Q^H*sub(C)	sub(C)*Q^H	(PCUNMQL only)

where Q is a real/complex orthogonal/unitary distributed matrix defined as the product of k elementary reflectors H_i

$$Q = H_k \cdots H_2 H_1$$

as returned by PSGEQLF/PCGEQLF. Q is of order m if SIDE = 'L' and of order n if SIDE = 'R'.

Arguments

SIDE   (global input) CHARACTER*1
        = 'L':   apply Q or Q^H from the Left;
        = 'R':   apply Q or Q^H from the Right.

TRANS   (global input) CHARACTER*1
        = 'N':   No transpose, apply Q
        = 'T':   Transpose, apply Q^T (PSORMQL only)
        = 'C':   Conjugate transpose, apply Q^H (PCUNMQL only)

M   (global input) INTEGER
        The number of rows to be operated on, i.e., the number of rows of the distributed submatrix sub(C). M $\geq$ 0.

N   (global input) INTEGER
        The number of columns to be operated on, i.e, the number of columns of the distributed submatrix sub(C). N $\geq$ 0.

K   (global input) INTEGER
        The number of elementary reflectors whose product defines the matrix Q.
        If SIDE = 'L', M $\geq$ K $\geq$ 0;
        if SIDE = 'R', N $\geq$ K $\geq$ 0.

A   (local input) REAL/COMPLEX pointer into the local memory to an array of dimension (LLD_A,LOCc(JA+K−1)).
        On entry, the jth column must contain the vector which defines the elementary reflector H_j, JA $\leq$ j $\leq$ JA+K−1, as returned by PSGEQLF/PCGEQLF in the K columns of its distributed matrix argument A(IA:*,JA:JA+K−1). A(IA:*,JA:JA+K−1) is modified by the routine but restored on exit.
        If SIDE = 'L', LLD_A $\geq$ max(1, LOCr(IA+M−1));
        if SIDE = 'R', LLD_A $\geq$ max(1, LOCr(IA+N−1)).

IA   (global input) INTEGER
        The row index in the global array A indicating the first row of sub(A).

JA   (global input) INTEGER
        The column index in the global array A indicating the first column of sub(A).

DESCA   (global and local input) INTEGER array, dimension (DLEN_).
        The array descriptor for the distributed matrix A.

TAU   (local input) REAL/COMPLEX array, dimension LOCc(JA+N−1).
        This array contains the scalar factors TAU(j) of the elementary reflectors H_j as returned by PSGEQLF/PCGEQLF. TAU is tied to the distributed matrix A.

C   (local input/local output) REAL/COMPLEX pointer into the local memory to an array of dimension (LLD_C,LOCc(JC+N−1)).
        On entry, the local pieces of the distributed matrix sub(C).
        On exit, sub(C) is overwritten by Q*sub(C) or Q^H*sub(C) or sub(C)*Q or sub(C)*Q^H.

IC   (global input) INTEGER
        The row index in the global array C indicating the first row of sub(C).

JC   (global input) INTEGER
        The column index in the global array C indicating the first column of sub(C).

DESCC   (global and local input) INTEGER array, dimension (DLEN_).
        The array descriptor for the distributed matrix C.

WORK   (local workspace/local output) REAL/COMPLEX array, dimension (LWORK)
        On exit, WORK(1) returns the minimal and optimal LWORK.

Purpose

PSORMQR/PCUNMQR overwrites the general real/complex m-by-n distributed matrix sub(C) = C(IC:IC+M−1,JC:JC+N−1) with

	SIDE = 'L'	SIDE = 'R'	
TRANS = 'N':	Q*sub(C)	sub(C)*Q	
TRANS = 'T':	Q^T*sub(C)	sub(C)*Q^T	(PSORMQR only)
TRANS = 'C':	Q^H*sub(C)	sub(C)*Q^H	(PCUNMQR only)

where Q is a real/complex orthogonal/unitary distributed matrix defined as the product of k elementary reflectors H_i

$$Q = H_1 H_2 \cdots H_k$$

as returned by PSGEQRF/PCGEQRF. Q is of order m if SIDE = 'L' and of order n if SIDE = 'R'.

Arguments

SIDE   (global input) CHARACTER*1
    = 'L':   apply Q or Q^H from the Left;
    = 'R':   apply Q or Q^H from the Right.

TRANS   (global input) CHARACTER*1
    = 'N':   No transpose, apply Q
    = 'T':   Transpose, apply Q^T (PSORMQR only)
    = 'C':   Conjugate transpose, apply Q^H (PCUNMQR only)

M   (global input) INTEGER
    The number of rows to be operated on, i.e., the number of rows of the distributed submatrix sub(C). M ≥ 0.

N   (global input) INTEGER
    The number of columns to be operated on, i.e., the number of columns of the distributed submatrix sub(C). N ≥ 0.

K   (global input) INTEGER
    The number of elementary reflectors whose product defines the matrix Q.
    If SIDE = 'L', M ≥ K ≥ 0;
    if SIDE = 'R', N ≥ K ≥ 0.

A   (local input) REAL/COMPLEX pointer into the local memory to an array of dimension (LLD_A,LOCc(JA+K−1)).
    On entry, the jth column must contain the vector which defines the elementary reflector H_j, JA ≤ j ≤ JA+K−1, as returned by PSGEQRF/PCGEQRF in the K columns of its distributed matrix argument A(IA:*,JA:JA+K−1). A(IA:*,JA:JA+K−1) is modified by the routine but restored on exit.
    If SIDE = 'L', LLD_A ≥ max(1, LOCr(IA+M−1));
    if SIDE = 'R', LLD_A ≥ max(1, LOCr(IA+N−1)).

LWORK   (local or global input) INTEGER
    The dimension of the array WORK.
    LWORK is local input and must be at least
    IF SIDE = 'L',
       LWORK ≥ max((NB_A*(NB_A−1))/2, (NqC0 + MpC0)*NB_A) + NB_A * NB_A
    ELSE IF SIDE = 'R',
       LWORK ≥ max((NB_A*(NB_A−1))/2, (NqC0 + max(NpA0 + NUMROC(NUMROC(N+ICOFFC, NB_A, 0, 0, NPCOL), NB_A, 0, 0, LCMQ), MpC0))*NB_A) + NB_A * NB_A
    END IF

    where LCMQ = LCM / NPCOL with LCM = ILCM(NPROW, NPCOL),
    IROFFA = mod(IA−1, MB_A), ICOFFA = mod(JA−1, NB_A),
    IAROW = INDXG2P(IA, MB_A, MYROW, RSRC_A, NPROW),
    NpA0 = NUMROC(N+IROFFA, MB_A, MYROW, IAROW, NPROW),
    IROFFC = mod(IC−1, MB_C), ICOFFC = mod(JC−1, NB_C),
    ICROW = INDXG2P(IC, MB_C, MYROW, RSRC_C, NPROW),
    ICCOL = INDXG2P(JC, NB_C, MYCOL, CSRC_C, NPCOL),
    MpC0 = NUMROC(M+IROFFC, MB_C, MYROW, ICROW, NPROW),
    NqC0 = NUMROC(N+ICOFFC, NB_C, MYCOL, ICCOL, NPCOL),

    ILCM, INDXG2P and NUMROC are ScaLAPACK tool functions; MYROW, MYCOL, NPROW and NPCOL can be determined by calling the subroutine BLACS_GRIDINFO.

    If LWORK = −1, then LWORK is global input and a workspace query is assumed; the routine only calculates the minimum and optimal size for all work arrays. Each of these values is returned in the first entry of the corresponding work array, and no error message is issued by PXERBLA.

INFO   (global output) INTEGER
    = 0:   successful exit
    < 0:   If the ith argument is an array and the j-entry had an illegal value, then INFO = −(i*100+j), if the ith argument is a scalar and had an illegal value, then INFO = −i.

PSORMQR/PCUNMQR

```
SUBROUTINE PSORMQR( SIDE, TRANS, M, N, K, A, IA, JA, DESCA, TAU,
$                   C, IC, JC, DESCC, WORK, LWORK, INFO )
     CHARACTER         SIDE, TRANS
     INTEGER           IA, IC, INFO, JA, JC, K, LWORK, M, N
     INTEGER           DESCA( * ), DESCC( * )
     REAL              A( * ), C( * ), TAU( * ), WORK( * )

SUBROUTINE PCUNMQR( SIDE, TRANS, M, N, K, A, IA, JA, DESCA, TAU,
$                   C, IC, JC, DESCC, WORK, LWORK, INFO )
     CHARACTER         SIDE, TRANS
     INTEGER           IA, IC, INFO, JA, JC, K, LWORK, M, N
     INTEGER           DESCA( * ), DESCC( * )
     COMPLEX           A( * ), C( * ), TAU( * ), WORK( * )
```

the subroutine BLACS_GRIDINFO.
If LWORK = −1, then LWORK is global input and a workspace query
is assumed; the routine only calculates the minimum and optimal size
for all work arrays. Each of these values is returned in the first entry of the corresponding work array, and no error message is issued by
PXERBLA.

INFO   (global output) INTEGER
= 0:   successful exit
< 0:   If the ith argument is an array and the j-entry had an illegal
     value, then INFO = −(i*100+j); if the ith argument is a scalar
     and had an illegal value, then INFO = −i.

PSORMRQ/PCUNMRQ

```
SUBROUTINE PSORMRQ( SIDE, TRANS, M, N, K, A, IA, JA, DESCA, TAU,
$                    C, IC, JC, DESCC, WORK, LWORK, INFO )
     CHARACTER        SIDE, TRANS
     INTEGER          IA, IC, INFO, JA, JC, K, LWORK, M, N
     INTEGER          DESCA( * ), DESCC( * )
     REAL             A( * ), C( * ), TAU( * ), WORK( * )

SUBROUTINE PCUNMRQ( SIDE, TRANS, M, N, K, A, IA, JA, DESCA, TAU,
$                    C, IC, JC, DESCC, WORK, LWORK, INFO )
     CHARACTER        SIDE, TRANS
     INTEGER          IA, IC, INFO, JA, JC, K, LWORK, M, N
     INTEGER          DESCA( * ), DESCC( * )
     COMPLEX          A( * ), C( * ), TAU( * ), WORK( * )
```

Purpose

PSORMRQ/PCUNMRQ overwrites the general real/complex m-by-n distributed
matrix sub(C) = C(IC:IC+M−1,JC:JC+N−1) with

	SIDE = 'L'	SIDE = 'R'
TRANS = 'N':	Q*sub(C)	sub(C)*Q	
TRANS = 'T':	Q^T*sub(C)	sub(C)*Q^T	*(PSORMRQ only)*
TRANS = 'C':	Q^H*sub(C)	sub(C)*Q^H	*(PCUNMRQ only)*

where Q is a real/complex orthogonal/unitary distributed matrix defined as the
product of k elementary reflectors H_i

$$Q = H_1 H_2 \cdots H_k \quad (PSORMRQ)$$

$$Q = H_1^H H_2^H \cdots H_k^H \quad (PCUNMRQ)$$

as returned by PSGERQF/PCGERQF. Q is of order m if SIDE = 'L' and of order
n if SIDE = 'R'.

IA   (global input) INTEGER
The row index in the global array A indicating the first row of sub(A).

JA   (global input) INTEGER
The column index in the global array A indicating the first column of
sub(A).

DESCA   (global and local input) INTEGER array, dimension (DLEN_).
The array descriptor for the distributed matrix A.

TAU   (local input) REAL/COMPLEX array, dimension LOCc(JA+K−1).
This array contains the scalar factors TAU(j) of the elementary reflectors H_j as returned by PSGEQRF/PCGEQRF. TAU is tied to the distributed matrix A.

C   (local input/local output) REAL/COMPLEX pointer into the local
memory to an array of dimension (LLD_C,LOCc(JC+N−1)).
On entry, the local pieces of the distributed matrix sub(C).
On exit, sub(C) is overwritten by Q*sub(C) or Q^H*sub(C) or
sub(C)*Q^H or sub(C)*Q.

IC   (global input) INTEGER
The row index in the global array C indicating the first row of sub(C).

JC   (global input) INTEGER
The column index in the global array C indicating the first column of
sub(C).

DESCC   (global and local input) INTEGER array, dimension (DLEN_).
The array descriptor for the distributed matrix C.

WORK   (local workspace/local output) REAL/COMPLEX array, dimension
(LWORK)
On exit, WORK(1) returns the minimal and optimal LWORK.

LWORK   (local or global input) INTEGER.
The dimension of the array WORK.
LWORK is local input and must be at least
IF SIDE = 'L',
  LWORK ≥ max((NB_A*(NB_A−1))/2, (NqC0 + MpC0)*NB_A) +
    NB_A * NB_A
ELSE IF SIDE = 'R',
  LWORK ≥ max((NB_A*(NB_A−1))/2, (NqC0 + max(NpA0 +
    NUMROC(NUMROC(N+ICOFFC, NB_A, 0, 0, NPCOL),
    NB_A, 0, 0, LCMQ), MpC0))*NB_A) + NB_A * NB_A
END IF
where LCMQ = LCM / NPCOL with LCM = ILCM(NPROW, NPCOL),
IROFFA = mod(IA−1, MB_A), ICOFFA = mod(JA−1, NB_A),
IAROW = INDXG2P(IA, MB_A, MYROW, RSRC_A, NPROW),
NpA0 = NUMROC(N+IROFFA, MB_A, MYROW, IAROW, NPROW),
IROFFC = mod(IC−1, MB_C), ICOFFC = mod(JC−1, NB_C),
ICROW = INDXG2P(IC, MB_C, MYROW, RSRC_C, NPROW),
ICCOL = INDXG2P(JC, NB_C, MYCOL, CSRC_C, NPCOL),
MpC0 = NUMROC(M+IROFFC, MB_C, MYROW, ICROW, NPROW),
NqC0 = NUMROC(N+ICOFFC, NB_C, MYCOL, ICCOL, NPCOL),
ILCM, INDXG2P and NUMROC are ScaLAPACK tool functions; MY-
ROW, MYCOL, NPROW and NPCOL can be determined by calling

Arguments

SIDE (global input) CHARACTER*1
= 'L': apply Q or Q^H from the Left;
= 'R': apply Q or Q^H from the Right.

TRANS (global input) CHARACTER*1
= 'N': No transpose, apply Q
= 'T': Transpose, apply Q^T (PSORMRQ only)
= 'C': Conjugate transpose, apply Q^H (PCUNMRQ only)

M (global input) INTEGER
The number of rows to be operated on, i.e., the number of rows of the distributed submatrix sub(C). M ≥ 0.

N (global input) INTEGER
The number of columns to be operated on, i.e., the number of columns of the distributed submatrix sub(C). N ≥ 0.

K (global input) INTEGER
The number of elementary reflectors whose product defines the matrix Q.
If SIDE = 'L', M ≥ K ≥ 0;
if SIDE = 'R', N ≥ K ≥ 0.

A (local input) REAL/COMPLEX pointer into the local memory to an array of dimension (LLD_A,LOCc(JA+M−1)) if SIDE='L', and (LLD_A,LOCc(JA+N−1)) if SIDE='R', where LLD_A ≥ max(1,LOCr(IA+K−1)).
On entry, the ith row must contain the vector which defines the elementary reflector H_i, IA ≤ i ≤ IA+K−1, as returned by PSGERQF/PCGERQF in the K rows of its distributed matrix argument A(IA:IA+K−1,JA:*). A(IA:IA+K−1,JA:*) is modified by the routine but restored on exit.

IA (global input) INTEGER
The row index in the global array A indicating the first row of sub(A).

JA (global input) INTEGER
The column index in the global array A indicating the first column of sub(A).

DESCA (global and local input) INTEGER array, dimension (DLEN_).
The array descriptor for the distributed matrix A.

TAU (local input) REAL/COMPLEX array, dimension LOCc(IA+K−1).
This array contains the scalar factors TAU(i) of the elementary reflectors H_i, as returned by PSGERQF/PCGERQF. TAU is tied to the distributed matrix A.

C (local input/local output) REAL/COMPLEX pointer into the local memory to an array of dimension (LLD_C,LOCc(JC+N−1)).
On entry, sub(C) is the local pieces of the distributed matrix sub(C).
On exit, sub(C) is overwritten by Q*sub(C) or Q^H*sub(C) or sub(C)*Q^H or sub(C)*Q.

IC (global input) INTEGER
The row index in the global array C indicating the first row of sub(C).

JC (global input) INTEGER
The column index in the global array C indicating the first column of sub(C).

DESCC (global and local input) INTEGER array, dimension (DLEN_).
The array descriptor for the distributed matrix C.

WORK (local workspace/local output) REAL/COMPLEX array, dimension (LWORK)
On exit, WORK(1) returns the minimal and optimal LWORK.

LWORK (local or global input) INTEGER
The dimension of the array WORK.
LWORK is local input and must be at least
IF SIDE = 'L',
  LWORK ≥ max((MB_A*(MB_A−1))/2, (MpC0 + max(MqA0 + NUMROC(NUMROC(M+IROFFC, MB_A, 0, 0, NPROW), MB_A, 0, 0, LCMP), NqC0))*MB_A) + MB_A * MB_A
ELSE IF SIDE = 'R',
  LWORK ≥ max((MB_A*(MB_A−1))/2, (MpC0 + NqC0)*MB_A) + MB_A * MB_A
END IF

where LCMP = LCM / NPROW with LCM = ILCM(NPROW, NPCOL),
IROFFA = mod(IA−1, MB_A), ICOFFA = mod(JA−1, NB_A),
IACOL = INDXG2P(JA, NB_A, MYCOL, CSRC_A, NPCOL),
MqA0 = NUMROC(M+ICOFFA, NB_A, MYCOL, IACOL, NPCOL),
IROFFC = mod(IC−1, MB_C), ICOFFC = mod(JC−1, NB_C),
ICROW = INDXG2P(IC, MB_C, MYROW, RSRC_C, NPROW),
ICCOL = INDXG2P(JC, NB_C, MYCOL, CSRC_C, NPCOL),
MpC0 = NUMROC(M+IROFFC, MB_C, MYROW, ICROW, NPROW),
NqC0 = NUMROC(N+ICOFFC, NB_C, MYCOL, ICCOL, NPCOL),
ILCM, INDXG2P and NUMROC are ScaLAPACK tool functions; MY-ROW, MYCOL, NPROW and NPCOL can be determined by calling the subroutine BLACS_GRIDINFO.

If LWORK = −1, then LWORK is global input and a workspace query is assumed; the routine only calculates the minimum and optimal size for all work arrays. Each of these values is returned in the first entry of the corresponding work array, and no error message is issued by PXERBLA.

INFO (global output) INTEGER
= 0: successful exit
< 0: If the ith argument is an array and the j-entry had an illegal value, then INFO = −(i*100+j), if the ith argument is a scalar and had an illegal value, then INFO = −i.

PSORMRZ/PCUNMRZ

```
SUBROUTINE PSORMRZ( SIDE, TRANS, M, N, K, L, A, IA, JA, DESCA,
$                   TAU, C, IC, JC, DESCC, WORK, LWORK, INFO )
      CHARACTER       SIDE, TRANS
      INTEGER         IA, IC, INFO, JA, JC, K, L, LWORK, M, N
      INTEGER         DESCA( * ), DESCC( * )
      REAL            A( * ), C( * ), TAU( * ), WORK( * )

SUBROUTINE PCUNMRZ( SIDE, TRANS, M, N, K, L, A, IA, JA, DESCA,
$                   TAU, C, IC, JC, DESCC, WORK, LWORK, INFO )
      CHARACTER       SIDE, TRANS
      INTEGER         IA, IC, INFO, JA, JC, K, L, LWORK, M, N
      INTEGER         DESCA( * ), DESCC( * )
      COMPLEX         A( * ), C( * ), TAU( * ), WORK( * )
```

Purpose

PSORMRZ/PCUNMRZ overwrites the general real/complex m-by-n distributed matrix sub(C) = C(IC:IC+M−1,JC:JC+N−1) with

	SIDE = 'L'	SIDE = 'R'	
TRANS = 'N':	Q*sub(C)	sub(C)*Q	
TRANS = 'T':	Q^T*sub(C)	sub(C)*Q^T	(PSORMRZ only)
TRANS = 'C':	Q^H*sub(C)	sub(C)*Q^H	(PCUNMRZ only)

where Q is a real/complex orthogonal/unitary distributed matrix defined as the product of k elementary reflectors H_i

$Q = H_1 H_2 \cdots H_k$ (PSORMRZ)

$Q = H_1{}^H H_2{}^H \cdots H_k{}^H$ (PCUNMRZ)

as returned by PSTZRZF/PCTZRZF. Q is of order m if SIDE = 'L' and of order n if SIDE = 'R'.

Arguments

SIDE   (global input) CHARACTER*1
         = 'L': apply Q or Q^H from the Left;
         = 'R': apply Q or Q^H from the Right.

TRANS   (global input) CHARACTER*1
         = 'N': No transpose, apply Q
         = 'T': Transpose, apply Q^T (PSORMRZ only)
         = 'C': Conjugate transpose, apply Q^H (PCUNMRZ only)

M   (global input) INTEGER
         The number of rows to be operated on, i.e., the number of rows of the
         distributed submatrix sub(C). M ≥ 0.

N   (global input) INTEGER
         The number of columns to be operated on, i.e., the number of columns
         of the distributed submatrix sub(C). N ≥ 0.

K   (global input) INTEGER
         The number of elementary reflectors whose product defines the matrix
         Q.
         If SIDE = 'L', M ≥ K ≥ 0;
         if SIDE = 'R', N ≥ K ≥ 0.

L   (global input) INTEGER
         The columns of the distributed submatrix sub(A) containing the mean-
         ingful part of the Householder reflectors.
         If SIDE = 'L', M ≥ L ≥ 0, if SIDE = 'R', N ≥ L ≥ 0.

A   (local input) REAL/COMPLEX pointer into the local memory
         to an array of dimension (LLD_A,LOCc(JA+M−1)) if SIDE='L',
         and (LLD_A,LOCc(JA+N−1)) if SIDE='R', where LLD_A ≥
         max(1,LOCr(IA+K−1)).
         On entry, the ith row must contain the vector which defines
         the elementary reflector H_i, IA ≤ i ≤ IA+K−1, as returned by
         PSTZRZF/PCTZRZF in the K rows of its distributed matrix argument
         A(IA:IA+K−1,JA:*). A(IA:IA+K−1,JA:*) is modified by the routine
         but restored on exit.

IA   (global input) INTEGER
         The row index in the global array A indicating the first row of sub(A).

JA   (global input) INTEGER
         The column index in the global array A indicating the first column of
         sub(A).

DESCA   (global and local input) INTEGER array, dimension (DLEN_).
         The array descriptor for the distributed matrix A.

TAU   (local input) REAL/COMPLEX array, dimension LOCc(IA+K-1).
         This array contains the scalar factors TAU(i) of the elementary reflectors
         H_i as returned by PSTZRZF/PCTZRZF. TAU is tied to the distributed
         matrix A.

C   (local input/local output) REAL/COMPLEX pointer into the local
         memory to an array of dimension (LLD_C,LOCc(JC+N−1)).
         On entry, the local pieces of the distributed matrix sub(C).
         On exit, sub(C) is overwritten by Q*sub(C) or Q^H*sub(C) or
         sub(C)*Q^H or sub(C)*Q.

IC   (global input) INTEGER
         The row index in the global array C indicating the first row of sub(C).

JC   (global input) INTEGER
         The column index in the global array C indicating the first column of
         sub(C).

DESCC   (global and local input) INTEGER array, dimension (DLEN_).
         The array descriptor for the distributed matrix C.

SUBROUTINE PCUNMTR(SIDE, UPLO, TRANS, M, N, A, IA, JA, DESCA,
$ TAU, C, IC, JC, DESCC, WORK, LWORK, INFO)
        CHARACTER   SIDE, TRANS, UPLO
        INTEGER   IA, IC, INFO, JA, JC, LWORK, M, N
        INTEGER   DESCA(*), DESCC(*)
        COMPLEX   A(*), C(*), TAU(*), WORK(*)

Purpose

PSORMTR/PCUNMTR overwrites the general real/complex m-by-n distributed matrix sub(C) = C(IC:IC+M−1,JC:JC+N−1) with

	SIDE = 'L'	SIDE = 'R'
TRANS = 'N':	Q*sub(C)	sub(C)*Q
TRANS = 'T':	Q^T*sub(C)	sub(C)*Q^T (PSORMTR only)
TRANS = 'C':	Q^H*sub(C)	sub(C)*Q^H (PCUNMTR only)

where Q is a real/complex orthogonal/unitary distributed matrix of order nq, with nq = m if SIDE = 'L' and nq = n if SIDE = 'R'. Q is defined as the product of nq−1 elementary reflectors, as returned by PSSYTRD/PCHETRD:

if UPLO = 'U', $Q = H_{nq-1} \cdots H_2 H_1$;

if UPLO = 'L', $Q = H_1 H_2 \cdots H_{nq-1}$.

Arguments

SIDE   (global input) CHARACTER*1
        = 'L': apply Q or Q^H from the Left;
        = 'R': apply Q or Q^H from the Right.

UPLO   (global input) CHARACTER*1
        = 'U': Upper triangle of A(IA:*,JA:*) contains elementary reflectors from PSSYTRD/PCHETRD;
        = 'L': Lower triangle of A(IA:*,JA:*) contains elementary reflectors from PSSYTRD/PCHETRD.

TRANS   (global input) CHARACTER*1
        = 'N': No transpose, apply Q
        = 'T': Transpose, apply Q^T (PSORMTR only)
        = 'C': Conjugate transpose, apply Q^H (PCUNMTR only)

M   (global input) INTEGER
        The number of rows to be operated on, i.e., the number of rows of the distributed submatrix sub(C). M ≥ 0.

N   (global input) INTEGER
        The number of columns to be operated on, i.e., the number of columns of the distributed submatrix sub(C). N ≥ 0.

A   (local input) REAL/COMPLEX pointer into the local memory to an array of dimension (LLD_A,LOCc(JA+M−1)) if SIDE=L, or (LLD_A,LOCc(JA+N−1)) if SIDE = 'R'. The vectors which define the elementary reflectors, as returned by

WORK   (local workspace/local output) REAL/COMPLEX array, dimension (LWORK)
        On exit, WORK(1) returns the minimal and optimal LWORK.

LWORK   (local or global input) INTEGER
        The dimension of the array WORK.
        LWORK is local input and must be at least
        IF SIDE = 'L',
          LWORK ≥ max((MB_A*(MB_A−1))/2, (MpC0 + max(MqA0 + NUMROC(NUMROC(M+IROFFC, MB_A, 0, 0, NPROW), MB_A, 0, 0, LCMP), NqC0))*MB_A) + MB_A * MB_A
        ELSE IF SIDE = 'R',
          LWORK ≥ max((MB_A*(MB_A−1))/2, (MpC0 + NqC0)*MB_A) + MB_A * MB_A
        END IF
        where LCMP = LCM / NPROW with LCM = ILCM(NPROW, NPCOL),
        IROFFA = mod(IA−1, MB_A), ICOFFA = mod(JA−1, NB_A),
        IACOL = INDXG2P(JA, NB_A, MYCOL, CSRC_A, NPCOL),
        MqA0 = NUMROC(M+ICOFFA, NB_A, MYCOL, IACOL, NPCOL),
        IROFFC = mod(IC−1, MB_C), ICOFFC = mod(JC−1, NB_C),
        ICROW = INDXG2P(IC, MB_C, MYROW, RSRC_C, NPROW),
        ICCOL = INDXG2P(JC, NB_C, MYCOL, CSRC_C, NPCOL),
        MpC0 = NUMROC(M+IROFFC, MB_C, MYROW, ICROW, NPROW),
        NqC0 = NUMROC(N+ICOFFC, NB_C, MYCOL, ICCOL, NPCOL),
        ILCM, INDXG2P and NUMROC are ScaLAPACK tool functions; MYROW, MYCOL, NPROW and NPCOL can be determined by calling the subroutine BLACS_GRIDINFO.
        If LWORK = −1, then LWORK is global input and a workspace query is assumed; the routine only calculates the minimum and optimal size for all work arrays. Each of these values is returned in the first entry of the corresponding work array, and no error message is issued by PXERBLA.

INFO   (global output) INTEGER
        = 0: successful exit
        < 0: If the ith argument is an array and the j-entry had an illegal value, then INFO = −(i*100+j), if the ith argument is a scalar and had an illegal value, then INFO = −i.

PSORMTR/PCUNMTR

SUBROUTINE PSORMTR(SIDE, UPLO, TRANS, M, N, A, IA, JA, DESCA,
$ TAU, C, IC, JC, DESCC, WORK, LWORK, INFO)
        CHARACTER   SIDE, TRANS, UPLO
        INTEGER   IA, IC, INFO, JA, JC, LWORK, M, N
        INTEGER   DESCA(*), DESCC(*)
        REAL   A(*), C(*), TAU(*), WORK(*)

PSSYTRD/PCHETRD.
If SIDE = 'L', LLD_A $\geq$ max(1,LOCr(IA+M−1));
if SIDE = 'R', LLD_A $\geq$ max(1,LOCr(IA+N−1)).

IA   (global input) INTEGER
      The row index in the global array A indicating the first row of sub(A).

JA   (global input) INTEGER
      The column index in the global array A indicating the first column of sub(A).

DESCA (global and local input) INTEGER array, dimension (DLEN_).
      The array descriptor for the distributed matrix A.

TAU   (local input) REAL/COMPLEX array, dimension LTAU,
      where if SIDE = 'L' and UPLO = 'U', LTAU = LOCc(M_A),
      if SIDE = 'L' and UPLO = 'L', LTAU = LOCc(JA+M−2),
      if SIDE = 'R' and UPLO = 'U', LTAU = LOCc(N_A),
      if SIDE = 'R' and UPLO = 'L', LTAU = LOCc(JA+N−2).
      TAU(i) must contain the scalar factor of the elementary reflector H_i,
      as returned by PSSYTRD/PCHETRD. TAU is tied to the distributed
      matrix A.

C   (local input/local output) REAL/COMPLEX pointer into the local
      memory to an array of dimension (LLD_C,LOCc(JC+N−1)).
      On entry, the local pieces of the distributed matrix sub(C).
      On exit, sub(C) is overwritten by Q*sub(C) or Q^H*sub(C) or
      sub(C)*Q^H or sub(C)*Q.

IC   (global input) INTEGER
      The row index in the global array C indicating the first row of sub(C).

JC   (global input) INTEGER
      The column index in the global array C indicating the first column of sub(C).

DESCC (global and local input) INTEGER array, dimension (DLEN_).
      The array descriptor for the distributed matrix C.

WORK (local workspace/local output) REAL/COMPLEX array, dimension
      (LWORK)
      On exit, WORK(1) returns the minimal and optimal LWORK.

LWORK (local or global input) INTEGER
      The dimension of the array WORK.
      LWORK is local input and must be at least
      IF UPLO = 'U',
        IAA = IA, JAA = JA+1, ICC = IC, JCC = JC;
      ELSE IF UPLO = 'L',
        IAA = IA+1, JAA = JA;
        if SIDE = 'L',
          ICC = IC+1; JCC = JC;
        else
          ICC = IC; JCC = JC+1;
        end if
      END IF

      IF SIDE = 'L',

MI = M-1; NI = N;
LWORK $\geq$ max((NB_A*(NB_A−1))/2, (NqC0 + MpC0)*NB_A) +
              NB_A * NB_A
ELSE IF SIDE = 'R',
  MI = M; NI = N−1;
  LWORK $\geq$ max((NB_A*(NB_A−1))/2, (NqC0 + max(NpA0 +
              NUMROC(N1+ICOFFC, NB_A, 0, 0, NPCOL),
              NB_A, 0, 0, LCMQ), MpC0))*NB_A) + NB_A * NB_A
END IF
where LCMQ = LCM / NPCOL with LCM = ICLM(NPROW, NPCOL),
IROFFA = mod(IAA−1, MB_A), ICOFFA = mod(JAA−1, NB_A),
IAROW = INDXG2P(IAA, MB_A, MYROW, RSRC_A, NPROW),
NpA0 = NUMROC(N1+IROFFA, MB_A, MYROW, IAROW, NPROW),
IROFFC = mod(ICC−1, MB_C), ICOFFC = mod(JCC−1, NB_C),
ICROW = INDXG2P(ICC, MB_C, MYROW, RSRC_C, NPROW),
ICCOL = INDXG2P(JCC, NB_C, MYCOL, CSRC_C, NPCOL),
MpC0 = NUMROC(MI+IROFFC, MB_C, MYROW, ICROW, NPROW),
NqC0 = NUMROC(N1+ICOFFC, NB_C, MYCOL, ICCOL, NPCOL),
ILCM, INDXG2P and NUMROC are ScaLAPACK tool functions; MY-
ROW, MYCOL, NPROW and NPCOL can be determined by calling
the subroutine BLACS_GRIDINFO.
If LWORK = −1, then LWORK is global input and a workspace query
is assumed; the routine only calculates the minimum and optimal size
for all work arrays. Each of these values is returned in the first en-
try of the corresponding work array, and no error message is issued by
PXERBLA.

INFO (global output) INTEGER
      = 0: successful exit
      < 0: If the ith argument is an array and the j-entry had an illegal
            value, then INFO = −(i*100+j), if the ith argument is a scalar
            and had an illegal value, then INFO = −i.

PSPBSV/PCPBSV

```
SUBROUTINE PSPBSV( UPLO, N, BW, NRHS, A, JA, DESCA, B, IB, DESCB,
$                  WORK, LWORK, INFO )
   CHARACTER        UPLO
   INTEGER          BW, IB, INFO, JA, LWORK, N, NRHS
   INTEGER          DESCA( * ), DESCB( * )
   REAL             A( * ), B( * ), WORK( * )

SUBROUTINE PCPBSV( UPLO, N, BW, NRHS, A, JA, DESCA, B, IB, DESCB,
$                  WORK, LWORK, INFO )
   CHARACTER        UPLO
   INTEGER          BW, IB, INFO, JA, LWORK, N, NRHS
   INTEGER          DESCA( * ), DESCB( * )
   COMPLEX          A( * ), B( * ), WORK( * )
```

Purpose

PSPBSV/PCPBSV computes the solution to a real/complex system of linear equations

$$A(1:N, JA:JA+N-1)* X = B(IB:IB+N-1, 1:NRHS)$$

where $A(1:N, JA:JA+N-1)$ is an n-by-n symmetric positive definite band distributed matrix and X and B are n-by-nrhs distributed matrices.

The Cholesky decomposition is used to factor $A(1:N, JA:JA+N-1)$ as
$A(1:N, JA:JA+N-1) = P*U^H*U*P^T$, if UPLO = 'U', or
$A(1:N, JA:JA+N-1) = P*L*L^H*P^T$, if UPLO = 'L',
where P is a permutation matrix and U and L are banded upper and lower triangular matrices respectively. The factored form of A is then used to solve the system of equations
$A(1:N, JA:JA+N-1)* X = B(IB:IB+N-1, 1:NRHS)$.

Arguments

UPLO   (global input) CHARACTER*1
        = 'U':   Upper triangle of $A(1:N, JA:JA+N-1)$ is stored;
        = 'L':   Lower triangle of $A(1:N, JA:JA+N-1)$ is stored.

N   (global input) INTEGER
        The number of rows and columns to be operated on, i.e., the order of the distributed submatrix $A(1:N, JA:JA+N-1)$. $N \geq 0$.

BW   (global input) INTEGER
        The number of superdiagonals of the distributed matrix if UPLO = 'U', or the number of subdiagonals if UPLO = 'L'. $BW \geq 0$.

NRHS   (global input) INTEGER
        The number of right hand sides, i.e., the number of columns of the distributed submatrix $B(IB:IB+N-1, 1:NRHS)$. $NRHS \geq 0$.

A   (local input/local output) REAL/COMPLEX pointer into the local memory to an array of dimension (LLD_A, LOCc(JA+N−1)).
        On entry, this array contains the local pieces of the upper or lower triangle of the symmetric band distributed matrix A.
        On exit, if INFO = 0, the permuted triangular factor U or L from the Cholesky factorization $A(1:N, JA:JA+N-1) = P*U^H*U*P^T$ or $A(1:N, JA:JA+N-1) = P*L*L^H*P^T$ of the band matrix A, in the same storage format as A. Note, the resulting factorization is *not* the same factorization as returned from LAPACK. Additional permutations are performed on the matrix for the sake of parallelism.

JA   (global input) INTEGER
        The index in the global array A that points to the start of the matrix to be operated on (which may be either all of A or a submatrix of A).

DESCA   (global and local input) INTEGER array, dimension (DLEN_).
        The array descriptor for the distributed matrix A.
        If DESCA(DTYPE_)=501 then DLEN_ $\geq$ 7,
        else if DESCA(DTYPE_)=1 then DLEN_ $\geq$ 9.

B   (local input/local output) REAL/COMPLEX pointer into the local memory to an array of dimension (LLD_B, LOCc(NRHS)).
        On entry, this array contains the the local pieces of the n-by-nrhs right hand side matrix B(IB:IB+N−1, 1:NRHS). On exit, if INFO = 0, this array contains the local pieces of the n-by-nrhs solution distributed matrix X.

IB   (global input) INTEGER
        The row index in the global array B indicating the first row of sub(B).

DESCB   (global and local input) INTEGER array, dimension (DLEN_).
        The array descriptor for the distributed matrix B.
        If DESCB(DTYPE_)=502 then DLEN_ $\geq$ 7,
        else if DESCB(DTYPE_)=1 then DLEN_ $\geq$ 9.

WORK   (local workspace/local output) REAL/COMPLEX array, dimension (LWORK)
        On exit, WORK(1) returns the minimal and optimal LWORK.

LWORK   (local or global input) INTEGER
        The dimension of the array WORK.
        LWORK is local input and must be at least
        LWORK $\geq$ (NB+2*BW)*BW+max((BW*NRHS),BW*BW).
        If LWORK = −1, then LWORK is global input and a workspace query is assumed; the routine only calculates the minimum and optimal size for all work arrays. Each of these values is returned in the first entry of the corresponding work array, and no error message is issued by PXERBLA.

INFO   (global output) INTEGER
        = 0:   successful exit
        < 0:   If the ith argument is an array and the j-entry had an illegal value, then INFO = −(i*100+j), if the ith argument is a scalar and had an illegal value, then INFO = −i.
        > 0:   If INFO = K $\leq$ NPROCS, the submatrix stored on processor INFO−NPROCS and factored locally was not positive definite, and the factorization was not completed.
                If INFO = K > NPROCS, the submatrix stored on processor INFO−NPROCS representing interactions with other processors was not nonsingular, and the factorization was not completed.

PSPBTRF/PCPBTRF

```
SUBROUTINE PSPBTRF( UPLO, N, BW, A, JA, DESCA, AF, LAF, WORK,
$                    LWORK, INFO )
CHARACTER        UPLO
INTEGER          BW, INFO, JA, LAF, LWORK, N
INTEGER          DESCA( * )
REAL             A( * ), AF( * ), WORK( * )

SUBROUTINE PCPBTRF( UPLO, N, BW, A, JA, DESCA, AF, LAF, WORK,
$                    LWORK, INFO )
CHARACTER        UPLO
INTEGER          BW, INFO, JA, LAF, LWORK, N
INTEGER          DESCA( * )
COMPLEX          A( * ), AF( * ), WORK( * )
```

Purpose

PSPBTRF/PCPBTRF computes the Cholesky factorization of an n-by-n real/complex symmetric/Hermitian positive definite banded distributed matrix $A(1:N, JA:JA+N-1)$.

The resulting factorization is *not* the same factorization as returned from LAPACK. Additional permutations are performed on the matrix for the sake of parallelism.

The factorization has the form

$A(1:N, JA:JA+N-1) = P*U^H*U*P^T$, if UPLO = 'U', or
$A(1:N, JA:JA+N-1) = P*L*L^H*P^T$, if UPLO = 'L',

where P is a permutation matrix and U and L are banded upper and lower triangular matrices respectively.

Arguments

UPLO   (global input) CHARACTER*1
= 'U':   Upper triangle of $A(1:N, JA:JA+N-1)$ is stored;
= 'L':   Lower triangle of $A(1:N, JA:JA+N-1)$ is stored.

N   (global input) INTEGER
The number of rows and columns to be operated on, i.e., the order of the distributed submatrix $A(1:N, JA:JA+N-1)$. $N \geq 0$.

BW   (global input) INTEGER
The number of superdiagonals of the distributed matrix if UPLO = 'U', or the number of subdiagonals if UPLO = 'L'. $BW \geq 0$.

A   (local input/local output) REAL/COMPLEX pointer into the local memory to an array of dimension (LLD_A, LOCc(JA+N-1)).
On entry, this array contains the local pieces of the upper or lower triangle of the symmetric band distributed matrix $A(1:N, JA:JA+N-1)$ to be factored.
On exit, if INFO = 0, the permuted triangular factor U or L from the Cholesky factorization $A(1:N, JA:JA+N-1) = P*U^H*U*P^T$ or $A(1:N, JA:JA+N-1) = P*L*L^H*P^T$ of the band matrix A. Note, the resulting

factorization is *not* the same factorization as returned from LAPACK. Additional permutations are performed on the matrix for the sake of parallelism.

JA   (global input) INTEGER
The index in the global array A that points to the start of the matrix to be operated on (which may be either all of A or a submatrix of A).

DESCA   (global and local input) INTEGER array, dimension (DLEN_)
The array descriptor for the distributed matrix A.
If DESCA(DTYPE_)=501 then DLEN_$\geq$7,
else if DESCA(DTYPE_)=1 then DLEN_$\geq$9.

AF   (local output) REAL/COMPLEX array, dimension (LAF)
Auxiliary Fillin space. Fillin is created during the factorization routine PSPBTRF/PCPBTRF and this is stored in AF. If a linear system is to be solved using PSPBTRS/PCPBTRS after the factorization routine, AF *must not be altered*.

LAF   (local input) INTEGER
The dimension of the array AF.
$LAF \geq (NB+2*BW)*BW$.
If LAF is not large enough, an error code will be returned and the minimum acceptable size will be returned in AF(1).

WORK   (local workspace/local output) REAL/COMPLEX array, dimension (LWORK)
On exit, WORK(1) returns the minimal and optimal LWORK.

LWORK   (local or global input) INTEGER
The dimension of the array WORK.
LWORK is local input and must be at least
$LWORK \geq BW*BW$.
If LWORK = −1, then LWORK is global input and a workspace query is assumed; the routine only calculates the minimum and optimal size for all work arrays. Each of these values is returned in the first entry of the corresponding work array, and no error message is issued by PXERBLA.

INFO   (global output) INTEGER
= 0:   successful exit
< 0:   If the ith argument had an illegal value, then INFO = −i.
If the ith argument is an array and the j-entry had an illegal value, then INFO = −(i*100+j), if the ith argument is a scalar and had an illegal value, then INFO = −i.
> 0:   If INFO = K $\leq$ NPROCS, the submatrix stored on processor INFO−NPROCS and factored locally was not positive definite, and the factorization was not completed.
If INFO = K > NPROCS, the submatrix stored on processor INFO−NPROCS representing interactions with other processors was not nonsingular, and the factorization was not completed.

PSPBTRS/PCPBTRS

```
SUBROUTINE PSPBTRS( UPLO, N, BW, NRHS, A, JA, DESCA, B, IB, DESCB,
$                   AF, LAF, WORK, LWORK, INFO )

CHARACTER          UPLO
INTEGER            BW, IB, INFO, JA, LAF, LWORK, N, NRHS
INTEGER            DESCA( * ), DESCB( * )
REAL               A( * ), AF( * ), B( * ), WORK( * )

SUBROUTINE PCPBTRS( UPLO, N, BW, NRHS, A, JA, DESCA, B, IB, DESCB,
$                   AF, LAF, WORK, LWORK, INFO )

CHARACTER          UPLO
INTEGER            BW, IB, INFO, JA, LAF, LWORK, N, NRHS
INTEGER            DESCA( * ), DESCB( * )
COMPLEX            A( * ), AF( * ), B( * ), WORK( * )
```

Purpose

PSPBTRS/PCPBTRS solves a system of linear equations $A(1:N, JA:JA+N-1)*X = B(IB:IB+N-1, 1:NRHS)$ with a symmetric positive definite banded distributed matrix $A(1:N, JA:JA+N-1)$ using the Cholesky factorization $A(1:N, JA:JA+N-1) = P*U^H*U*P^T$, or $A(1:N, JA:JA+N-1) = P*L*L^H*P^T$ computed by PSPBTRF/PCPBTRF.

Arguments

UPLO (global input) CHARACTER*1
= 'U': Upper triangle of $A(1:N, JA:JA+N-1)$ is stored;
= 'L': Lower triangle of $A(1:N, JA:JA+N-1)$ is stored.

N (global input) INTEGER
The number of rows and columns to be operated on, i.e., the order of the distributed submatrix $A(1:N, JA:JA+N-1)$. $N \geq 0$.

BW (global input) INTEGER
The number of superdiagonals of the distributed matrix if UPLO = 'U', or the number of subdiagonals if UPLO = 'L'. $BW \geq 0$.

NRHS (global input) INTEGER
The number of right hand sides, i.e., the number of columns of the distributed submatrix $B(IB:IB+N-1, 1:NRHS)$. $NRHS \geq 0$.

A (local input) REAL/COMPLEX pointer into the local memory to an array of dimension (LLD_A, LOCc(JA+N-1)).
The permuted triangular factor U or L from the Cholesky factorization $A(1:N, JA:JA+N-1) = P*U^H*U*P^T$ or $A(1:N, JA:JA+N-1) = P*L*L^H*P^T$ of the band matrix A, as returned by PSPBTRF/PCPBTRF.

JA (global input) INTEGER
The index in the global array A that points to the start of the matrix to be operated on (which may be either all of A or a submatrix of A).

DESCA (global and local input) INTEGER array, dimension (DLEN_).
The array descriptor for the distributed matrix A.
If DESCA(DTYPE_)=501 then DLEN_ $\geq$ 7,
else if DESCA(DTYPE_)=1 then DLEN_ $\geq$ 9.

B (local input/local output) REAL/COMPLEX pointer into the local memory to an array of dimension (LLD_B, LOCc(NRHS)).
On entry, this array contains the the local pieces of the n-by-nrhs right hand side distributed matrix $B(IB:IB+N-1, 1:NRHS)$.
On exit, if INFO = 0, this array contains the local pieces of the n-by-nrhs solution distributed matrix X.

IB (global input) INTEGER
The row index in the global array B indicating the first row of sub(B).

DESCB (global and local input) INTEGER array, dimension (DLEN_).
The array descriptor for the distributed matrix B.
If DESCB(DTYPE_)=502 then DLEN_ $\geq$ 7,
else if DESCB(DTYPE_)=1 then DLEN_ $\geq$ 9.

AF (local input) REAL/COMPLEX array, dimension (LAF)
Auxiliary Fillin space. Fillin is created during the factorization routine PSDBTRF/PCDBTRF and this is stored in AF.

LAF (local input) INTEGER
The dimension of the array AF.
$LAF \geq NRHS*BW$.
If LAF is not large enough, an error code will be returned and the minimum acceptable size will be returned in AF(1).

WORK (local workspace/local output) REAL/COMPLEX array, dimension (LWORK)
On exit, WORK(1) returns the minimal and optimal LWORK.

LWORK (local or global input) INTEGER
The dimension of the array WORK.
LWORK is local input and must be at least $LWORK \geq BW*BW$.
If LWORK = -1, then LWORK is global input and a workspace query is assumed; the routine only calculates the minimum and optimal size for all work arrays. Each of these values is returned in the first entry of the corresponding work array, and no error message is issued by PXERBLA.

INFO (global output) INTEGER
= 0: successful exit
< 0: If the ith argument is an array and the j-entry had an illegal value, then INFO = $-(i*100+j)$, if the ith argument is a scalar and had an illegal value, then INFO = $-i$.

PSPOCON/PCPOCON

```
SUBROUTINE PSPOCON( UPLO, N, A, IA, JA, DESCA, ANORM, RCOND, WORK,
$                   LWORK, IWORK, LIWORK, INFO )

    CHARACTER        UPLO
    INTEGER          IA, INFO, JA, LIWORK, LWORK, N
    REAL             ANORM, RCOND
    INTEGER          DESCA( * ), IWORK( * )
    REAL             A( * ), WORK( * )

SUBROUTINE PCPOCON( UPLO, N, A, IA, JA, DESCA, ANORM, RCOND, WORK,
$                   LWORK, RWORK, LRWORK, INFO )

    CHARACTER        UPLO
    INTEGER          IA, INFO, JA, LRWORK, LWORK, N
    REAL             ANORM, RCOND
    INTEGER          DESCA( * )
    REAL             RWORK( * )
    COMPLEX          A( * ), WORK( * )
```

Purpose

PSPOCON/PCPOCON estimates the reciprocal of the condition number (in the 1-norm) of a real/complex symmetric/Hermitian positive definite distributed matrix using the Cholesky factorization $A = U^H*U$ or $A = L*L^H$ computed by PSPOTRF/PCPOTRF.

An estimate is obtained for $\|A(IA:IA+N-1, JA:JA+N-1)^{-1}\|$, and the reciprocal of the condition number is computed as

$$RCOND = \frac{1}{(\|A(IA:IA+N-1,JA:JA+N-1)\|*\|A(IA:IA+N-1,JA:JA+N-1)^{-1}\|)}.$$

Arguments

UPLO (global input) CHARACTER*1
      Specifies whether the factor stored in $A(IA:IA+N-1,JA:JA+N-1)$ is upper or lower triangular.
      = 'U': Upper triangular;
      = 'L': Lower triangular.

N (global input) INTEGER
   The order of the distributed matrix $A(IA:IA+N-1,JA:JA+N-1)$. $N \geq 0$.

A (local input) REAL/COMPLEX pointer into the local memory to an array of dimension (LLD_A, LOCc(JA+N−1)).
   On entry, this array contains the local pieces of the factors L or U from the Cholesky factorization $A(IA:IA+N-1,JA:JA+N-1) = U^H*U$ or $L*L^H$, as computed by PSPOTRF/PCPOTRF.

IA (global input) INTEGER
    The row index in the global array A indicating the first row of sub(A).

JA (global input) INTEGER
    The column index in the global array A indicating the first column of

DESCA (global and local input) INTEGER array, dimension (DLEN_). The array descriptor for the distributed matrix A.

ANORM (global input) REAL
       The 1-norm (or infinity-norm) of the symmetric/Hermitian distributed matrix $A(IA:IA+N-1,JA:JA+N-1)$.

RCOND (global output) REAL
       The reciprocal of the condition number of the distributed matrix $A(IA:IA+N-1,JA:JA+N-1)$, computed as $RCOND =$
       $$\frac{1}{(\|A(IA:IA+N-1,JA:JA+N-1)\|*\|A(IA:IA+N-1,JA:JA+N-1)^{-1}\|)}.$$

WORK (local workspace/local output) REAL/COMPLEX array, dimension (LWORK)
      On exit, WORK(1) returns the minimal and optimal LWORK.

LWORK (local or global input) INTEGER
       The dimension of the array WORK.
       LWORK is local input and must be at least
       PSPOCON
       $LWORK \geq 2*LOCr(N+mod(IA-1,MB\_A)) + 2*LOCc(N+mod(JA-1,NB\_A)) + max(2, max(NB\_A*CEIL(NPROW-1,NPCOL), LOCc(N+mod(JA-1,NB\_A)) + NB\_A*CEIL(NPCOL-1,NPROW)))$.
       PCPOCON
       $LWORK \geq 2*LOCr(N+mod(IA-1,MB\_A)) + max(2, max(NB\_A*max(1,CEIL(P-1,Q)), LOCc(N+mod(JA-1,NB\_A)) + NB\_A*max(1,CEIL(Q-1,P))))$.

       If LWORK = −1, then LWORK is global input and a workspace query is assumed; the routine only calculates the minimum and optimal size for all work arrays. Each of these values is returned in the first entry of the corresponding work array, and no error message is issued by PXERBLA.

IWORK (local workspace/local output) INTEGER array, dimension (LIWORK)
       On exit, IWORK(1) returns the minimal and optimal LIWORK.

LIWORK PSPOCON only (local or global input) INTEGER
        The dimension of the array IWORK.
        LIWORK is local input and must be at least
        $LIWORK \geq LOCr(N+mod(IA-1,MB\_A))$.
        If LIWORK = −1, then LIWORK is global input and a workspace query is assumed; the routine only calculates the minimum and optimal size for all work arrays. Each of these values is returned in the first entry of the corresponding work array, and no error message is issued by PXERBLA.

RWORK PCPOCON only (local workspace/local output) REAL array, dimension (LRWORK)
       On exit, RWORK(1) returns the minimal and optimal LRWORK.

LRWORK PCPOCON only (local or global input) INTEGER
        The dimension of the array RWORK.

LRWORK is local input and must be at least
$LRWORK \geq 2*LOCc(N+mod(JA-1,NB\_A))$.
If LRWORK = −1, then LRWORK is global input and a workspace query is assumed; the routine only calculates the minimum and optimal size for all work arrays. Each of these values is returned in the first entry of the corresponding work array, and no error message is issued by PXERBLA.

INFO   (global output) INTEGER
= 0: successful exit
< 0: If the ith argument is an array and the j-entry had an illegal value, then INFO = $-(i*100+j)$, if the ith argument is a scalar and had an illegal value, then INFO = $-i$.

PSPOEQU/PCPOEQU

```
SUBROUTINE PSPOEQU( N, A, IA, JA, DESCA, SR, SC, SCOND, AMAX,
$                    INFO )
INTEGER              IA, INFO, JA, N
REAL                 AMAX, SCOND
INTEGER              DESCA( * )
REAL                 A( * ), SC( * ), SR( * )

SUBROUTINE PCPOEQU( N, A, IA, JA, DESCA, SR, SC, SCOND, AMAX,
$                    INFO )
INTEGER              IA, INFO, JA, N
REAL                 AMAX, SCOND
INTEGER              DESCA( * )
REAL                 SC( * ), SR( * )
COMPLEX              A( * )
```

Purpose

PSPOEQU/PCPOEQU computes row and column scalings intended to equilibrate a symmetric/Hermitian positive definite matrix sub(A) = A(IA:IA+N−1,JA:JA+N−1) and reduce its condition number (with respect to the two-norm). SR and SC contain the scale factors, $S(i) = 1/\sqrt{(A(i,i))}$, chosen so that the scaled distributed matrix B with elements B(i,j) = S(i)*A(i,j)*S(j) has ones on the diagonal. This choice of SR and SC puts the condition number of B within a factor N of the smallest possible condition number over all possible diagonal scalings.

Arguments

N   (global input) INTEGER
The number of rows and columns to be operated on, i.e., the order of the distributed submatrix sub(A). N ≥ 0.

A   (local input) REAL/COMPLEX pointer into the local memory to an array of local dimension (LLD_A, LOCc(JA+N−1)).
The n-by-n symmetric/Hermitian positive definite distributed matrix sub(A) whose scaling factors are to be computed. Only the diagonal elements of sub(A) are referenced.

IA   (global input) INTEGER
The row index in the global array A indicating the first row of sub(A).

JA   (global input) INTEGER
The column index in the global array A indicating the first column of sub(A).

DESCA   (global and local input) INTEGER array, dimension (DLEN_).
The array descriptor for the distributed matrix A.

SR   (local output) REAL array, dimension LOCr(M_A)
If INFO = 0, SR(IA:IA+N−1) contains the row scale factors for sub(A). SR is aligned with the distributed matrix A, and replicated across every process column. SR is tied to the distributed matrix A.

SC   (local output) REAL array, dimension LOCc(N_A)
If INFO = 0, SC(JA:JA+N−1) contains the column scale factors for sub(A). SC is aligned with the distributed matrix A(IA:IA+M−1,JA:JA+N−1). SC is replicated down every process row. SC is tied to the distributed matrix A.

SCOND   (global output) REAL
If INFO = 0, SCOND contains the ratio of the smallest SR(i) (or SC(j)) to the largest SR(i) (or SC(j)), with IA ≤ i ≤ IA+N−1 and JA ≤ j ≤ JA+N−1. If SCOND ≥ 0.1 and AMAX is neither too large nor too small, it is not worth scaling by SR (or SC).

AMAX   (global output) REAL
Absolute value of largest matrix element. If AMAX is very close to overflow or very close to underflow, the matrix should be scaled.

INFO   (global output) INTEGER
= 0: successful exit
< 0: If the ith argument is an array and the j-entry had an illegal value, then INFO = $-(i*100+j)$, if the ith argument is a scalar and had an illegal value, then INFO = $-i$.
> 0: If INFO = K, the Kth diagonal entry of sub(A) is nonpositive.

PSPORFS/PCPORFS

```
SUBROUTINE PSPORFS( UPLO, N, NRHS, A, IA, JA, DESCA, AF, IAF, JAF,
$                   DESCAF, B, IB, JB, DESCB, X, IX, JX, DESCX,
$                   FERR, BERR, WORK, LWORK, IWORK, LIWORK, INFO )
   CHARACTER         UPLO
   INTEGER           IA, IAF, IB, INFO, IX, JA, JAF, JB, JX,
$                    LIWORK, LWORK, N, NRHS
   INTEGER           DESCA( * ), DESCAF( * ), DESCB( * ),
$                    DESCX( * ), IWORK( * )
   REAL              A( * ), AF( * ), B( * ),
$                    BERR( * ), FERR( * ), WORK( * ), X( * )

SUBROUTINE PCPORFS( UPLO, N, NRHS, A, IA, JA, DESCA, AF, IAF, JAF,
$                   DESCAF, B, IB, JB, DESCB, X, IX, JX, DESCX,
$                   FERR, BERR, WORK, LWORK, RWORK, LRWORK, INFO )
   CHARACTER         UPLO
   INTEGER           IA, IAF, IB, INFO, IX, JA, JAF, JB, JX,
$                    LRWORK, LWORK, N, NRHS
   INTEGER           DESCA( * ), DESCAF( * ), DESCB( * ),
$                    DESCX( * )
   COMPLEX           A( * ), AF( * ), B( * ),
$                    BERR( * ), FERR( * ), WORK( * ), X( * )
   REAL              RWORK( * )
```

Purpose

PSPORFS/PCPORFS improves the computed solution to a system of linear equations when the coefficient matrix is symmetric/Hermitian positive definite and provides error bounds and backward error estimates for the solutions.

Arguments

UPLO   (global input) CHARACTER*1
Specifies whether the upper or lower triangular part of the symmetric/Hermitian matrix sub(A) is stored.
= 'U':   Upper triangular;
= 'L':   Lower triangular.

N   (global input) INTEGER
The order of the matrix sub(A). $N \geq 0$.

NRHS   (global input) INTEGER
The number of right hand sides, i.e., the number of columns of the matrices sub(B) and sub(X). $NRHS \geq 0$.

A   (local input) REAL/COMPLEX pointer into the local memory to an array of local dimension (LLD_A,LOCc(JA+N−1)).
This array contains the local pieces of the n-by-n symmetric/Hermitian distributed matrix sub(A) to be factored. If UPLO = 'U', the leading n-by-n upper triangular part of sub(A) contains the upper triangular part of the matrix, and its strictly lower triangular part is not referenced.

If UPLO = 'L', the leading n-by-n lower triangular part of sub(A) contains the lower triangular part of the distributed matrix, and its strictly upper triangular part is not referenced.

IA   (global input) INTEGER
The row index in the global array A indicating the first row of sub(A).

JA   (global input) INTEGER
The column index in the global array A indicating the first column of sub(A).

DESCA   (global and local input) INTEGER array, dimension (DLEN_).
The array descriptor for the distributed matrix A.

AF   (input) REAL/COMPLEX pointer into the local memory to an array of local dimension (LLD_AF,LOCc(JA+N−1)).
On entry, this array contains the factors L or U from the Cholesky factorization sub(A) = $L*L^H$ or U^H*U, as computed by PSPOTRF/PCPOTRF.

IAF   (global input) INTEGER
The row index in the global array AF indicating the first row of sub(AF).

JAF   (global input) INTEGER
The column index in the global array AF indicating the first column of sub(AF).

DESCAF   (global and local input) INTEGER array, dimension (DLEN_).
The array descriptor for the distributed matrix AF.

B   (local input) REAL/COMPLEX pointer into the local memory to an array of local dimension (LLD_B, LOCc(JB+NRHS−1)).
On entry, this array contains the the local pieces of the right hand sides sub(B).

IB   (global input) INTEGER
The row index in the global array B indicating the first row of sub(B).

JB   (global input) INTEGER
The column index in the global array B indicating the first column of sub(B).

DESCB   (global and local input) INTEGER array, dimension (DLEN_).
The array descriptor for the distributed matrix B.

X   (local input/local output) REAL/COMPLEX pointer into the local memory to an array of local dimension (LLD_X, LOCc(JX+NRHS−1)).
On entry, this array contains the the local pieces of the solution vectors sub(X).
On exit, it contains the improved solution vectors.

IX   (global input) INTEGER
The row index in the global array X indicating the first row of sub(X).

JX   (global input) INTEGER
          The column index in the global array X indicating the first column of
          sub(X).

DESCX   (global and local input) INTEGER array, dimension (DLEN_).
          The array descriptor for the distributed matrix X.

FERR   (local output) REAL array, dimension LOCc(JB+NRHS−1).
          The estimated forward error bound for each solution vector of sub(X).
          If XTRUE is the true solution corresponding to sub(X), FERR is
          an estimated upper bound for the magnitude of the largest element in
          (sub(X) − XTRUE) divided by the magnitude of the largest element
          in sub(X). The estimate is as reliable as the estimate for RCOND, and
          is almost always a slight overestimate of the true error. This array is
          tied to the distributed matrix X.

BERR   (local output) REAL array, dimension (NRHS)
          The componentwise relative backward error of each solution vector X(j)
          (i.e., the smallest relative change in any element of A or B that makes
          X(j) an exact solution).

WORK   (local workspace/local output) REAL/COMPLEX array, dimension
          (LWORK)
          On exit, WORK(1) returns the minimal and optimal LWORK.

LWORK   (local input or global output) INTEGER
          The dimension of the array WORK.
          LWORK is local input and must be at least
          PSPORFS
          LWORK ≥ 3*LOCr(N + mod(IA−1, MB_A)).
          PCPORFS
          LWORK ≥ 2*LOCr(N + mod(IA−1, MB_A)).
          If LWORK = −1, then LWORK is global input and a workspace query is
          assumed; the routine only calculates the minimum and optimal size for
          all work arrays. Each of these values is returned in the first entry of the
          corresponding work array, and no error message is issued by PXERBLA.

IWORK   PSPORFS only (local workspace/local output) INTEGER array, di-
          mension (LIWORK)
          On exit, IWORK(1) returns the minimal and optimal LIWORK.

LIWORK   PSPORFS only (local or global input) INTEGER
          The dimension of the array IWORK.
          LIWORK is local input and must be at least
          LWORK ≥ LOCr(N + mod(IB−1, MB_B)).
          If LIWORK = −1, then LIWORK is global input and a workspace
          query is assumed; the routine only calculates the minimum and optimal
          size for all work arrays. Each of these values is returned in the first
          entry of the corresponding work array, and no error message is issued
          by PXERBLA.

RWORK   PCPORFS only (local workspace/local output) REAL array, dimension
          (LRWORK)
          On exit, RWORK(1) returns the minimal and optimal LRWORK.

LRWORK   PCPORFS only (local or global input) INTEGER
          The dimension of the array RWORK.
          LRWORK is local input and must be at least
          LRWORK ≥ LOCr(N + mod(IB−1, MB_B)).
          If LRWORK = −1, then LRWORK is global input and a workspace
          query is assumed; the routine only calculates the minimum and optimal
          size for all work arrays. Each of these values is returned in the first
          entry of the corresponding work array, and no error message is issued
          by PXERBLA.

INFO   (global output) INTEGER
          = 0: successful exit
          < 0: If the ith argument is an array and the j-entry had an illegal
                value, then INFO = −(i*100+j), if the ith argument is a scalar
                and had an illegal value, then INFO = −i.

PSPOSV/PCPOSV

```
SUBROUTINE PSPOSV( UPLO, N, NRHS, A, IA, JA, DESCA, B, IB, JB,
$                  DESCB, INFO )
     CHARACTER      UPLO
     INTEGER        IA, IB, INFO, JA, JB, N, NRHS
     INTEGER        DESCA( * ), DESCB( * )
     REAL           A( * ), B( * )

SUBROUTINE PCPOSV( UPLO, N, NRHS, A, IA, JA, DESCA, B, IB, JB,
$                  DESCB, INFO )
     CHARACTER      UPLO
     INTEGER        IA, IB, INFO, JA, JB, N, NRHS
     INTEGER        DESCA( * ), DESCB( * )
     COMPLEX        A( * ), B( * )
```

Purpose

PSPOSV/PCPOSV computes the solution to a real/complex system of linear equations sub(A)*X = sub(B), where sub(A) denotes A(IA:IA+N−1,JA:JA+N−1)
and is an n-by-n symmetric/Hermitian distributed positive definite matrix and X
and sub(B) denoting B(IB:IB+N−1,JB:JB+NRHS−1) are n-by-nrhs distributed
matrices.

The Cholesky decomposition is used to factor sub(A) as sub(A) = U^H*U, if
UPLO = 'U', or sub(A) = L*L^H, if UPLO = 'L', where U is an upper triangular
matrix and L is a lower triangular matrix. The factored form of sub(A) is then
used to solve the system of equations.

Arguments

UPLO   (global input) CHARACTER*1
          = 'U': Upper triangle of sub(A) is stored;

> 0:   If INFO = K, the leading minor of order K, A(IA:IA+K−1,JA:JA+K−1) is not positive definite, and the factorization could not be completed, and the solution has not been computed.

PSPOSVX/PCPOSVX

```
SUBROUTINE PSPOSVX( FACT, UPLO, N, NRHS, A, IA, JA, DESCA, AF,
$                   IAF, JAF, DESCAF, EQUED, SR, SC, B, IB, JB,
$                   DESCB, X, IX, JX, DESCX, RCOND, FERR, BERR,
$                   WORK, LWORK, IWORK, LIWORK, INFO )
CHARACTER          EQUED, FACT, UPLO
INTEGER            IA, IAF, IB, INFO, IX, JA, JAF, JB, JX, LIWORK,
$                   LWORK, N, NRHS
REAL               RCOND
INTEGER            DESCA( * ), DESCAF( * ), DESCB( * ),
$                   DESCX( * ), IWORK( * )
REAL               A( * ), AF( * ),
$                   B( * ), BERR( * ), FERR( * ),
$                   SC( * ), SR( * ), WORK( * ), X( * )

SUBROUTINE PCPOSVX( FACT, UPLO, N, NRHS, A, IA, JA, DESCA, AF,
$                   IAF, JAF, DESCAF, EQUED, SR, SC, B, IB, JB,
$                   DESCB, X, IX, JX, DESCX, RCOND, FERR, BERR,
$                   WORK, LWORK, RWORK, LRWORK, INFO )
CHARACTER          EQUED, FACT, UPLO
INTEGER            IA, IAF, IB, INFO, IX, JA, JAF, JB, JX, LRWORK,
$                   LWORK, N, NRHS
REAL               RCOND
INTEGER            DESCA( * ), DESCAF( * ), DESCB( * ), DESCX( * ),
REAL               BERR( * ), FERR( * ), SC( * ),
$                   SR( * ), RWORK( * )
COMPLEX            A( * ), AF( * ),
$                   B( * ), WORK( * ), X( * )
```

Purpose

PSPOSVX/PCPOSVX uses the Cholesky factorization $A = U^H*U$ or $A = L*L^H$ to compute the solution to a real/complex system of linear equations $A(IA:IA+N−1,JA:JA+N−1)*X = B(IB:IB+N−1,JB:JB+NRHS−1)$, where $A(IA:IA+N−1,JA:JA+N−1)$ is an n-by-n matrix and X and B(IB:IB+N−1,JB:JB+NRHS−1) are n-by-nrhs matrices.

Error bounds on the solution and a condition estimate are also provided. In the following comments Y denotes $Y(IY:IY+M−1,JY:JY+K−1)$ a m-by-k matrix where Y can be A, AF, B and X.

= 'L':   Lower triangle of sub(A) is stored.

N   (global input) INTEGER
The number of rows and columns to be operated on, i.e., the order of the distributed submatrix sub(A). N ≥ 0.

NRHS   (global input) INTEGER
The number of right hand sides, i.e., the number of columns of the distributed submatrix sub(B). NRHS ≥ 0.

A   (local input/local output) REAL/COMPLEX pointer into the local memory to an array of dimension (LLD_A, LOCc(JA+N−1)). On entry, this array contains the local pieces of the n-by-n symmetric/Hermitian distributed matrix sub(A) to be factored. If UPLO = 'U', the leading n-by-n upper triangular part of sub(A) contains the upper triangular part of the matrix, and its strictly lower triangular part is not referenced. If UPLO = 'L', the leading n-by-n lower triangular part of sub(A) contains the lower triangular part of the distributed matrix, and its strictly upper triangular part is not referenced. On exit, if INFO = 0, this array contains the local pieces of the factor U or L from the Cholesky factorization zation sub(A) = U^H*U or $L*L^H$.

IA   (global input) INTEGER
The row index in the global array A indicating the first row of sub(A).

JA   (global input) INTEGER
The column index in the global array A indicating the first column of sub(A).

DESCA   (global and local input) INTEGER array, dimension (DLEN_).
The array descriptor for the distributed matrix A.

B   (local input/local output) REAL/COMPLEX pointer into the local memory to an array of dimension (LLD_B,LOC(JB+NRHS−1)). On entry, the local pieces of the right hand sides distributed matrix sub(B). On exit, if INFO = 0, sub(B) is overwritten with the solution distributed matrix X.

IB   (global input) INTEGER
The row index in the global array B indicating the first row of sub(B).

JB   (global input) INTEGER
The column index in the global array B indicating the first column of sub(B).

DESCB   (global and local input) INTEGER array, dimension (DLEN_).
The array descriptor for the distributed matrix B.

INFO   (global output) INTEGER
= 0:   successful exit
< 0:   If the ith argument is an array and the j-entry had an illegal value, then INFO = −(i*100+j), if the ith argument is a scalar and had an illegal value, then INFO = −i.

Description

The following steps are performed:

1. If FACT = 'E', real scaling factors are computed to equilibrate the system:

$$\text{diag(SR)}*A*\text{diag(SC)}*(\text{diag(SC)})^{-1}*X = \text{diag(SR)}*B$$

Whether or not the system will be equilibrated depends on the scaling of the matrix A, but if equilibration is used, A is overwritten by diag(SR)*A*diag(SC) and B by diag(SR)*B.

2. If FACT = 'N' or 'E', the Cholesky decomposition is used to factor the matrix A (after equilibration if FACT = 'E') as
$A = U^H*U$, if UPLO = 'U', or
$A = L*L^H$, if UPLO = 'L',
where U is an upper triangular matrix and L is a lower triangular matrix.

3. If the leading i-by-i principal minor is not positive definite, then the routine returns with INFO = i. Otherwise, the factored form of A is used to estimate the condition number of the matrix A. If the reciprocal of the condition number is less than relative machine precision, INFO = N+1 is returned as a warning, but the routine still goes on to solve for X and compute error bounds as described below.

4. The system of equations is solved for X using the factored form of A.

5. Iterative refinement is applied to improve the computed solution matrix and calculate error bounds and backward error estimates for it.

6. If equilibration was used, the matrix X is premultiplied by diag(S) so that it solves the original system before equilibration.

Arguments

FACT   (global input) CHARACTER*1
Specifies whether or not the factored form of the matrix A is supplied on entry, and if not, whether the matrix A should be equilibrated before it is factored.
= 'F': On entry, AF contains the factored form of A. If EQUED = 'Y', the matrix A has been equilibrated with scaling factors given by S. A and AF will not be modified.
= 'N': The matrix A will be copied to AF and factored.
= 'E': The matrix A will be equilibrated if necessary, then copied to AF and factored.

UPLO   (global input) CHARACTER*1
= 'U': Upper triangle of A is stored;
= 'L': Lower triangle of A is stored.

N   (global input) INTEGER
The number of rows and columns to be operated on, i.e., the order of the distributed submatrix A(IA:IA+N−1,JA:JA+N−1). N ≥ 0.

NRHS   (global input) INTEGER
The number of right hand sides, i.e., the number of columns of the distributed submatrices B and X. NRHS ≥ 0.

A   (input/output) REAL/COMPLEX pointer into the local memory to an array of local dimension (LLD_A, LOCc(JA+N−1)).
On entry, the symmetric/Hermitian matrix A, except if FACT = 'F' and EQUED = 'Y', then A must contain the equilibrated matrix diag(SR)*A*diag(SC). If UPLO = 'U', the leading n-by-n upper triangular part of A contains the upper triangular part of the matrix A, and the strictly lower triangular part of A is not referenced. If UPLO = 'L', the leading n-by-n lower triangular part of A contains the lower triangular part of the matrix A, and the strictly upper triangular part of A is not referenced. A is not modified if FACT = 'F' or 'N', or if FACT = 'E' and EQUED = 'N' on exit.
On exit, if FACT = 'E' and EQUED = 'Y', A is overwritten by diag(SR)*A*diag(SC).

IA   (global input) INTEGER
The row index in the global array A indicating the first row of sub(A).

JA   (global input) INTEGER
The column index in the global array A indicating the first column of sub(A).

DESCA   (global and local input) INTEGER array, dimension (DLEN_).
The array descriptor for the distributed matrix A.

AF   (local input or local output) REAL/COMPLEX pointer into the local memory to an array of local dimension (LLD_AF, LOCc(JA+N−1)).
If FACT = 'F', then AF is an input argument and on entry contains the triangular factor U or L from the Cholesky factorization $A = U^H*U$ or $A = L*L^H$, in the same storage format as A. If EQUED ≠ 'N', then AF is the factored form of the equilibrated matrix diag(SR)*A*diag(SC).
If FACT = 'N', then AF is an output argument and on exit returns the triangular factor U or L from the Cholesky factorization $A = U^H*U$ or $A = L*L^H$ of the original matrix A.
If FACT = 'E', then AF is an output argument and on exit returns the triangular factor U or L from the Cholesky factorization $A = U^H*U$ or $A = L*L^H$ of the equilibrated matrix A (see the description of A for the form of the equilibrated matrix).

IAF   (global and local input) INTEGER
The row index in the global array AF indicating the first row of sub(AF).

JAF   (global and local input) INTEGER
The column index in the global array AF indicating the first column of sub(AF).

DESCAF (global and local input) INTEGER array, dimension (DLEN_).
The array descriptor for the distributed matrix AF.

RCOND
(global output) REAL
The estimate of the reciprocal condition number of the matrix A after equilibration (if done). If RCOND is less than the relative machine precision (in particular, if RCOND = 0), the matrix is singular to working precision. This condition is indicated by a return code of INFO > 0.

FERR
(local output) REAL array, dimension (LOCc(N_B))
The estimated forward error bounds for each solution vector X(j) (the jth column of the solution matrix X(IX:IX+N−1,JX:JX+NRHS−1). If XTRUE is the true solution, FERR(j) bounds the magnitude of the largest entry in (X(j) − XTRUE) divided by the magnitude of the largest entry in X(j)). The estimate is as reliable as the estimate for RCOND, and is almost always a slight overestimate of the true error. FERR is replicated in every process row, and is aligned with the matrices B and X.

BERR
(local output) REAL array, dimension (LOCc(N_B))
The componentwise relative backward error of each solution vector X(j) (i.e., the smallest relative change in any element of A or B that makes X(j) an exact solution).

WORK
(local workspace/local output) REAL/COMPLEX array, dimension (LWORK)
On exit, WORK(1) returns the minimal and optimal LWORK.

LWORK
(local or global input) INTEGER
The dimension of the array WORK.
LWORK is local input and must be at least
PSPOSVX
LWORK = max(PSPOCON(LWORK), PSPORFS(LWORK)) + LOCr(N_A).
PCPOSVX
LWORK = max(PCPOCON(LWORK), PCPORFS(LWORK)) + LOCr(N_A).
If LWORK = −1, then LWORK is global input and a workspace query is assumed; the routine only calculates the minimum and optimal size for all work arrays. Each of these values is returned in the first entry of the corresponding work array, and no error message is issued by PXERBLA.

IWORK
PSPOSVX only (local workspace/local output) INTEGER array, dimension (LIWORK)
On exit, IWORK(1) returns the minimal and optimal LIWORK.

LIWORK
PSPOSVX only (local or global input) INTEGER
The dimension of the array IWORK.
LIWORK is local input and must be at least
LIWORK = DESCA(LLD_).
If LIWORK = −1, then LIWORK is global input and a workspace query is assumed; the routine only calculates the minimum and optimal size for all work arrays. Each of these values is returned in the first entry of the corresponding work array, and no error message is issued by PXERBLA.

RWORK
PCPOSVX only (local workspace/local output) REAL array, dimension (LRWORK)

EQUED
(global input or global output) CHARACTER*1
Specifies the form of equilibration that was done.
= 'N': No equilibration (always true if FACT = 'N').
= 'Y': Equilibration was done, i.e., A has been replaced by diag(SR)*A*diag(SC).
EQUED is an input argument if FACT = 'F'; otherwise, it is an output argument.

SR
(local input or local output) REAL array, dimension (LLD_A)
The scale factors for A distributed across process rows; not accessed if EQUED = 'N'. SR is an input variable if FACT = 'F'; otherwise, SR is an output variable. If FACT = 'F' and EQUED = 'Y', each element of SR must be positive.

SC
(local input or local output) REAL array, dimension (LOC(N_A))
The scale factors for A distributed across process columns; not accessed if EQUED = 'N'. SC is an input variable if FACT = 'F'; otherwise, SC is an output variable. If FACT = 'F' and EQUED = 'Y', each element of SC must be positive.

B
(local input/local output) REAL/COMPLEX pointer into the local memory to an array of local dimension (LLD_B, LOCc(JB+NRHS−1)).
On entry, the n-by-nrhs right-hand side matrix B.
On exit, if EQUED = 'N', B is not modified; if TRANS = 'N' and EQUED = 'R' or 'B', B is overwritten by diag(SR)*B; if TRANS = 'T' or 'C' and EQUED = 'C' or 'B', B is overwritten by diag(SC)*B.

IB
(global input) INTEGER
The row index in the global array B indicating the first row of sub(B).

JB
(global input) INTEGER
The column index in the global array B indicating the first column of sub(B).

DESCB
(global and local input) INTEGER array, dimension (DLEN_).
The array descriptor for the distributed matrix B.

X
(local output) REAL/COMPLEX pointer into the local memory to an array of local dimension (LLD_X, LOCc(JX+NRHS−1)).
If INFO = 0, the n-by-nrhs solution matrix X to the original system of equations. Note that A and B are modified on exit if EQUED ≠ 'N', and the solution to the equilibrated system is diag(SC)⁻¹*X if TRANS = 'N' and EQUED = 'C' or 'B', or diag(SR)⁻¹*X if TRANS = 'T' or 'C' and EQUED = 'R' or 'B'.

IX
(global input) INTEGER
The row index in the global array X indicating the first row of sub(X).

JX
(global input) INTEGER
The column index in the global array X indicating the first column of sub(X).

DESCX
(global and local input) INTEGER array, dimension (DLEN_).
The array descriptor for the distributed matrix X.

On exit, RWORK(1) returns the minimal and optimal LRWORK.

LRWORK *PCPOSVX only* (local or global input) INTEGER
The dimension of the array RWORK.
LRWORK is local input and must be at least

LRWORK = 2*LOCc(N_A).

If LRWORK = −1, then LRWORK is global input and a workspace query is assumed; the routine only calculates the minimum and optimal size for all work arrays. Each of these values is returned in the first entry of the corresponding work array, and no error message is issued by PXERBLA.

INFO   (global output) INTEGER
= 0:   successful exit
< 0:   If the ith argument is an array and the j-entry had an illegal value, then INFO = −(i*100+j), if the ith argument is a scalar and had an illegal value, then INFO = −i,
> 0:   if INFO = i, and i is
≤ N: the leading minor of order i of A is not positive definite, so the factorization could not be completed, and the solution and error bounds could not be computed. RCOND = 0 is returned.
= N+1: U is nonsingular, but RCOND is less than relative machine precision, meaning that the matrix is singular to working precision. Nevertheless, the solution and error bounds are computed because there are a number of situations where the computed solution can be more accurate than the value of RCOND would suggest.

PSPOTRF/PCPOTRF

```
SUBROUTINE PSPOTRF( UPLO, N, A, IA, JA, DESCA, INFO )
CHARACTER        UPLO
INTEGER          IA, INFO, JA, N
INTEGER          DESCA( * )
REAL             A( * )
SUBROUTINE PCPOTRF( UPLO, N, A, IA, JA, DESCA, INFO )
CHARACTER        UPLO
INTEGER          IA, INFO, JA, N
INTEGER          DESCA( * )
COMPLEX          A( * )
```

Purpose

PSPOTRF/PCPOTRF computes the Cholesky factorization of a real/complex symmetric/Hermitian positive definite distributed matrix sub(A) denoted A(IA:IA+N−1, JA:JA+N−1).

The factorization has the form sub(A) = U^H*U, if UPLO = 'U', or sub(A) = L*L^H, if UPLO = 'L', where U is an upper triangular matrix and L is lower triangular.

Arguments

UPLO   (global input) CHARACTER*1
= 'U': Upper triangle of sub(A) is stored;
= 'L': Lower triangle of sub(A) is stored.

N   (global input) INTEGER
The number of rows and columns to be operated on, i.e., the order of the distributed submatrix sub(A). N ≥ 0.

A   (local input/local output) REAL/COMPLEX pointer into the local memory to an array of dimension (LLD_A, LOCc(JA+N−1)).
On entry, this array contains the local pieces of the n-by-n symmetric/Hermitian distributed matrix sub(A) to be factored.
If UPLO = 'U', the leading n-by-n upper triangular part of sub(A) contains the upper triangular part of the matrix, and its strictly lower triangular part is not referenced.
If UPLO = 'L', the leading n-by-n lower triangular part of sub(A) contains the lower triangular part of the distributed matrix, and its strictly upper triangular part is not referenced.
On exit, if UPLO = 'U', the upper triangular part of the distributed matrix contains the Cholesky factor U, if UPLO = 'L', the lower triangular part of the distributed matrix contains the Cholesky factor L.

IA   (global input) INTEGER
The row index in the global array A indicating the first row of sub(A).

JA   (global input) INTEGER
The column index in the global array A indicating the first column of sub(A).

DESCA   (global and local input) INTEGER array, dimension (DLEN_).
The array descriptor for the distributed matrix A.

INFO   (global output) INTEGER
= 0:   successful exit
< 0:   If the ith argument is an array and the j-entry had an illegal value, then INFO = −(i*100+j), if the ith argument is a scalar and had an illegal value, then INFO = −i.
> 0:   If INFO = K, the leading minor of order K, A(IA:IA+K−1,JA:JA+K−1) is not positive definite, and the factorization could not be completed.

PSPOTRI/PCPOTRI

```
SUBROUTINE PSPOTRI( UPLO, N, A, IA, JA, DESCA, INFO )
        CHARACTER       UPLO
        INTEGER         IA, INFO, JA, N
        INTEGER         DESCA( * )
        REAL            A( * )

SUBROUTINE PCPOTRI( UPLO, N, A, IA, JA, DESCA, INFO )
        CHARACTER       UPLO
        INTEGER         IA, INFO, JA, N
        INTEGER         DESCA( * )
        COMPLEX         A( * )
```

Purpose

PSPOTRI/PCPOTRI computes the inverse of a real/complex symmetric/-Hermitian positive definite distributed matrix

sub(A) = A(IA:IA+N−1,JA:JA+N−1)

using the Cholesky factorization sub(A) = U^H*U or sub(A) = $L*L^H$ computed by PSPOTRF/PCPOTRF.

Arguments

UPLO   (global input) CHARACTER*1
        = 'U': Upper triangle of sub(A) is stored;
        = 'L': Lower triangle of sub(A) is stored.

N   (global input) INTEGER
        The number of rows and columns to be operated on, i.e., the order of
        the distributed submatrix sub(A). N ≥ 0.

A   (local input/local output) REAL/COMPLEX pointer into the local
        memory to an array of dimension (LLD_A, LOCc(JA+N−1)).
        On entry, the local pieces of the triangular factor U or L from the
        Cholesky factorization of the distributed matrix sub(A) = U^H*U or
        $L*L^H$, as computed by PSPOTRF/PCPOTRF.
        On exit, the local pieces of the upper or lower triangle of the (symmet-
        ric)/(Hermitian) inverse of sub(A), overwriting the input factor U or
        L.

IA   (global input) INTEGER
        The row index in the global array A indicating the first row of sub(A).

JA   (global input) INTEGER
        The column index in the global array A indicating the first column of
        sub(A).

DESCA   (global and local input) INTEGER array, dimension (DLEN_).
        The array descriptor for the distributed matrix A.

INFO   (global output) INTEGER
        = 0: successful exit

< 0:   If the ith argument is an array and the j-entry had an illegal
        value, then INFO = −(i*100+j), if the ith argument is a scalar
        and had an illegal value, then INFO = −i.

> 0:   If INFO = i, the (i,i) element of the factor U or L is zero, and
        the inverse could not be computed.

PSPOTRS/PCPOTRS

```
SUBROUTINE PSPOTRS( UPLO, N, NRHS, A, IA, JA, DESCA, B, IB, JB,
     $                    DESCB, INFO )
        CHARACTER       UPLO
        INTEGER         IA, IB, INFO, JA, JB, N, NRHS
        INTEGER         DESCA( * ), DESCB( * )
        REAL            A( * ), B( * )

SUBROUTINE PCPOTRS( UPLO, N, NRHS, A, IA, JA, DESCA, B, IB, JB,
     $                    DESCB, INFO )
        CHARACTER       UPLO
        INTEGER         IA, IB, INFO, JA, JB, N, NRHS
        INTEGER         DESCA( * ), DESCB( * )
        COMPLEX         A( * ), B( * )
```

Purpose

PSPOTRS/PCPOTRS solves a system of linear equations sub(A)*X = sub(B)
where sub(A) denotes A(IA:IA+N−1,JA:JA+N−1) and is an n-by-n symmet-
ric/Hermitian positive definite distributed matrix using the Cholesky factorization
sub(A) = U^H*U or $L*L^H$ computed by PSPOTRF/PCPOTRF, and sub(B)
denotes the distributed matrix B(IB:IB+N−1,JB:JB+NRHS−1).

Arguments

UPLO   (global input) CHARACTER*1
        = 'U': Upper triangle of sub(A) is stored;
        = 'L': Lower triangle of sub(A) is stored.

N   (global input) INTEGER
        The number of rows and columns to be operated on, i.e., the order of
        the distributed submatrix sub(A). N ≥ 0.

NRHS   (global input) INTEGER
        The number of right hand sides, i.e., the number of columns of the
        distributed submatrix sub(B). NRHS ≥ 0.

A   (local input) REAL/COMPLEX pointer into local memory to an array
        of dimension (LLD_A, LOCc(JA+N−1)).
        On entry, this array contains the factors L or U from the
        Cholesky factorization sub(A) = $L*L^H$ or U^H*U, as computed by
        PSPOTRF/PCPOTRF.

IA   (global input) INTEGER
The row index in the global array A indicating the first row of sub(A).

JA   (global input) INTEGER
The column index in the global array A indicating the first column of sub(A).

DESCA   (global and local input) INTEGER array, dimension (DLEN_).
The array descriptor for the distributed matrix A.

B   (local input/local output) REAL/COMPLEX pointer into the local memory to an array of local dimension (LLD_B,LOCc(JB+NRHS−1)).
On entry, this array contains the the local pieces of the right hand sides sub(B).
On exit, this array contains the local pieces of the solution distributed sub(B).

IB   (global input) INTEGER
The row index in the global array B indicating the first row of sub(B).

JB   (global input) INTEGER
The column index in the global array B indicating the first column of sub(B).

DESCB   (global and local input) INTEGER array, dimension (DLEN_).
The array descriptor for the distributed matrix B.

INFO   (global output) INTEGER
$= 0$:   successful exit
< 0:   If the ith argument is an array and the j-entry had an illegal value, then INFO $= -(i*100+j)$; if the ith argument is a scalar and had an illegal value, then INFO $= -i$.

PSPTSV/PCPTSV

```
SUBROUTINE PSPTSV( N, NRHS, D, E, JA, DESCA, B, IB, DESCB, WORK,
$                  LWORK, INFO )
INTEGER           IB, INFO, JA, LWORK, N, NRHS
INTEGER           DESCA( * ), DESCB( * )
REAL              B( * ), D( * ), E( * ), WORK( * )

SUBROUTINE PCPTSV( N, NRHS, D, E, JA, DESCA, B, IB, DESCB, WORK,
$                  LWORK, INFO )
INTEGER           IB, INFO, JA, LWORK, N, NRHS
INTEGER           DESCA( * ), DESCB( * )
COMPLEX           B( * ), E( * ), WORK( * )
REAL              D( * )
```

Purpose

PSPTSV/PCPTSV computes the solution to a real/complex system of linear equations $A(1:N, JA:JA+N-1)*X = B(IB:IB+N-1, 1:NRHS)$, where A(1:N, JA:JA+N−1) is an n-by-n symmetric positive definite tridiagonal distributed matrix, and X and B are n-by-nrhs distributed matrices.

A is factored as $A(1:N, JA:JA+N-1) = P*L*D*L^H*P^T$, where P is a permutation matrix, and the factored form of A is then used to solve the system of equations.

Arguments

N   (global input) INTEGER
The number of rows and columns to be operated on, i.e., the order of the distributed submatrix A(1:N, JA:JA+N−1). $N \geq 0$.

NRHS   (global input) INTEGER
The number of right hand sides, i.e., the number of columns of the distributed submatrix B(IB:IB+N−1, 1:NRHS). $NRHS \geq 0$.

D   (local input/local output) REAL/COMPLEX pointer into the local memory to an array of dimension (DESCA(NB_)).
On entry, the local part of the global vector storing the main diagonal of the distributed matrix A.
On exit, details of the factorization.

E   (local input/local output) REAL/COMPLEX pointer into the local memory to an array of dimension (DESCA(NB_)).
On entry, the local part of the global vector storing the upper diagonal of the distributed matrix A.
On exit, details of the factorization.

JA   (global input) INTEGER
The index in the global array A that points to the start of the matrix to be operated on (which may be either all of A or a submatrix of A).

DESCA   (global and local input) INTEGER array, dimension (DLEN_).
The array descriptor for the distributed matrix A.
If DESCA(DTYPE_)=501 or 502 then DLEN_$\geq$7,
else if DESCA(DTYPE_)=1 then DLEN_$\geq$9.

B   (local input/local output) REAL/COMPLEX pointer into the local memory to an array of dimension (LLD_B, LOCc(NRHS)).
On entry, this array contains the the local pieces of the right hand side distributed matrix B(IB:IB+N−1, 1:NRHS).
On exit, this array contains the local pieces of the solution distributed matrix X.

IB   (global input) INTEGER
The row index in the global array B that points to the first row of the matrix to be operated on (which may be either all of B or a submatrix of B).

DESCB   (global and local input) INTEGER array, dimension (DLEN_).
The array descriptor for the distributed matrix B.
If DESCB(DTYPE_)=502 then DLEN_$\geq$7,
else if DESCB(DTYPE_)=1 then DLEN_$\geq$9.

WORK   (local workspace/local output) REAL/COMPLEX array, dimension (LWORK)
On exit, WORK(1) returns the minimal and optimal LWORK.

LWORK   (local or global input) INTEGER
The dimension of the array WORK.
LWORK is local input and must be at least
LWORK $\geq$ (12*NPCOL + 3*NB) +max((10+2*min(100,NRHS))*NPCOL + 4*NRHS, 8*NPCOL).

If LWORK = -1, then LWORK is global input and a workspace query is assumed; the routine only calculates the minimum and optimal size for all work arrays. Each of these values is returned in the first entry of the corresponding work array, and no error message is issued by PXERBLA.

INFO   (global output) INTEGER
= 0: successful exit
< 0: If the ith argument is an array and the j-entry had an illegal value, then INFO = $-(i*100+j)$, if the ith argument is a scalar and had an illegal value, then INFO = $-i$.
> 0: If INFO = K $\leq$ NPROCS, the submatrix stored on processor INFO$-$NPROCS and factored locally was not positive definite, and the factorization was not completed.
If INFO = K > NPROCS, the submatrix stored on processor INFO$-$NPROCS representing interactions with other processors was not nonsingular, and the factorization was not completed.

PSPTTRF/PCPTTRF

```
SUBROUTINE PSPTTRF( N, D, E, JA, DESCA, AF, LAF, WORK, LWORK,
$                   INFO )
INTEGER             INFO, JA, LAF, LWORK, N
INTEGER             DESCA( * )
REAL                AF( * ), D( * ), E( * ), WORK( * )

SUBROUTINE PCPTTRF( N, D, E, JA, DESCA, AF, LAF, WORK, LWORK,
$                   INFO )
INTEGER             INFO, JA, LAF, LWORK, N
INTEGER             DESCA( * )
COMPLEX             AF( * ), E( * ), WORK( * )
REAL                D( * )
```

Purpose

PSPTTRF/PCPTTRF computes the Cholesky factorization of an n-by-n real/complex tridiagonal symmetric/Hermitian positive definite distributed matrix A(1:N, JA:JA+N$-$1). The resulting factorization is *not* the same factorization as returned from LAPACK. Additional permutations are performed on the matrix for the sake of parallelism.

The factorization has the form
A(1:N, JA:JA+N$-$1) = $P*L*D*L^H*P^T$, or
A(1:N, JA:JA+N$-$1) = $P*U^H*D*U*P^T$, or
where P is a permutation matrix, and U is a tridiagonal upper triangular matrix and L is tridiagonal lower triangular.

Arguments

N   (global input) INTEGER
The number of rows and columns to be operated on, i.e., the order of the distributed submatrix A(1:N, JA:JA+N$-$1). N $\geq$ 0.

D   (local input/local output) REAL/COMPLEX pointer into the local memory to an array of dimension (DESCA(NB_)).
On entry, the local part of the global vector storing the main diagonal of the distributed matrix A.
On exit, details of the factorization.

E   (local input/local output) REAL/COMPLEX pointer into the local memory to an array of dimension (DESCA(NB_)).
On entry, the local part of the global vector storing the upper diagonal of the distributed matrix A.
On exit, details of the factorization.

JA   (global input) INTEGER
The index in the global array A that points to the start of the matrix to be operated on (which may be either all of A or a submatrix of A).

DESCA   (global and local input) INTEGER array, dimension (DLEN_)
The array descriptor for the distributed matrix A.
If DESCA(DTYPE_)=501 or 502 then DLEN_$\geq$7,
else if DESCA(DTYPE_)=1 then DLEN_$\geq$9.

AF   (local output) REAL/COMPLEX array of dimension (LAF)
Auxiliary Fillin space. Fillin is created during the factorization routine PSPTTRF/PCPTTRF and this is stored in AF. If a linear system is to be solved using PSPTTRS/PCPTTRS after the factorization routine, AF *must not be altered*.

LAF   (local input) INTEGER
The dimension of the array AF.
LAF $\geq$ NB+2.
If LAF is not large enough, an error code will be returned and the minimum acceptable size will be returned in AF(1).

WORK   (local workspace/local output) REAL/COMPLEX array, dimension (LWORK)
On exit, WORK(1) returns the minimal and optimal LWORK.

LWORK   (local or global input) INTEGER
The dimension of the array WORK.
LWORK is local input and must be at least
LWORK $\geq$ 8*NPCOL.
If LWORK = -1, then LWORK is global input and a workspace query is

assumed; the routine only calculates the minimum and optimal size for all work arrays. Each of these values is returned in the first entry of the corresponding work array, and no error message is issued by PXERBLA.

INFO (global output) INTEGER
= 0: successful exit
< 0: If the ith argument is an array and the j-entry had an illegal value, then INFO = $-(i*100+j)$, if the ith argument is a scalar and had an illegal value, then INFO = $-i$.
> 0: If INFO = K $\leq$ NPROCS, the submatrix stored on processor INFO−NPROCS and factored locally was not positive definite, and the factorization was not completed.
If INFO = K > NPROCS, the submatrix stored on processor INFO−NPROCS representing interactions with other processors was not nonsingular, and the factorization was not completed.

PSPTTRS/PCPTTRS

```
SUBROUTINE PSPTTRS( N, NRHS, D, E, JA, DESCA, B, IB, DESCB, AF,
$                    LAF, WORK, LWORK, INFO )
INTEGER           IB, INFO, JA, LAF, LWORK, N, NRHS
INTEGER           DESCA( * ), DESCB( * )
REAL              AF( * ), B( * ), D( * ), E( * ), WORK( * )

SUBROUTINE PCPTTRS( UPLO, N, NRHS, D, E, JA, DESCA, B, IB, DESCB,
$                    AF, LAF, WORK, LWORK, INFO )
CHARACTER         UPLO
INTEGER           IB, INFO, JA, LAF, LWORK, N, NRHS
INTEGER           DESCA( * ), DESCB( * )
COMPLEX           AF( * ), B( * ), E( * ), WORK( * )
REAL              D( * )
```

Purpose

PSPTTRS/PCPTTRS solves a system of linear equations A(1:N, JA:JA+N−1)*X = B(IB:IB+N−1, 1:NRHS) with a symmetric/Hermitian positive definite tridiagonal distributed matrix A using the factorization
A(1:N, JA:JA+N−1) = $P*L*D*L^H*P^T$, or
A(1:N, JA:JA+N−1) = $P*U^H*D*U*P^T$
computed by PSPTTRF/PCPTTRF.

Arguments

UPLO   *PCPTTRS only* (global input) CHARACTER*1
= 'U': Upper triangle of A(1:N, JA:JA+N−1) is stored;
= 'L': Lower triangle of A(1:N, JA:JA+N−1) is stored.

N   (global input) INTEGER
The number of rows and columns to be operated on, i.e., the order of the distributed submatrix A(1:N, JA:JA+N−1). N $\geq$ 0.

NRHS   (global input) INTEGER
The number of right hand sides, i.e., the number of columns of the distributed submatrix B(IB:IB+N−1, 1:NRHS). NRHS $\geq$ 0.

D   (local input) REAL/COMPLEX pointer into the local memory to an array of dimension (DESCA(NB_)).
Details of the factorization as returned by PSPTTRF/PCPTTRF.

E   (local input) REAL/COMPLEX pointer into the local memory to an array of dimension (DESCA(NB_)).
Details of the factorization as returned by PSPTTRF/PCPTTRF.

JA   (global input) INTEGER
The index in the global array A that points to the start of the matrix to be operated on (which may be either all of A or a submatrix of A).

DESCA (global and local input) INTEGER array, dimension (DLEN_).
The array descriptor for the distributed matrix A.
If DESCA(DTYPE_)=501 or 502 then DLEN_$\geq$7,
else if DESCA(DTYPE_)=1 then DLEN_$\geq$9.

B   (local input/local output) REAL/COMPLEX pointer into the local memory to an array of dimension (LLD_B, LOCc(NRHS)).
On entry, this array contains the the local pieces of the right hand sides B(IB:IB+N−1, 1:NRHS).
On exit, this array contains the local pieces of the solution distributed matrix X.

IB   (global input) INTEGER
The row index in the global array B that points to the first row of the matrix to be operated on (which may be either all of B or a submatrix of B).

DESCB (global and local input) INTEGER array, dimension (DLEN_).
The array descriptor for the distributed matrix B.
If DESCB(DTYPE_)=502 then DLEN_$\geq$7,
else if DESCB(DTYPE_)=1 then DLEN_$\geq$9.

AF   (local input) REAL/COMPLEX array of dimension (LAF)
Auxiliary Fillin space. Fillin is created during the factorization routine PSPTTRF/PCPTTRF and this is stored in AF.

LAF   (local input) INTEGER
The dimension of the array AF.
LAF $\geq$ NB+2.
If LAF is not large enough, an error code will be returned and the minimum acceptable size will be returned in AF(1).

WORK   (local workspace/local output) REAL/COMPLEX array, dimension (LWORK)
On exit, WORK(1) returns the minimal and optimal LWORK.

LWORK (local or global input) INTEGER
The dimension of the array WORK.

LWORK is local input and must be at least

LWORK ≥ (10+2*min(100,NRHS))*NPCOL + 4*NRHS.

If LWORK = −1, then LWORK is global input and a workspace query is assumed; the routine only calculates the minimum and optimal size for all work arrays. Each of these values is returned in the first entry of the corresponding work array, and no error message is issued by PXERBLA.

INFO   (global output) INTEGER
= 0:   successful exit
< 0:   If the ith argument is an array and the j-entry had an illegal value, then INFO = −(i*100+j), if the ith argument is a scalar and had an illegal value, then INFO = −i.

PSSTEBZ

```
SUBROUTINE PSSTEBZ( ICTXT, RANGE, ORDER, N, VL, VU, IL, IU,
$                   ABSTOL, D, E, M, NSPLIT, W, IBLOCK, ISPLIT,
$                   WORK, LWORK, IWORK, LIWORK, INFO )
   CHARACTER    ORDER, RANGE
   INTEGER      ICTXT, IL, INFO, IU, LIWORK, LWORK, M, N,
$               NSPLIT
   REAL         ABSTOL, VL, VU
   INTEGER      IBLOCK( * ), ISPLIT( * ), IWORK( * )
   REAL         D( * ), E( * ), W( * ), WORK( * )
```

Purpose

PSSTEBZ computes the eigenvalues of a symmetric tridiagonal matrix in parallel. The user may ask for all eigenvalues, all eigenvalues in the interval [VL, VU], or the eigenvalues indexed IL through IU. A static partitioning of work is done at the beginning of PSSTEBZ which results in all processes finding an (almost) equal number of eigenvalues.

To avoid overflow, the matrix must be scaled so that its largest entry is no greater than overflow$^{1/2}$*underflow$^{1/4}$ in absolute value, and for greatest accuracy, it should not be much smaller than that.

NOTE : It is assumed that the user is on an IEEE machine. If the user is not on an IEEE machine, set the compile time flag NOIEEE to 1 (in SLmake.inc). The features of IEEE arithmetic that are needed for the "fast" Sturm Count are : (a) infinity arithmetic (b) the sign bit of a double precision floating point number is assumed be in the 32nd or 64th bit position (c) the sign of negative zero.

Arguments

ICTXT   (global input) INTEGER
The BLACS context handle.

RANGE   (global input) CHARACTER*1
Specifies which eigenvalues are to be found.
= 'A':   ("All") all eigenvalues will be found.
= 'V':   ("Value") all eigenvalues in the interval [VL, VU] will be found.
= 'I':   ("Index") the ILth through IUth eigenvalues (of the entire matrix) will be found.

ORDER   (global input) CHARACTER*1
Specifies the order in which the eigenvalues and their block numbers are stored in W and IBLOCK.
= 'B':   ("By Block") the eigenvalues will be grouped by split-off block (see IBLOCK, ISPLIT) and ordered from smallest to largest within the block.
= 'E':   ("Entire matrix") the eigenvalues for the entire matrix will be ordered from smallest to largest.

N   (global input) INTEGER
The order of the tridiagonal matrix T. N ≥ 0.

VL, VU   (global input) REAL
If RANGE='V', the lower and upper bounds of the interval to be searched for eigenvalues. Eigenvalues less than or equal to VL, or greater than VU, will not be returned. VL < VU.
Not referenced if RANGE = 'A' or 'I'.

IL, IU   (global input) INTEGER
If RANGE='I', the indices (in ascending order) of the smallest and largest eigenvalues to be returned.
1 ≤ IL ≤ IU ≤ N, if N > 0; IL = 1 and IU = 0 if N = 0.
Not referenced if RANGE = 'A' or 'V'.

ABSTOL   (global input) REAL
The absolute tolerance for the eigenvalues. An eigenvalue (or cluster) is considered to be located if it has been determined to lie in an interval whose width is ABSTOL or less. If ABSTOL is less than or equal to zero, then EPS*||T||$_1$ will be used in its place, where EPS is the relative machine precision.
Eigenvalues will be computed most accurately when ABSTOL is set to the underflow threshold SLAMCH('U'), not zero. Note : If eigenvectors are desired later by inverse iteration (PSSTEIN), ABSTOL should be set to 2*PSLAMCH('S').

D   (global input) REAL array, dimension (N)
The n diagonal elements of the tridiagonal matrix T.

E   (global input) REAL array, dimension (N−1)
The (n−1) off-diagonal elements of the tridiagonal matrix T.

M   (global output) INTEGER
The actual number of eigenvalues found (0 ≤ M ≤ N). (See also the description of INFO = 2.)

NSPLIT   (global output) INTEGER
The number of diagonal blocks in the matrix T. (1 ≤ NSPLIT ≤ N).

W (global output) REAL array, dimension (N)
On exit, the first M elements of W will contain the eigenvalues.

IBLOCK (global output) INTEGER array, dimension (N)
At each row/column j where E(j) is zero or small, the matrix T is considered to split into a block diagonal matrix. On exit, if INFO = 0, IBLOCK(i) specifies to which block (from 1 to NSPLIT) the eigenvalue W(i) belongs.
NOTE: in the (theoretically impossible) event that bisection does not converge for some or all eigenvalues, INFO is set to 1 and the ones for which it did not are identified by a negative block number.

ISPLIT (global output) INTEGER array, dimension (N)
The splitting points, at which T breaks up into submatrices. The first submatrix consists of rows/columns 1 to ISPLIT(1), the second of rows/columns ISPLIT(1)+1 through ISPLIT(2), etc., and the NSPLITth consists of rows/columns ISPLIT(NSPLIT-1)+1 through ISPLIT(NSPLIT)=N. (Only the first NSPLIT elements will actually be used, but since the user cannot know a priori what value NSPLIT will have, N words must be reserved for ISPLIT.)

WORK (local workspace/local output) REAL array, dimension (LWORK)
On exit, WORK(1) returns the minimal and optimal LWORK.

LWORK (local or global input) INTEGER
The dimension of the array WORK.
LWORK is local input and must be at least
LWORK $\geq$ max(5*N, 7).
If LWORK = -1, then LWORK is global input and a workspace query is assumed; the routine only calculates the minimum and optimal size for all work arrays. Each of these values is returned in the first entry of the corresponding work array, and no error message is issued by PXERBLA.

IWORK (local workspace/local output) INTEGER array, dimension (LIWORK)
On exit, IWORK(1) returns the minimal and optimal LIWORK.

LIWORK (local or global input) INTEGER
The dimension of the array IWORK.
LIWORK is local input and must be at least
LIWORK $\geq$ max(4*N, 14, NPROCS).
If LIWORK = -1, then LIWORK is global input and a workspace query is assumed; the routine only calculates the minimum and optimal size for all work arrays. Each of these values is returned in the first entry of the corresponding work array, and no error message is issued by PXERBLA.

INFO (global output) INTEGER
= 0: successful exit
< 0: If the ith argument is an array and the j-entry had an illegal value, then INFO = $-(i*100+j)$, if the ith argument is a scalar and had an illegal value, then INFO = $-i$.
> 0: some or all of the eigenvalues failed to converge or were not computed:

= 1: Bisection failed to converge for some eigenvalues; these eigenvalues are flagged by a negative block number. The effect is that the eigenvalues may not be as accurate as the absolute and relative tolerances. This is generally caused by arithmetic which is less accurate than PSLAMCH says.

= 2: There is a mismatch between the number of eigenvalues output and the number desired.

= 3: RANGE='i', and the Gershgorin interval initially used was incorrect. No eigenvalues were computed. Probable cause: your machine has sloppy floating point arithmetic.
Cure: Increase the PARAMETER "FUDGE", recompile, and try again.

PSSTEIN/PCSTEIN

```
SUBROUTINE PSSTEIN( N, D, E, M, W, IBLOCK, ISPLIT, ORFAC, Z, IZ,
$                    JZ, DESCZ, WORK, LWORK, IWORK, LIWORK, IFAIL,
$                    ICLUSTR, GAP, INFO )

INTEGER      INFO, IZ, JZ, LIWORK, LWORK, M, N
REAL         ORFAC
INTEGER      DESCZ( * ), IBLOCK( * ), ICLUSTR( * ),
$            IFAIL( * ), ISPLIT( * ), IWORK( * )
REAL         D( * ), E( * ), GAP( * ), W( * ), WORK( * ),
$            Z( * )

SUBROUTINE PCSTEIN( N, D, E, M, W, IBLOCK, ISPLIT, ORFAC, Z, IZ,
$                    JZ, DESCZ, WORK, LWORK, IWORK, LIWORK, IFAIL,
$                    ICLUSTR, GAP, INFO )

INTEGER      INFO, IZ, JZ, LIWORK, LWORK, M, N
REAL         ORFAC
INTEGER      DESCZ( * ), IBLOCK( * ), ICLUSTR( * ),
$            IFAIL( * ), ISPLIT( * ), IWORK( * )
REAL         D( * ), E( * ), GAP( * ), W( * ), WORK( * )
COMPLEX      Z( * )
```

Purpose

PSSTEIN/PCSTEIN computes the eigenvectors of a symmetric tridiagonal matrix in parallel, using inverse iteration. The eigenvectors found correspond to user specified eigenvalues. PSSTEIN does not orthogonalize vectors that are on different processes. The extent of orthogonalization is controlled by the input parameter LWORK. Eigenvectors that are to be orthogonalized are computed by the same process. PSSTEIN/PCSTEIN decides on the allocation of work among the processes and then calls SSTEIN2 (modified LAPACK routine) on each individual process. If insufficient workspace is allocated, the expected orthogonalization may not be done.

Note : If the eigenvectors obtained are not orthogonal, increase LWORK and run the code again.

In the Argument section, P = NPROW*NPCOL is the total number of processes.

Arguments

N   (global input) INTEGER
The order of the tridiagonal matrix T. $N \geq 0$.

D   (global input) REAL array, dimension (N)
The n diagonal elements of the tridiagonal matrix T.

E   (global input) REAL array, dimension (N−1)
The (n−1) subdiagonal elements of the tridiagonal matrix T.

M   (global input) INTEGER
The total number of eigenvectors to be found. $0 \leq M \leq N$.

W   (global input) REAL array, dimension (N)
The first M elements of W contain the eigenvalues for which eigenvectors are to be computed. The eigenvalues should be grouped by split-off block and ordered from smallest to largest within the block. (The output array W from SSTEBZ with ORDER = 'B' is expected here.) This array should be replicated on all processes. On output, the first M elements contain the input eigenvalues in ascending order.
Note : To obtain orthogonal vectors, it is best if eigenvalues are computed to highest accuracy (this can be done by setting ABSTOL to the underflow threshold = SLAMCH('U') — ABSTOL is an input parameter to PSSTEBZ)

IBLOCK   (global input) INTEGER array, dimension (N)
The submatrix indices associated with the corresponding eigenvalues in W; IBLOCK(i)=1 if eigenvalue W(i) belongs to the first submatrix from the top, =2 if W(i) belongs to the second submatrix, etc. (The output array IBLOCK from SSTEBZ is expected here.)

ISPLIT   (global input) INTEGER array, dimension (N)
The splitting points, at which T breaks up into submatrices. The first submatrix consists of rows/columns 1 to ISPLIT(1), the second of rows/columns ISPLIT(1)+1 through ISPLIT(2), etc. (The output array ISPLIT from SSTEBZ is expected here.)

ORFAC   (global input) REAL
ORFAC specifies which eigenvectors should be orthogonalized. Eigenvectors that correspond to eigenvalues which are within ORFAC*$\|T\|_1$ of each other are to be orthogonalized. However, if the workspace is insufficient (see LWORK), this tolerance may be decreased until all eigenvectors to be orthogonalized can be stored in one process. No orthogonalization will be done if ORFAC equals zero. A default value of 10^{-3} is used if ORFAC is negative. ORFAC should be identical on all processes.

Z   (local output) REAL/COMPLEX array, dimension (LLD_Z, LOCc(JZ+N−1))
Z contains the computed eigenvectors associated with the specified eigenvalues. Any vector which fails to converge is set to its current iterate after MAXITS iterations (See SSTEIN2).
On output, Z is distributed across the P processes in block cyclic format.

IZ   (global input) INTEGER
The row index in the global array Z indicating the first row of sub(Z).

JZ   (global input) INTEGER
The column index in the global array Z indicating the first column of sub(Z).

DESCZ   (global and local input) INTEGER array, dimension (DLEN_).
The array descriptor for the distributed matrix Z.

WORK   (local workspace/local output) REAL array, dimension (LWORK)
On exit, WORK(1) gives a lower bound on the workspace (LWORK) that guarantees the user desired orthogonalization (see ORFAC). Note that this may overestimate the minimum workspace needed.

LWORK   (local or global input) INTEGER
The dimension of the array WORK.
LWORK controls the extent of orthogonalization which can be done. The number of eigenvectors for which storage is allocated on each process is NVEC = $\lfloor$(LWORK − max(5*N,NP00*MQ00))/N$\rfloor$. Eigenvectors corresponding to eigenvalue clusters of size NVEC − $\lceil M/P \rceil$ + 1 are guaranteed to be orthogonal (the orthogonality is similar to that obtained from SSTEIN2). Note : LWORK must be no smaller than: max(5*N,NP00*MQ00) + $\lceil M/P \rceil$*N, and should have the same input value on all processes. It is the minimum value of LWORK input on different processes that is significant.

IWORK   (local workspace/local output) INTEGER array, dimension (3*N+P+1)
On exit, IWORK(1) contains the amount of integer workspace required, and IWORK(2) through IWORK(P+2) indicate the eigenvectors computed by each process. Process I computes eigenvectors indexed IWORK(I+2)+1 through IWORK(I+3).

LIWORK   (local or global input) INTEGER
The dimension of the array IWORK.
LIWORK is local input and must be at least
$LIWORK \geq 3*N + P + 1$.
If LIWORK = −1, then LIWORK is global input and a workspace query is assumed; the routine only calculates the minimum and optimal size for all work arrays. Each of these values is returned in the first entry of the corresponding work array, and no error message is issued by PXERBLA.

IFAIL   (global output) INTEGER array, dimension (M)
On normal exit, all elements of IFAIL are zero. If one or more eigenvectors fail to converge after MAXITS iterations (as in SSTEIN/CSTEIN),

In its present form, PSSYEV assumes a homogeneous system and makes no checks for consistency of the eigenvalues or eigenvectors across the different processes. Because of this, it is possible that a heterogeneous system may return incorrect results without any error messages.

In the Argument section, NP = the number of rows local to a given process. NQ = the number of columns local to a given process.

Arguments

JOBZ   (global input) CHARACTER*1
        = 'N': Compute eigenvalues only;
        = 'V': Compute eigenvalues and eigenvectors.

UPLO   (global input) CHARACTER*1
        = 'U': Upper triangle of A is stored;
        = 'L': Lower triangle of A is stored.

N   (global input) INTEGER
        The number of rows and columns of the matrix A. $N \geq 0$.

A   (local input/local workspace) REAL array, dimension (LLD_A, LOCc(JA+N−1))
        On entry, the symmetric matrix A. If UPLO = 'U', only the upper triangular part of A is used to define the elements of the symmetric matrix. If UPLO = 'L', only the lower triangular part of A is used to define the elements of the symmetric matrix.
        On exit, the lower triangle (if UPLO='L') or the upper triangle (if UPLO='U') of A, including the diagonal, is destroyed.

IA   (global input) INTEGER
        The row index in the global array A indicating the first row of sub(A).

JA   (global input) INTEGER
        The column index in the global array A indicating the first column of sub(A).

DESCA   (global and local input) INTEGER array, dimension (DLEN_).
        The array descriptor for the distributed matrix A.

W   (global output) REAL array, dimension (N)
        If INFO = 0, the eigenvalues in ascending order.

Z   (local output) REAL array, dimension (LLD_Z, LOCc(JZ+N−1))
        If JOBZ = 'V', then on normal exit the first M columns of Z contain the orthonormal eigenvectors of the matrix corresponding to the selected eigenvalues. If JOBZ = 'N', then Z is not referenced.

IZ   (global input) INTEGER
        The row index in the global array Z indicating the first row of sub(Z).

JZ   (global input) INTEGER
        The column index in the global array Z indicating the first column of sub(Z).

---

then INFO > 0 is returned. If mod(INFO,M+1)>0, then for I=1 to mod(INFO,M+1), the eigenvector corresponding to the eigenvalue W(IFAIL(I)) failed to converge (W refers to the array of eigenvalues on output).

ICLUSTR (global output) INTEGER array, dimension (2*P)
        This output array contains indices of eigenvectors corresponding to a cluster of eigenvalues that could not be orthogonalized due to insufficient workspace (see LWORK, ORFAC and INFO). Eigenvectors corresponding to clusters of eigenvalues indexed ICLUSTR(2*I−1) to ICLUSTR(2*I), I = 1 to INFO/(M+1), could not be orthogonalized due to lack of workspace. Hence the eigenvectors corresponding to these clusters may not be orthogonal. ICLUSTR is a zero terminated array − (ICLUSTR(2*K)≠0 .AND. ICLUSTR(2*K+1) = 0) if and only if K is the number of clusters.

GAP   (global output) REAL array, dimension (P)
        This output array contains the gap between eigenvalues whose eigenvectors could not be orthogonalized. The INFO/M output values in this array correspond to the INFO/(M+1) clusters indicated by the array ICLUSTR. As a result, the dot product between eigenvectors corresponding to the Ith cluster may be as high as (O(n)*macheps) / GAP(I).

INFO   (global output) INTEGER
        = 0: successful exit
        < 0: If the ith argument is an array and the j-entry had an illegal value, then INFO = −(i*100+j), if the ith argument is a scalar and had an illegal value, then INFO = −i.
        > 0: if mod(INFO,M+1) = I, then I eigenvectors failed to converge in MAXITS iterations. Their indices are stored in the array IFAIL. if INFO/(M+1) = I, then eigenvectors corresponding to I clusters of eigenvalues could not be orthogonalized due to insufficient workspace. The indices of the clusters are stored in the array ICLUSTR.

PSSYEV

```
SUBROUTINE PSSYEV( JOBZ, UPLO, N, A, IA, JA, DESCA, W, Z, IZ, JZ,
$                  DESCZ, WORK, LWORK, INFO )
    CHARACTER    JOBZ, UPLO
    INTEGER      IA, INFO, IZ, JA, JZ, LWORK, N
    INTEGER      DESCA( * ), DESCZ( * )
    REAL         A( * ), W( * ), WORK( * ), Z( * )
```

Purpose

PSSYEV computes selected eigenvalues and, optionally, eigenvectors of a real symmetric matrix A by calling the recommended sequence of ScaLAPACK routines.

PSSYEVX/PCHEEVX

```
SUBROUTINE PSSYEVX( JOBZ, RANGE, UPLO, N, A, IA, JA, DESCA, VL,
                    VU, IL, IU, ABSTOL, M, NZ, W, ORFAC, Z, IZ,
                    JZ, DESCZ, WORK, LWORK, IWORK, LIWORK, IFAIL,
                    ICLUSTR, GAP, INFO )

CHARACTER           JOBZ, RANGE, UPLO
INTEGER             IA, IL, INFO, IU, IZ, JA, JZ, LIWORK, LWORK, M,
    $               N, NZ
REAL                ABSTOL, ORFAC, VL, VU
INTEGER             DESCA( * ), DESCZ( * ), ICLUSTR( * ),
    $               IFAIL( * ), IWORK( * )
REAL                A( * ), GAP( * ), W( * ), WORK( * ), Z( * )


SUBROUTINE PCHEEVX( JOBZ, RANGE, UPLO, N, A, IA, JA, DESCA, VL,
                    VU, IL, IU, ABSTOL, M, NZ, W, ORFAC, Z, IZ,
                    JZ, DESCZ, WORK, LWORK, RWORK, LRWORK, IWORK,
                    LIWORK, IFAIL, ICLUSTR, GAP, INFO )

CHARACTER           JOBZ, RANGE, UPLO
INTEGER             IA, IL, INFO, IU, IZ, JA, JZ, LIWORK, LRWORK,
    $               LWORK, M, N, NZ
REAL                ABSTOL, ORFAC, VL, VU
INTEGER             DESCA( * ), DESCZ( * ), ICLUSTR( * ),
    $               IFAIL( * ), IWORK( * )
REAL                GAP( * ), RWORK( * ), W( * )
COMPLEX             A( * ), WORK( * ), Z( * )
```

Purpose

PSSYEVX/PCHEEVX computes selected eigenvalues and, optionally, eigenvectors of a real/complex symmetric/Hermitian matrix A by calling the recommended sequence of ScaLAPACK routines. Eigenvalues/vectors can be selected by specifying a range of values or a range of indices for the desired eigenvalues.

In the Argument section, NP = the number of rows local to a given process. NQ = the number of columns local to a given process.

Arguments

JOBZ   (global input) CHARACTER*1
        = 'N':   Compute eigenvalues only;
        = 'V':   Compute eigenvalues and eigenvectors.

RANGE   (global input) CHARACTER*1
        = 'A':   all eigenvalues will be found.
        = 'V':   all eigenvalues in the interval [VL,VU] will be found.
        = 'I':   the ILth through IU^{th} eigenvalues will be found.

UPLO   (global input) CHARACTER*1
        = 'U':   Upper triangle of A is stored;
        = 'L':   Lower triangle of A is stored.

DESCZ   (global and local input) INTEGER array, dimension (DLEN_).
        The array descriptor for the distributed matrix Z.

WORK   (local workspace/local output) REAL array, dimension (LWORK)
        If JOBZ='N' WORK(1) = minimal=optimal amount of workspace.
        If JOBZ='V' WORK(1) = minimal workspace required to generate all
        the eigenvectors.

LWORK   (local or global input) INTEGER
        The length of the array WORK.
        If no eigenvectors are requested (JOBZ = 'N') then
        LWORK $\geq$ 5*N + 2*NP0 + MQ0 + NB*NN. If eigenvectors are requested
        (JOBZ = 'V') then the amount of workspace required to guarantee
        that all eigenvectors are computed is:
        LWORK $\geq$ 5*N + max(2*NP0 + MQ0 + NB*NN, 2*NN−2) + N*LDC.
        Variable definitions:
        NB = DESCA(MB_) = DESCA(NB_) = DESCZ(MB_) = DESCZ(NB_)
        NN = max(N, NB, 2)
        DESCA(RSRC_) = DESCA(RSRC_) = 0
        DESCZ(RSRC_) = DESCZ(CSRC_) = 0
        NP0 = NUMROC(NN, NB, 0, 0, NPROW)
        MQ0 = NUMROC(max(N, NB, 2), NB, 0, 0, NPCOL)
        NRC = NUMROC(N, NB, MYPROWC, 0, NPROCS)
        LDC = max(1, NRC)

        With MYPROWC defined when a new context is created as:
        CALL BLACS.GET(DESCA CTXT_), 0, CONTEXTC)
        CALL BLACS.GRIDINIT(CONTEXTC, 'R', NPROCS, 1)
        CALL BLACS.GRIDINFO(CONTEXTC, NPROWC, NPCOLC, MYPROWC,
        MYPCOLC)
        If LWORK = −1, the LWORK is global input and a workspace query is
        assumed; the routine only calculates the minimum size for the WORK
        array. The required workspace is returned as the first element of WORK
        and no error message is issued by PXERBLA.

INFO   (global output) INTEGER
        = 0:   successful exit
        < 0:   If the ith argument is an array and the j-entry had an illegal
               value, then INFO = −(i*100+j), if the ith argument is a scalar
               and had an illegal value, then INFO = −i.
        > 0:   If INFO = 1 through N, the ith eigenvalue did not converge in
               SSTEQR2 after a total of 30*N iterations. If INFO = N+1, then
               PSSYEV has detected heterogeneity by finding that eigenvalues
               were not identical across the process grid. In this case, the accu-
               racy of the results from PSSYEV cannot be guaranteed.
```
```

N (global input) INTEGER
The number of rows and columns of the matrix A. $N \geq 0$.

A (local input/workspace) REAL/COMPLEX array, dimension (LLD_A, LOCc(JA+N−1))
On entry, the Hermitian matrix A. If UPLO = 'U', only the upper triangular part of A is used to define the elements of the Hermitian matrix. If UPLO = 'L', only the lower triangular part of A is used to define the elements of the Hermitian matrix.
On exit, the lower triangle (if UPLO='L') or the upper triangle (if UPLO='U') of A, including the diagonal, is destroyed.

IA (global input) INTEGER
The row index in the global array A indicating the first row of sub(A).

JA (global input) INTEGER
The column index in the global array A indicating the first column of sub(A).

DESCA (global and local input) INTEGER array, dimension (DLEN_).
The array descriptor for the distributed matrix A.

VL, VU (global input) REAL
If RANGE='V', the lower and upper bounds of the interval to be searched for eigenvalues. VL < VU.
Not referenced if RANGE = 'A' or 'I'.

IL, IU (global input) INTEGER
If RANGE='I', the indices (in ascending order) of the smallest and largest eigenvalues to be returned.
$1 \leq IL \leq IU \leq N$, if $N > 0$; $IL = 1$ and $IU = 0$ if $N = 0$.
Not referenced if RANGE = 'A' or 'V'.

ABSTOL (global input) REAL
The absolute error tolerance for the eigenvalues. An approximate eigenvalue is accepted as converged when it is determined to lie in an interval [a,b] of width less than or equal to ABSTOL + EPS*max(|a|,|b|), where EPS is the relative machine precision. If ABSTOL is less than or equal to zero, then EPS*$\|T\|_1$ will be used in its place, where T is the tridiagonal matrix obtained by reducing A to tridiagonal form. Eigenvalues will be computed most accurately when ABSTOL is set to twice the underflow threshold 2*SLAMCH('S'), not zero. If this routine returns with ((mod(INFO,2).NE.0) .OR. (mod(INFO/8,2).NE.0)), indicating that some eigenvalues or eigenvectors did not converge, try setting ABSTOL to 2*PSLAMCH('S').

M (global output) INTEGER
The total number of eigenvalues found. $0 \leq M \leq N$.

NZ (global output) INTEGER
Total number of eigenvectors computed. $0 \leq NZ \leq M$.
The number of columns of Z that are filled.
If JOBZ .NE. 'V', NZ is not referenced.
If JOBZ .EQ. 'V', NZ = M unless the user supplies insufficient space and PSSYEVX/PCHEEVX is not able to detect this before beginning computation. To get all the eigenvectors requested, the user must supply both sufficient space to hold the eigenvectors in Z ($M \leq$ DESCZ(N_)) and sufficient workspace to compute them. (See LWORK below.) PSSYEVX/PCHEEVX is always able to detect insufficient space without computation unless RANGE = 'V'.

W (global output) REAL array, dimension (N)
On normal exit, the first M entries contain the selected eigenvalues in ascending order.

ORFAC (global input) REAL
Specifies which eigenvectors should be reorthogonalized. Eigenvectors that correspond to eigenvalues which are within tol=ORFAC*norm(A) of each other are to be reorthogonalized. However, if the workspace is insufficient (see LWORK), tol may be decreased until all eigenvectors to be reorthogonalized can be stored in one process. No reorthogonalization will be done if ORFAC equals zero. A default value of 10^{-3} is used if ORFAC is negative. ORFAC should be identical on all processes.

Z (local output) REAL/COMPLEX array, dimension (LLD_Z, LOCc(JZ+N−1))
If JOBZ = 'V', then on normal exit the first M columns of Z contain the orthonormal eigenvectors of the matrix corresponding to the selected eigenvalues. If an eigenvector fails to converge, then that column of Z contains the latest approximation to the eigenvector, and the index of the eigenvector is returned in IFAIL.
If JOBZ = 'N', then Z is not referenced.

IZ (global input) INTEGER
The row index in the global array Z indicating the first row of sub(Z).

JZ (global input) INTEGER
The column index in the global array Z indicating the first column of sub(Z).

DESCZ (global and local input) INTEGER array, dimension (DLEN_).
The array descriptor for the distributed matrix Z.

WORK (local workspace/local output) REAL/COMPLEX array, dimension (LWORK)
On exit, WORK(1) returns the workspace needed to guarantee completion, but not orthogonality of the eigenvectors. If the input parameters are incorrect, WORK(1) may also be incorrect.
Later we will modify this so if enough workspace is given to complete the request, WORK(1) will return the amount of workspace needed to guarantee orthogonality. This is described in great detail below:
if INFO≥0, then
if JOBZ='N' WORK(1) = minimal=optimal amount of workspace
if JOBZ='V' WORK(1) = minimal workspace required to guarantee orthogonal eigenvectors on the given input matrix with the given ORFAC. (In version 1.0 WORK(1) = minimal workspace required to compute eigenvalues.)

if INFO < 0 then

if JOBZ='N' WORK(1) = minimal=optimal amount of workspace

if JOBZ='V' then

if RANGE='A' or RANGE='I' then

WORK(1) = minimal workspace required to compute all eigenvectors (no guarantee on orthogonality)

if RANGE='V' then

WORK(1) = minimal workspace required to compute N.Z = DESCZ(N_) eigenvectors (no guarantee on orthogonality.) (In version 1.0 WORK(1) = minimal workspace required to compute eigenvalues.)

LWORK   (local or global input) INTEGER
The dimension of the array WORK.
PSSYEVX
If no eigenvectors are requested (JOBZ = 'N') then

$$LWORK \geq 5*N + \max(5*NN, NB*(NP0 + 1))$$

If eigenvectors are requested (JOBZ = 'V') then the amount of workspace required to guarantee that all eigenvectors are computed is:

$$LWORK \geq 5*N + \max(5*NN, NP0*MQ0 + 2*NB*NB) + ICEIL(NEIG, NPROW*NPCOL)*NN$$

PCHEEVX
If only eigenvalues are requested:

$$LWORK \geq N + (NP0 + MQ0 + NB) * NB$$

If eigenvectors are requested:

$$LWORK \geq N + \max(NB * (NP0 + 1), 3).$$

The computed eigenvectors may not be orthogonal if the minimal workspace is supplied and ORFAC is too small. If you want to guarantee orthogonality (at the cost of potentially poor performance) you should add the following to LWORK:

$$(CLUSTERSIZE-1)*N$$

where CLUSTERSIZE is the number of eigenvalues in the largest cluster, where a cluster is defined as a set of close eigenvalues:

$$W(K),\ldots,W(K+CLUSTERSIZE-1)$$
$$W(J+1) \leq W(J) + ORFAC*2*norm(A)$$

Variable definitions:

NEIG = number of eigenvectors requested

NB = DESCA(MB_) = DESCA(NB_) = DESCZ(MB_) = DESCZ(NB_)

NN = max(N, NB, 2)

DESCA(RSRC_) = DESCA(NB_) = DESCZ(NB_) = DESCZ(RSRC_) = DESCZ(CSRC_) = 0

NP0 = NUMROC(NN, NB, 0, 0, NPROW)

MQ0 = NUMROC(max(NEIG, NB, 2), NB, 0, 0, NPCOL)

ICEIL(X, Y) is a ScaLAPACK function returning ceiling(X/Y)

When LWORK is too small:

If LWORK is too small to guarantee orthogonality, PSSYEVX/PCHEEVX attempts to maintain orthogonality in the clusters with the smallest spacing between the eigenvalues. If LWORK is too small to compute all the eigenvectors requested, no computation is performed and INFO=-23 is returned. Note that when RANGE='V', PSSYEVX/PCHEEVX does not know how many eigenvectors are requested until the eigenvalues are computed. Therefore, when RANGE='V' and as long as LWORK is large enough to allow PSSYEVX/PCHEEVX to compute the eigenvalues, PSSYEVX/PCHEEVX will compute the eigenvalues and as many eigenvectors as it can.

Relationship between workspace, orthogonality & performance:
If $CLUSTERSIZE \geq \frac{N}{\sqrt{NPROW*NPCOL}}$, then providing enough space to compute all the eigenvectors orthogonally will cause serious degradation in performance. In the limit (i.e. CLUSTERSIZE = N−1) PSSTEIN/PCSTEIN will perform no better than SSTEIN/CSTEIN on 1 processor. For $CLUSTERSIZE = \frac{N}{\sqrt{NPROW*NPCOL}}$ reorthogonalizing all eigenvectors will increase the total execution time by a factor of 2 or more. For $CLUSTERSIZE > \frac{N}{\sqrt{NPROW*NPCOL}}$ execution time will grow as the square of the cluster size, all other factors remaining equal and assuming enough workspace. Less workspace means less reorthogonalization but faster execution.
If LWORK = −1, then LWORK is global input and a workspace query is assumed; the routine only calculates the minimum and optimal size for all work arrays. Each of these values is returned in the first entry of the corresponding work array, and no error message is issued by PXERBLA.

RWORK   *PCHEEVX only* (local workspace/local output) REAL array, dimension (LRWORK)
On exit, RWORK(1) returns the workspace needed to guarantee completion, but not orthogonality of the eigenvectors.

LRWORK   *PCHEEVX only* (local or global input) INTEGER
The dimension of the array RWORK.
If no eigenvectors are requested (JOBZ = 'N') then

$$LRWORK \geq 5 * NN + 4 * N$$

If eigenvectors are requested (JOBZ = 'V') then the amount of workspace required to guarantee that all eigenvectors are computed is:

$$LRWORK \geq 4*N + \max(5*NN, NP0 * MQ0) + ICEIL(NEIG, NPROW*NPCOL)*NN$$

The computed eigenvectors may not be orthogonal if the minimal workspace is supplied and ORFAC is too small. If you want to guarantee orthogonality (at the cost of potentially poor performance) you should add the following to LRWORK:

$$(CLUSTERSIZE-1)*N$$

where CLUSTERSIZE is the number of eigenvalues in the largest clus-

ter, where a cluster is defined as a set of close eigenvalues:
$W(K),\ldots,W(K+CLUSTERSIZE-1)$
$W(J+1) \leq W(J) + ORFAC*2*norm(A)$

Variable definitions:

NEIG = number of eigenvectors requested
NB = DESCA(MB_) = DESCA(NB_) = DESCZ(MB_) = DESCZ(NB_)
NN = max(N, NB, 2)
DESCA(RSRC_) = DESCA(NB_) = DESCZ(RSRC_) = DESCZ(CSRC_) = 0
NP0 = NUMROC(NN, NB, 0, 0, NPROW)
MQ0 = NUMROC(max(NEIG, NB, 2), NB, 0, 0, NPCOL)

ICEIL(X, Y) is a ScaLAPACK function returning ceiling(X/Y)

When LRWORK is too small:

If LRWORK is too small to guarantee orthogonality, PCHEEVX attempts to maintain orthogonality in the clusters with the smallest spacing between the eigenvalues. If LRWORK is too small to compute all the eigenvectors requested, no computation is performed and INFO=−25 is returned. Note that when RANGE='V', PCHEEVX does not know how many eigenvectors are requested until the eigenvalues are computed. Therefore, when RANGE='V' and as long as LRWORK is large enough to allow PCHEEVX to compute the eigenvalues, PCHEEVX will compute the eigenvalues and as many eigenvectors as it can.

Relationship between workspace, orthogonality & performance:
If $CLUSTERSIZE \geq \dfrac{N}{\sqrt{NPROW*NPCOL}}$, then providing enough space to compute all the eigenvectors orthogonally will cause serious degradation in performance. In the limit (i.e. CLUSTERSIZE = N−1) PCSTEIN will perform no better than CSTEIN on 1 processor. For $CLUSTERSIZE = \dfrac{N}{\sqrt{NPROW*NPCOL}}$ reorthogonalizing all eigenvectors will increase the total execution time by a factor of 2 or more. For $CLUSTERSIZE > \dfrac{N}{\sqrt{NPROW*NPCOL}}$ execution time will grow as the square of the cluster size, all other factors remaining equal and assuming enough workspace. Less workspace means less reorthogonalization but faster execution.

If LRWORK = −1, then LRWORK is global input and a workspace query is assumed; the routine only calculates the minimum and optimal size for all work arrays. Each of these values is returned in the first entry of the corresponding work array, and no error message is issued by PXERBLA.

IWORK   (local workspace/local output) INTEGER array, dimension (LIWORK)
On exit, IWORK(1) contains the amount of integer workspace required.
If the input parameters are incorrect, IWORK(1) may also be incorrect.

LIWORK   (local or global input) INTEGER
The dimension of the array IWORK.
$LIWORK \geq 6 * NNP$, where $NNP = max(N, NPROW*NPCOL + 1, 4)$.

If LIWORK = −1, then LIWORK is global input and a workspace query is assumed; the routine only calculates the minimum and optimal size for all work arrays. Each of these values is returned in the first entry of the corresponding work array, and no error message is issued by PXERBLA.

IFAIL   (global output) INTEGER array, dimension (N)
If JOBZ = 'V', then on normal exit, the first M elements of IFAIL are zero. If (mod(INFO,2).NE.0) on exit, then IFAIL contains the indices of the eigenvectors that failed to converge. If JOBZ = 'N', then IFAIL is not referenced.

ICLUSTR   (global output) INTEGER array, dimension (2*NPROW*NPCOL)
This array contains indices of eigenvalues corresponding to a cluster of eigenvalues that could not be reorthogonalized due to insufficient workspace (see LWORK, ORFAC and INFO). Eigenvectors corresponding to clusters of eigenvalues indexed ICLUSTR(2*I−1) to ICLUSTR(2*I), could not be reorthogonalized due to lack of workspace. Hence the eigenvectors corresponding to these clusters may not be orthogonal. ICLUSTR() is a zero terminated array. (ICLUSTR(2*K).NE.0 .AND. ICLUSTR(2*K+1).EQ.0) if and only if K is the number of clusters. ICLUSTR is not referenced if JOBZ = 'N'.

GAP   (global output) REAL array, dimension (NPROW*NPCOL)
This array contains the gap between eigenvalues whose eigenvectors could not be reorthogonalized. The output values in this array correspond to the clusters indicated by the array ICLUSTR. As a result, the dot product between eigenvectors corresponding to the Ith cluster may be as high as (C * n) / GAP(I) where C is a small constant.

INFO   (global output) INTEGER
= 0:   successful exit
< 0:   If the ith argument is an array and the j-entry had an illegal value, then INFO = −(i*100+j), if the ith argument is a scalar and had an illegal value, then INFO = −i.
> 0:   if (mod(INFO,2).NE.0), then one or more eigenvectors failed to converge. Their indices are stored in IFAIL. Ensure ABSTOL=2.0*PSLAMCH('U').
if (mod(INFO/2,2).NE.0),then eigenvectors corresponding to one or more clusters of eigenvalues could not be reorthogonalized because of insufficient workspace. The indices of the clusters are stored in the array ICLUSTR.
if (mod(INFO/4,2).NE.0), then space limit prevented PSSYEVX/PCHEEVX from computing all of the eigenvectors between VL and VU. The number of eigenvectors computed is returned in NZ.
if (mod(INFO/8,2).NE.0), then PSSTEBZ failed to compute eigenvalues. Ensure ABSTOL=2.0*PSLAMCH('U').

PSSYGST/PCHEGST

```
SUBROUTINE PSSYGST( IBTYPE, UPLO, N, A, IA, JA, DESCA, B, IB, JB,
$                    DESCB, SCALE, INFO )
CHARACTER       UPLO
INTEGER         IA, IB, IBTYPE, INFO, JA, JB, N
REAL            SCALE
INTEGER         DESCA( * ), DESCB( * )
REAL            A( * ), B( * )


SUBROUTINE PCHEGST( IBTYPE, UPLO, N, A, IA, JA, DESCA, B, IB, JB,
$                    DESCB, SCALE, INFO )
CHARACTER       UPLO
INTEGER         IA, IB, IBTYPE, INFO, JA, JB, N
REAL            SCALE
INTEGER         DESCA( * ), DESCB( * )
COMPLEX         A( * ), B( * )
```

Purpose

PSSYGST/PCHEGST reduces a real/complex symmetric/Hermitian definite generalized eigenproblem to standard form.

In the following sub(A) denotes A(IA:IA+N−1, JA:JA+N−1) and sub(B) denotes B(IB:IB+N−1, JB:JB+N−1).

If IBTYPE = 1, the problem is sub(A)*x = λ*sub(B)*x, and sub(A) is overwritten by $(U^H)^{-1}$*sub(A)*U^{-1} or L^{-1}*sub(A)*$(L^H)^{-1}$.

If IBTYPE = 2 or 3, the problem is sub(A)*sub(B)*x = λ*x or sub(B)*sub(A)*x = λ*x, and sub(A) is overwritten by U*sub(A)*U^H or L^H*sub(A)*L.

sub(B) must have been previously factorized as U^H*U or $L*L^H$ by PSPOTRF/PCPOTRF.

Arguments

IBTYPE   (global input) INTEGER
         = 1: compute $(U^H)^{-1}$*sub(A)*U^{-1} or L^{-1}*sub(A)*$(L^H)^{-1}$;
         = 2 or 3: compute U*sub(A)*U^H or L^H*sub(A)*L.

UPLO   (global input) CHARACTER
         = 'U': Upper triangle of sub(A) is stored and sub(B) is factored as U^H*U;
         = 'L': Lower triangle of sub(A) is stored and sub(B) is factored as $L*L^H$.

N   (global input) INTEGER
         The order of the matrices sub(A) and sub(B). N $\geq$ 0.

A   (local input/local output) REAL/COMPLEX pointer into the local memory to an array of dimension (LLD_A, LOCc(JA+N−1)).
         On entry, this array contains the local pieces of the n-by-n symmetric/Hermitian distributed matrix sub(A). If UPLO = 'U', the leading n-by-n upper triangular part of sub(A) contains the upper triangular part of the matrix, and its strictly lower triangular part is not referenced. If UPLO = 'L', the leading n-by-n lower triangular part of sub(A) contains the lower triangular part of the matrix, and its strictly upper triangular part is not referenced.
         On exit, if INFO = 0, the transformed matrix, stored in the same format as sub(A).

IA   (global input) INTEGER
         The row index in the global array A indicating the first row of sub(A).

JA   (global input) INTEGER
         The column index in the global array A indicating the first column of sub(A).

DESCA   (global and local input) INTEGER array, dimension (DLEN_).
         The array descriptor for the distributed matrix A.

B   (local input) REAL/COMPLEX pointer into the local memory to an array of dimension (LLD_B, LOCc(JB+N−1)).
         On entry, this array contains the local pieces of the triangular factor from the Cholesky factorization of sub(B), as returned by PSPOTRF/PCPOTRF.

IB   (global input) INTEGER
         The row index in the global array B indicating the first row of sub(B).

JB   (global input) INTEGER
         The column index in the global array B indicating the first column of sub(B).

DESCB   (global and local input) INTEGER array, dimension (DLEN_).
         The array descriptor for the distributed matrix B.

SCALE   (global output) REAL
         Amount by which the eigenvalues should be scaled to compensate for the scaling performed in this routine. At present, SCALE is always returned as 1.0, it is returned here to allow for future enhancement.

INFO   (global output) INTEGER
         = 0: successful exit
         < 0: If the ith argument is an array and the j-entry had an illegal value, then INFO = −(i*100+j), if the ith argument is a scalar and had an illegal value, then INFO = −i.

PSSYGVX/PCHEGVX

```
SUBROUTINE PSSYGVX( IBTYPE, JOBZ, RANGE, UPLO, N, A, IA, JA,
$                   DESCA, B, IB, JB, DESCB, VL, VU, IL, IU,
$                   ABSTOL, M, NZ, W, ORFAC, Z, IZ, JZ, DESCZ,
$                   WORK, LWORK, IWORK, LIWORK, IFAIL, ICLUSTR,
$                   GAP, INFO )
CHARACTER          JOBZ, RANGE, UPLO
INTEGER            IA, IB, IBTYPE, IL, INFO, IU, IZ, JA, JB, JZ,
$                  LIWORK, LWORK, M, N, NZ
REAL               ABSTOL, ORFAC, VL, VU
INTEGER            DESCA( * ), DESCB( * ), DESCZ( * ),
$                  ICLUSTR( * ), IFAIL( * ), IWORK( * )
REAL               A( * ), B( * ), GAP( * ), W( * ), WORK( * ),
$                  Z( * )

SUBROUTINE PCHEGVX( IBTYPE, JOBZ, RANGE, UPLO, N, A, IA, JA,
$                   DESCA, B, IB, JB, DESCB, VL, VU, IL, IU,
$                   ABSTOL, M, NZ, W, ORFAC, Z, IZ, JZ, DESCZ,
$                   WORK, LWORK, RWORK, LRWORK, IWORK, LIWORK,
$                   IFAIL, ICLUSTR, GAP, INFO )
CHARACTER          JOBZ, RANGE, UPLO
INTEGER            IA, IB, IBTYPE, IL, INFO, IU, IZ, JA, JB, JZ,
$                  LIWORK, LRWORK, LWORK, M, N, NZ
REAL               ABSTOL, ORFAC, VL, VU
INTEGER            DESCA( * ), DESCB( * ), DESCZ( * ),
$                  ICLUSTR( * ), IFAIL( * ), IWORK( * )
REAL               GAP( * ), RWORK( * ), W( * )
COMPLEX            A( * ), B( * ), WORK( * ), Z( * )
```

Purpose

PSSYGVX/PCHEGVX computes all the eigenvalues, and optionally, the eigenvectors of a real/complex generalized symmetric/Hermitian definite eigenproblem, of the form

$$\text{sub}(A)x = \lambda\,\text{sub}(B)x, \quad \text{sub}(A)\,\text{sub}(B)x = \lambda x, \quad \text{or} \quad \text{sub}(B)\,\text{sub}(A)x = \lambda x.$$

Here sub(A) denoting A(IA:IA+N−1, JA:JA+N−1) is assumed to be symmetric/Hermitian, and sub(B) denoting B(IB:IB+N−1, JB:JB+N−1) is assumed to be symmetric/Hermitian positive definite.

Arguments

IBTYPE   (input) INTEGER
Specifies the problem type to be solved:
= 1: $\text{sub}(A)x = \lambda\,\text{sub}(B)x$
= 2: $\text{sub}(A)\,\text{sub}(B)x = \lambda x$
= 3: $\text{sub}(B)\,\text{sub}(A)x = \lambda x$

JOBZ   (global input) CHARACTER*1
= 'N':   Compute eigenvalues only;
= 'V':   Compute eigenvalues and eigenvectors.

UPLO   (global input) CHARACTER*1
= 'U':   Upper triangles of sub(A) and sub(B) are stored;
= 'L':   Lower triangles of sub(A) and sub(B) are stored.

N   (global input) INTEGER
The order of the matrices sub(A) and sub(B). N ≥ 0.

A   (local input/local output) REAL/COMPLEX pointer into the local memory to an array of dimension (LLD_A, LOCc(JA+N−1)). On entry, this array contains the local pieces of the n-by-n symmetric/Hermitian distributed matrix sub(A). If UPLO = 'U', the leading n-by-n upper triangular part of sub(A) contains the upper triangular part of the matrix. If UPLO = 'L', the leading n-by-n lower triangular part of sub(A) contains the lower triangular part of the matrix.
On exit, if JOBZ = 'V', then if INFO = 0, sub(A) contains the distributed matrix Z of eigenvectors. The eigenvectors are normalized as follows:
if IBTYPE = 1 or 2, $Z^H * \text{sub}(B) * Z = I;$
if IBTYPE = 3, $Z^H * \text{sub}(B)^{-1} * Z = I.$
If JOBZ = 'N', then on exit the upper triangle (if UPLO='U') or the lower triangle (if UPLO='L') of A, including the diagonal, is destroyed.

IA   (global input) INTEGER
The row index in the global array A indicating the first row of sub(A).

JA   (global input) INTEGER
The column index in the global array A indicating the first column of sub(A).

DESCA   (global and local input) INTEGER array, dimension (DLEN_).
The array descriptor for the distributed matrix A.

B   (local input/local output) REAL/COMPLEX pointer into the local memory to an array of dimension (LLD_B, LOCc(JB+N−1)). On entry, this array contains the local pieces of the n-by-n symmetric/Hermitian distributed matrix sub(B). If UPLO = 'U', the leading n-by-n upper triangular part of sub(B) contains the upper triangular part of the matrix. If UPLO = 'L', the leading n-by-n lower triangular part of sub(B) contains the lower triangular part of the matrix.
On exit, if INFO ≤ N, the part of sub(B) containing the matrix is overwritten by the triangular factor U or L from the Cholesky factorization $\text{sub}(B) = U^H * U$ or $\text{sub}(B) = L * L^H.$

IB   (global input) INTEGER
The row index in the global array B indicating the first row of sub(B).

JB   (global input) INTEGER
The column index in the global array B indicating the first column of sub(B).

DESCB (global and local input) INTEGER array, dimension (DLEN_). The array descriptor for the distributed matrix B.

VL, VU (global input) REAL If RANGE='V', the lower and upper bounds of the interval to be searched for eigenvalues. VL < VU. Not referenced if RANGE = 'A' or 'I'.

IL, IU (global input) INTEGER If RANGE='I', the indices (in ascending order) of the smallest and largest eigenvalues to be returned. $1 \leq IL \leq IU \leq N$, if $N > 0$; IL = 1 and IU = 0 if N = 0. Not referenced if RANGE = 'A' or 'V'.

ABSTOL (global input) REAL If JOBZ='V', setting ABSTOL to PSLAMCH(CONTEXT, 'U') yields the most orthogonal eigenvectors.
The absolute error tolerance for the eigenvalues. An approximate eigenvalue is accepted as converged when it is determined to lie in an interval [a,b] of width less than or equal to ABSTOL + EPS*max(|a|,|b|), where EPS is the relative machine precision. If ABSTOL is less than or equal to zero, then EPS*||T||1 will be used in its place, where T is the tridiagonal matrix obtained by reducing A to tridiagonal form. Eigenvalues will be computed most accurately when ABSTOL is set to twice the underflow threshold 2*SLAMCH('S'), not zero. If this routine returns with ((mod(INFO,2).NE.0) .OR. (mod(INFO/8,2).NE.0)), indicating that some eigenvalues or eigenvectors did not converge, try setting ABSTOL to 2*PSLAMCH('S').

M (global output) INTEGER Total number of eigenvalues found. $0 \leq M \leq N$.

NZ (global output) INTEGER Total number of eigenvectors computed. $0 \leq NZ \leq M$. The number of columns of Z that are filled. If JOBZ $\neq$ 'V', NZ is not referenced. If JOBZ = 'V', NZ = M unless the user supplies insufficient space and PSSYGVX/PCHEGVX is not able to detect this before beginning computation. To get all the eigenvectors requested, the user must supply both sufficient space to hold the eigenvectors in Z (M $\leq$ DESCZ(N_)) and sufficient workspace to compute them. (See LWORK below.) PSSYGVX/PCHEGVX is always able to detect insufficient space without computation unless RANGE = 'V'.

W (global output) REAL array, dimension (N) If INFO = 0, the eigenvalues in ascending order.

ORFAC (global input) REAL Specifies which eigenvectors should be reorthogonalized. Eigenvectors that correspond to eigenvalues which are within tol=ORFAC*norm(A) of each other are to be reorthogonalized. However, if the workspace is insufficient (see LWORK), tol may be decreased until all eigenvectors to be reorthogonalized can be stored in one process. No reorthogonaliza-

tion will be done if ORFAC equals zero. A default value of 10^{-3} is used if ORFAC is negative. ORFAC should be identical on all processes.

Z (local output) COMPLEX array, dimension (LLD_Z, LOCc(JZ+N−1)) If JOBZ = 'V', then on normal exit the first M columns of Z contain the orthonormal eigenvectors of the matrix corresponding to the selected eigenvalues. If an eigenvector fails to converge, then that column of Z contains the latest approximation to the eigenvector, and the index of the eigenvector is returned in IFAIL. If JOBZ = 'N', then Z is not referenced.

IZ (global input) INTEGER The row index in the global array Z indicating the first row of sub(Z).

JZ (global input) INTEGER The column index in the global array Z indicating the first column of sub(Z).

DESCZ (global and local input) INTEGER array, dimension (DLEN_). The array descriptor for the distributed matrix Z.

WORK (local workspace/local output) REAL/COMPLEX array, dimension (LWORK) On output, WORK(1) returns the workspace needed to guarantee completion, but not orthogonality of the eigenvectors. If the input parameters are incorrect, WORK(1) may also be incorrect. Later we will modify this so if enough workspace is given to complete the request, WORK(1) will return the amount of workspace needed to guarantee orthogonality. This is described in great detail below:

if INFO $\geq$ 0, then
  if JOBZ='N' WORK(1) = minimal=optimal amount of workspace
  if JOBZ='V' WORK(1) = minimal workspace required to guarantee orthogonal eigenvectors on the given input matrix with the given ORFAC. (In version 1.0 WORK(1) = minimal workspace required to compute eigenvalues.)
if INFO < 0 then
  if JOBZ='N' WORK(1) = minimal=optimal amount of workspace
  if JOBZ='V' then
    if RANGE='A' or RANGE='I' then
      WORK(1) = minimal workspace required to compute all eigenvectors (no guarantee on orthogonality)
    if RANGE='V' then
      WORK(1) = minimal workspace required to compute N_Z = DESCZ(N_) eigenvectors (no guarantee on orthogonality.) (In version 1.0 WORK(1) = minimal workspace required to compute eigenvalues.)

LWORK (local or global input) INTEGER The dimension of the array WORK.
PSSYGVX
If no eigenvectors are requested (JOBZ = 'N') then
LWORK $\geq$ 5*N + max(5*NN, NB*(NP0 + 1))

If eigenvectors are requested (JOBZ = 'V') then the amount of workspace required to guarantee that all eigenvectors are computed is:

$$LWORK \geq 5*N + max(5*NN, NP0*MQ0 + 2*NB*NB) + ICEIL(NEIG, NPROW*NPCOL)*NN$$

PCHEGVX

If only eigenvalues are requested:
$$LWORK \geq N + (NP0 + MQ0 + NB) * NB$$
If eigenvectors are requested:
$$LWORK \geq N + max(NB * (NP0 + 1), 3)$$

The computed eigenvectors may not be orthogonal if the minimal workspace is supplied and ORFAC is too small. If you want to guarantee orthogonality (at the cost of potentially poor performance) you should add the following to LWORK: (CLUSTERSIZE-1)*N where a CLUSTERSIZE is the number of eigenvalues in the largest cluster, where a cluster is defined as a set of close eigenvalues:

$$W(K),...,W(K+CLUSTERSIZE-1)$$
$$W(J+1) \leq W(J) + ORFAC*2*norm(A)$$

Variable definitions:

NEIG = number of eigenvectors requested

NB = DESCA(MB_) = DESCA(NB_) = DESCZ(MB_) = DESCZ(NB_)

NN = max(N, NB, 2)

DESCA(RSRC_) = DESCA(NB_) = DESCZ(RSRC_) = DESCZ(CSRC_) = 0

NP0 = NUMROC(NN, NB, 0, 0, NPROW)

MQ0 = NUMROC(max(NEIG, NB, 2), NB, 0, 0, NPCOL)

ICEIL(X, Y) is a ScaLAPACK function returning ceiling(X/Y)

When LWORK is too small:

If LWORK is too small to guarantee orthogonality, PSSYGVX/PCHEGVX attempts to maintain orthogonality in the clusters with the smallest spacing between the eigenvalues. If LWORK is too small to compute all the eigenvectors requested, no computation is performed and INFO=-23 is returned. Note that when RANGE='V', PSSYGVX/PCHEGVX does not know how many eigenvectors are requested until the eigenvalues are computed. Therefore, when RANGE='V' and as long as LWORK is large enough to allow PSSYGVX/PCHEGVX to compute the eigenvalues, PSSYGVX/PCHEGVX will compute the eigenvalues and as many eigenvectors as it can.

Relationship between workspace, orthogonality & performance:

If $CLUSTERSIZE \geq \frac{N}{\sqrt{NPROW*NPCOL}}$, then providing enough space to compute all the eigenvectors orthogonally will cause serious degradation in performance. In the limit (i.e. CLUSTERSIZE = N−1) PSSTEIN will perform no better than SSTEIN on 1 processor. For

$CLUSTERSIZE = \frac{N}{\sqrt{NPROW*NPCOL}}$ reorthogonalizing all eigenvectors will increase the total execution time by a factor of 2 or more. For $CLUSTERSIZE > \frac{N}{\sqrt{NPROW*NPCOL}}$ execution time will grow as the square of the cluster size, all other factors remaining equal and assuming enough workspace. Less workspace means less reorthogonalization but faster execution.

If LWORK = −1, then LWORK is global input and a workspace query is assumed; the routine only calculates the minimum and optimal size for all work arrays. Each of these values is returned in the first entry of the corresponding work array, and no error message is issued by PXERBLA.

RWORK   *PCHEGVX only* (local workspace/local output) REAL array, dimension (LRWORK)

On exit, RWORK(1) returns the workspace needed to guarantee completion, but not orthogonality of the eigenvectors.

LRWORK   *PCHEGVX only* (local or global input) INTEGER

The dimension of the array RWORK.

If no eigenvectors are requested (JOBZ = 'N') then
$$LRWORK \geq 5 * NN + 4 * N$$

If eigenvectors are requested (JOBZ = 'V') then the amount of workspace required to guarantee that all eigenvalues are computed is:
$$LRWORK \geq 4*N + max(5*NN, NP0*MQ0) + ICEIL(NEIG, NPROW*NPCOL)*NN$$

The computed eigenvectors may not be orthogonal if the minimal workspace is supplied and ORFAC is too small. If you want to guarantee orthogonality (at the cost of potentially poor performance) you should add the following to LRWORK:

(CLUSTERSIZE−1)*N
where CLUSTERSIZE is the number of eigenvalues in the largest cluster, where a cluster is defined as a set of close eigenvalues:

$$W(K),...,W(K+CLUSTERSIZE-1)$$
$$W(J+1) \leq W(J) + ORFAC*2*norm(A)$$

Variable definitions:

NEIG = number of eigenvectors requested

NB = DESCA(MB_) = DESCA(NB_) = DESCZ(MB_) = DESCZ(NB_)

NN = max(N, NB, 2)

DESCA(RSRC_) = DESCA(NB_) = DESCZ(RSRC_) = DESCZ(CSRC_) = 0

NP0 = NUMROC(NN, NB, 0, 0, NPROW)

MQ0 = NUMROC(max(NEIG, NB, 2), NB, 0, 0, NPCOL)

ICEIL(X, Y) is a ScaLAPACK function returning ceiling(X/Y)

When LRWORK is too small:

If LRWORK is too small to guarantee orthogonality, PCHEGVX attempts to maintain orthogonality in the clusters with the smallest spacing between the eigenvalues. If LRWORK is too small to compute all the eigenvectors requested, no computation is performed and INFO=−25 is returned. Note that when RANGE='V', PCHEGVX does not know how many eigenvectors are requested until the eigenvalues are computed. Therefore, when RANGE='V' and as long as LRWORK is large enough to allow PCHEGVX to compute the eigenvalues, PCHEGVX will compute the eigenvalues and as many eigenvectors as it can.

Relationship between workspace, orthogonality & performance:
If CLUSTERSIZE $\geq \frac{N}{\sqrt{NPROW*NPCOL}}$, then providing enough space to compute all the eigenvectors orthogonally will cause serious degradation in performance. In the limit (i.e. CLUSTERSIZE = N−1) PCSTEIN will perform no better than CSTEIN on 1 processor. For CLUSTERSIZE $= \frac{N}{\sqrt{NPROW*NPCOL}}$ reorthogonalizing all eigenvectors will increase the total execution time by a factor of 2 or more. For CLUSTERSIZE $> \frac{N}{\sqrt{NPROW*NPCOL}}$ execution time will grow as the square of the cluster size, all other factors remaining equal and assuming enough workspace. Less workspace means less reorthogonalization but faster execution.
If LRWORK = −1, then LRWORK is global input and a workspace query is assumed; the routine only calculates the minimum and optimal size for all work arrays. Each of these values is returned in the first entry of the corresponding work array, and no error message is issued by PXERBLA.

IWORK   (local workspace/local output) INTEGER array, dimension (LIWORK)
On return, IWORK(1) contains the amount of integer workspace required. If the input parameters are incorrect, IWORK(1) may also be incorrect.

LIWORK (local or global input) INTEGER.
The dimension of the array IWORK.
LIWORK ≥ 6 * NNP, where NNP = max(N, NPROW*NPCOL + 1, 4).
If LIWORK = −1, then LIWORK is global input and a workspace query is assumed; the routine only calculates the minimum and optimal size for all work arrays. Each of these values is returned in the first entry of the corresponding work array, and no error message is issued by PXERBLA.

IFAIL   (global output) INTEGER array, dimension (N)
IFAIL provides additional information when INFO .NE. 0 If (mod(INFO/16,2).NE.0) then IFAIL(1) indicates the order of the smallest minor which is not positive definite. If (mod(INFO/2).NE.0) on exit, then IFAIL contains the indices of the eigenvectors that failed to converge.
If neither of the above error conditions hold and JOBZ = 'V', then the first M elements of IFAIL are set to zero.

ICLUSTR (global output) INTEGER array, dimension (2*NPROW*NPCOL)
This array contains indices of eigenvectors corresponding to a cluster of eigenvalues that could not be reorthogonalized due to insufficient workspace (see LWORK, ORFAC and INFO). Eigenvectors corresponding to clusters of eigenvalues indexed ICLUSTR(2*I−1)
to ICLUSTR(2*I), could not be reorthogonalized due to lack of workspace. Hence the eigenvectors corresponding to these clusters may not be orthogonal. ICLUSTR() is a zero terminated array. (ICLUSTR(2*K).NE.0 .AND. ICLUSTR(2*K+1).EQ.0) if and only if K is the number of clusters. ICLUSTR is not referenced if JOBZ = 'N'.

GAP   (global output) REAL array, dimension (NPROW*NPCOL)
This array contains the gap between eigenvalues whose eigenvectors could not be reorthogonalized. The output values in this array correspond to the clusters indicated by the array ICLUSTR. As a result, the dot product between eigenvectors corresponding to the Ith cluster may be as high as (C * n) / GAP(I) where C is a small constant.

INFO   (global output) INTEGER
= 0:   successful exit
< 0:   If the ith argument is an array and the j-entry had an illegal value, then INFO = −(i*100+j), if the ith argument is a scalar and had an illegal value, then INFO = −i.
> 0:   if (mod(INFO,2).NE.0), then one or more eigenvectors failed to converge. Their indices are stored in IFAIL.
if (mod(INFO/2,2).NE.0),then eigenvectors corresponding to one or more clusters of eigenvalues could not be reorthogonalized because of insufficient workspace. The indices of the clusters are stored in the array ICLUSTR.
if (mod(INFO/4,2).NE.0), then space limit prevented PSSYEVX from computing all of the eigenvectors between VL and VU. The number of eigenvectors computed is returned in NZ.
if (mod(INFO/8,2).NE.0), then PSSTEBZ failed to compute eigenvalues.
if (mod(INFO/16,2).NE.0), then B was not positive definite. IFAIL(1) indicates the order of the smallest minor which is not positive definite.

PSSYTRD/PCHETRD

```
SUBROUTINE PSSYTRD( UPLO, N, A, IA, JA, DESCA, D, E, TAU, WORK,
$                   LWORK, INFO )
      CHARACTER         UPLO
      INTEGER           IA, INFO, JA, LWORK, N
      INTEGER           DESCA( * )
      REAL              A( * ), D( * ), E( * ), TAU( * ), WORK( * )
```

```
SUBROUTINE PCHETRD( UPLO, N, A, IA, JA, DESCA, D, E, TAU, WORK,
$                   LWORK, INFO )

    CHARACTER       UPLO
    INTEGER         IA, INFO, JA, LWORK, N
    INTEGER         DESCA( * )
    REAL            D( * ), E( * )
    COMPLEX         A( * ), TAU( * ), WORK( * )
```

Purpose

PSSYTRD/PCHETRD reduces a real/complex symmetric/Hermitian matrix sub(A) to real symmetric tridiagonal form T by an orthogonal/unitary similarity transformation:
$Q^H*\text{sub}(A)*Q = T$, where $\text{sub}(A) = A(IA:IA+N-1,JA:JA+N-1)$.

Arguments

UPLO   (global input) CHARACTER*1
        Specifies whether the upper or lower triangular part of the symmetric/Hermitian matrix sub(A) is stored:
        = 'U': Upper triangular;
        = 'L': Lower triangular.

N   (global input) INTEGER
        The number of rows and columns to be operated on, i.e., the order of the distributed submatrix sub(A). $N \geq 0$.

A   (local input/local output) REAL/COMPLEX pointer into the local memory to an array of dimension (LLD_A,LOCc(JA+N-1)).
        On entry, this array contains the local pieces of the symmetric/Hermitian distributed matrix sub(A).
        If UPLO = 'U', the leading n-by-n upper triangular part of sub(A) contains the upper triangular part of the matrix, and its strictly lower triangular part is not referenced.
        If UPLO = 'L', the leading n-by-n lower triangular part of sub(A) contains the lower triangular part of the matrix, and its strictly upper triangular part is not referenced.
        On exit, if UPLO = 'U', the diagonal and first superdiagonal of sub(A) are overwritten by the corresponding elements of the tridiagonal matrix T, and the elements above the first superdiagonal, with the array TAU, represent the orthogonal/unitary matrix Q as a product of elementary reflectors; if UPLO = 'L', the diagonal and first subdiagonal of sub(A) are overwritten by the corresponding elements of the tridiagonal matrix T, and the elements below the first subdiagonal, with the array TAU, represent the orthogonal/unitary matrix Q as a product of elementary reflectors.

IA   (global input) INTEGER
        The row index in the global array A indicating the first row of sub(A).

JA   (global input) INTEGER
        The column index in the global array A indicating the first column of sub(A).

DESCA   (global and local input) INTEGER array, dimension (DLEN_).
        The array descriptor for the distributed matrix A.

D   (local output) REAL array, dimension LOCc(JA+N-1)
        The diagonal elements of the tridiagonal matrix T: D(i) = A(i,i). D is tied to the distributed matrix A.

E   (local output) REAL array, dimension LOCc(JA+N-1)
        The off-diagonal elements of the tridiagonal matrix T: E(i) = A(i,i+1) if UPLO = 'U', E(i) = A(i+1,i) if UPLO = 'L'. E is tied to the distributed matrix A.

TAU   (local output) REAL/COMPLEX array, dimension LOCc(JA+N-1)
        The scalar factors of the elementary reflectors. TAU is tied to the distributed matrix A.

WORK   (local workspace/local output) REAL/COMPLEX array, dimension (LWORK)
        On exit, WORK(1) returns the minimal and optimal LWORK.

LWORK   (local or global input) INTEGER
        The dimension of the array WORK.
        LWORK is local input and must be at least
        $\text{LWORK} \geq \max(NB * (NP +1), 3 * NB)$.
        where NB = MB_A = NB_A,
        NP = NUMROC(N, NB, MYROW, IAROW, NPROW),
        IAROW = INDXG2P(IA, NB, MYROW, RSRC_A, NPROW).
        INDXG2P and NUMROC are ScaLAPACK tool functions; MYROW, MYCOL, NPROW and NPCOL can be determined by calling the subroutine BLACS_GRIDINFO.
        If LWORK = -1, then LWORK is global input and a workspace query is assumed; the routine only calculates the minimum and optimal size for all work arrays. Each of these values is returned in the first entry of the corresponding work array, and no error message is issued by PXERBLA.

INFO   (global output) INTEGER
        = 0: successful exit
        < 0: If the ith argument is an array and the j-entry had an illegal value, then INFO = -(i*100+j), if the ith argument is a scalar and had an illegal value, then INFO = -i.
```
```

PSTRCON/PCTRCON

```
SUBROUTINE PSTRCON( NORM, UPLO, DIAG, N, A, IA, JA, DESCA, RCOND,
$                   WORK, LWORK, IWORK, LIWORK, INFO )
    CHARACTER    DIAG, NORM, UPLO
    INTEGER      IA, JA, INFO, LIWORK, LWORK, N
    REAL         RCOND
    INTEGER      DESCA( * ), IWORK( * )
    REAL         A( * ), WORK( * )

SUBROUTINE PCTRCON( NORM, UPLO, DIAG, N, A, IA, JA, DESCA, RCOND,
$                   WORK, LWORK, RWORK, LRWORK, INFO )
    CHARACTER    DIAG, NORM, UPLO
    INTEGER      IA, JA, INFO, LRWORK, LWORK, N
    REAL         RCOND
    INTEGER      DESCA( * )
    REAL         RWORK( * )
    COMPLEX      A( * ), WORK( * )
```

Purpose

PSTRCON/PCTRCON estimates the reciprocal of the condition number of a triangular matrix A(IA:IA+N−1,JA:JA+N−1), in either the 1-norm or the infinity-norm.

The norm of A(IA:IA+N−1,JA:JA+N−1) is computed and an estimate is obtained for $\|A(IA:IA+N-1,JA:JA+N-1)^{-1}\|$, then the reciprocal of the condition number is computed as

$$RCOND = \frac{1}{\left(\|A(IA:IA+N-1,JA:JA+N-1)\| * \|A(IA:IA+N-1,JA:JA+N-1)^{-1}\|\right)}.$$

Arguments

NORM   (global input) CHARACTER*1
Specifies whether the 1-norm condition number or the infinity-norm condition number is required:
= '1' or 'O': 1-norm;
= 'I':   Infinity-norm.

UPLO   (global input) CHARACTER*1
= 'U':   A(IA:IA+N−1,JA:JA+N−1) is upper triangular;
= 'L':   A(IA:IA+N−1,JA:JA+N−1) is lower triangular.

DIAG   (global input) CHARACTER*1
= 'N':   A(IA:IA+N−1,JA:JA+N−1) is non-unit triangular;
= 'U':   A(IA:IA+N−1,JA:JA+N−1) is unit triangular.

N   (global input) INTEGER
The order of the distributed matrix A(IA:IA+N−1,JA:JA+N−1). N ≥ 0.

A   (local input) REAL/COMPLEX pointer into the local memory to an array of dimension (LLD_A, LOCc(JA+N−1)).
This array contains the local pieces of the triangular distributed matrix A(IA:IA+N−1,JA:JA+N−1).
If UPLO = 'U', the leading n-by-n upper triangular part of this distributed matrix contains the upper triangular matrix, and its strictly lower triangular part is not referenced.
If UPLO = 'L', the leading n-by-n lower triangular part of this distributed matrix contains the lower triangular matrix, and the strictly upper triangular part is not referenced.
If DIAG = 'U', the diagonal elements of A(IA:IA+N−1,JA:JA+N−1) are also not referenced and are assumed to be 1.

IA   (global input) INTEGER
The row index in the global array A indicating the first row of sub(A).

JA   (global input) INTEGER
The column index in the global array A indicating the first column of sub(A).

DESCA (global and local input) INTEGER array, dimension (DLEN_).
The array descriptor for the distributed matrix A.

RCOND (global output) REAL
The reciprocal of the condition number of the distributed matrix A(IA:IA+N−1,JA:JA+N−1), computed as
$$RCOND = \frac{1}{\left(\|A(IA:IA+N-1,JA:JA+N-1)\| * \|A(IA:IA+N-1,JA:JA+N-1)^{-1}\|\right)}.$$

WORK   (local workspace/local output) REAL/COMPLEX array, dimension (LWORK)
On exit, WORK(1) returns the minimal and optimal LWORK.

LWORK (local or global input) INTEGER.
The dimension of the array WORK.
LWORK is local input and must be at least

PSTRCON
LWORK ≥ 2*LOCr(N+mod(IA−1,MB_A)) + LOCc(N+mod(JA−1,NB_A)) + max(2, max(1, CEIL(NPROW−1,NPCOL)), LOCc(N+mod(JA−1,NB_A)) + NB_A*max(1, CEIL(NPCOL−1,NPROW))).

PCTRCON
LWORK ≥ 2*LOCr(N+mod(IA−1,MB_A)) + max(2, max(NB_A*CEIL(P−1,Q),LOCc(N+mod(JA−1,NB_A)) + NB_A*CEIL(Q−1,P))).

If LWORK = −1, then LWORK is global input and a workspace query is assumed; the routine only calculates the minimum and optimal size for all work arrays. Each of these values is returned in the first entry of the corresponding work array, and no error message is issued by PXERBLA.

IWORK *PSTRCON only* (local workspace/local output) INTEGER array, dimension (LIWORK)
On exit, IWORK(1) returns the minimal and optimal LIWORK.

LIWORK *PSTRCON only* (local or global input) INTEGER
The dimension of the array IWORK.
LIWORK is local input and must be at least
LIWORK ≥ LOCr(N+mod(IA−1,MB_A)).
If LIWORK = −1, then LIWORK is global input and a workspace

Purpose

PSTRRFS/PCTRRFS provides error bounds and backward error estimates for the solution to a system of linear equations with a triangular coefficient matrix.

The solution matrix X must be computed by PSTRTRS/PCTRTRS or some other means before entering this routine. PSTRRFS/PCTRRFS does not do iterative refinement because doing so cannot improve the backward error.

In the following comments, sub(A), sub(X) and sub(B) denote respectively A(IA:IA+N−1,JA:JA+N−1), X(IX:IX+N−1,JX:JX+NRHS−1) and B(IB:IB+N−1,JB:JB+NRHS−1).

Arguments

UPLO   (global input) CHARACTER*1
        = 'U':   sub(A) is upper triangular;
        = 'L':   sub(A) is lower triangular.

TRANS   (global input) CHARACTER*1
        Specifies the form of the system of equations:
        = 'N':   sub(A)*sub(X) = sub(B) (No transpose)
        = 'T':   sub(A)^T*sub(X) = sub(B) (Transpose)
        = 'C':   sub(A)^H*sub(X) = sub(B) (Conjugate transpose)

DIAG   (global input) CHARACTER*1
        = 'N':   sub(A) is non-unit triangular;
        = 'U':   sub(A) is unit triangular.

N   (global input) INTEGER
        The order of the matrix sub(A). N ≥ 0.

NRHS   (global input) INTEGER
        The number of right hand sides, i.e., the number of columns of the matrices sub(B) and sub(X). NRHS ≥ 0.

A   (local input) REAL/COMPLEX pointer into the local memory to an array of local dimension (LLD_A,LOCc(JA+N−1)).
        This array contains the local pieces of the original triangular distributed matrix sub(A).
        If UPLO = 'U', the leading n-by-n upper triangular part of sub(A) contains the upper triangular part of the matrix, and its strictly lower triangular part is not referenced.
        If UPLO = 'L', the leading n-by-n lower triangular part of sub(A) contains the lower triangular part of the distributed matrix, and its strictly upper triangular part is not referenced.
        If DIAG = 'U', the diagonal elements of sub(A) are also not referenced and are assumed to be 1.

IA   (global input) INTEGER
        The row index in the global array A indicating the first row of sub(A).

JA   (global input) INTEGER
        The column index in the global array A indicating the first column of sub(A).

query is assumed; the routine only calculates the minimum and optimal size for all work arrays. Each of these values is returned in the first entry of the corresponding work array, and no error message is issued by PXERBLA.

RWORK   PCTRCON only (local workspace/local output) REAL array, dimension (LRWORK).
        On exit, RWORK(1) returns the minimal and optimal LRWORK.

LRWORK PCTRCON only (local or global input) INTEGER
        The dimension of the array RWORK.
        LRWORK is local input and must be at least
        LRWORK ≥ LOCc(N+mod(JA−1,NB_A)).
        If LRWORK = −1, then LRWORK is global input and a workspace query is assumed; the routine only calculates the minimum and optimal size for all work arrays. Each of these values is returned in the first entry of the corresponding work array, and no error message is issued by PXERBLA.

INFO   (global output) INTEGER
        = 0:   successful exit
        < 0:   If the ith argument is an array and the j-entry had an illegal value, then INFO = −(i*100+j), if the ith argument is a scalar and had an illegal value, then INFO = −i.

PSTRRFS/PCTRRFS

```
SUBROUTINE PSTRRFS( UPLO, TRANS, DIAG, N, NRHS, A, IA, JA, DESCA,
     $                    B, IB, JB, DESCB, X, IX, JX, DESCX, FERR,
     $                    BERR, WORK, LWORK, IWORK, LIWORK, INFO )
       CHARACTER         DIAG, TRANS, UPLO
       INTEGER           INFO, IA, IB, IX, JA, JB, JX, LIWORK, LWORK,
     $                    N, NRHS
       INTEGER           DESCA( * ), DESCB( * ), DESCX( * ), IWORK( * )
       REAL              A( * ), B( * ), BERR( * ), FERR( * ),
     $                    WORK( * ), X( * )

SUBROUTINE PCTRRFS( UPLO, TRANS, DIAG, N, NRHS, A, IA, JA, DESCA,
     $                    B, IB, JB, DESCB, X, IX, JX, DESCX, FERR,
     $                    BERR, WORK, LWORK, RWORK, LRWORK, INFO )
       CHARACTER         DIAG, TRANS, UPLO
       INTEGER           INFO, IA, IB, IX, JA, JB, JX, LRWORK, LWORK,
     $                    N, NRHS
       INTEGER           DESCA( * ), DESCB( * ), DESCX( * )
       REAL              BERR( * ), FERR( * ), RWORK( * )
       COMPLEX           A( * ), B( * ), WORK( * ), X( * )
```

DESCA   (global and local input) INTEGER array, dimension (DLEN_). The array descriptor for the distributed matrix A.

B   (local input) REAL/COMPLEX pointer into the local memory to an array of local dimension (LLD_B, LOCc(JB+NRHS−1)). On entry, this array contains the the local pieces of the right hand sides sub(B).

IB   (global input) INTEGER The row index in the global array B indicating the first row of sub(B).

JB   (global input) INTEGER The column index in the global array B indicating the first column of sub(B).

DESCB   (global and local input) INTEGER array, dimension (DLEN_). The array descriptor for the distributed matrix B.

X   (local input) REAL/COMPLEX pointer into the local memory to an array of local dimension (LLDX, LOCc(JX+NRHS−1)). On entry, this array contains the the local pieces of the solution vectors sub(X).

IX   (global input) INTEGER The row index in the global array X indicating the first row of sub(X).

JX   (global input) INTEGER The column index in the global array X indicating the first column of sub(X).

DESCX   (global and local input) INTEGER array, dimension (DLEN_). The array descriptor for the distributed matrix X.

FERR   (local output) REAL array, dimension LOCc(JB+NRHS−1). The estimated forward error bounds for each solution vector of sub(X). If XTRUE is the true solution, FERR bounds the magnitude of the largest entry in (sub(X) − XTRUE) divided by the magnitude of the largest entry in sub(X). The estimate is as reliable as the estimate for RCOND, and is almost always a slight overestimate of the true error. This array is tied to the distributed matrix X.

BERR   (local output) REAL array, dimension LOCc(JB+NRHS−1). The componentwise relative backward error of each solution vector (i.e., the smallest relative change in any entry of sub(A) or sub(B) that makes sub(X) an exact solution). This array is tied to the distributed matrix X.

WORK   (local workspace/local output) REAL/COMPLEX array, dimension (LWORK) On exit, WORK(1) returns the minimal and optimal LWORK.

LWORK   (local or global input) INTEGER The dimension of the array WORK. LWORK is local input and must be at least
PSTRRFS
LWORK ≥ 3*LOCr(N + mod(IA−1, MB_A)).
PCTRRFS
LWORK ≥ 2*LOCr(N + mod(IA−1, MB_A)).
If LWORK = −1, then LWORK is global input and a workspace query is assumed; the routine only calculates the minimum and optimal size for all work arrays. Each of these values is returned in the first entry of the corresponding work array, and no error message is issued by PXERBLA.

IWORK   *PSTRRFS only* (local workspace/local output) INTEGER array, dimension (LIWORK) On exit, IWORK(1) returns the minimal and optimal LIWORK.

LIWORK   *PSTRRFS only* (local or global input) INTEGER The dimension of the array IWORK. LIWORK is local input and must be at least
LIWORK ≥ LOCr(N + mod(IB−1, MB_B)).
If LIWORK = −1, then LIWORK is global input and a workspace query is assumed; the routine only calculates the minimum and optimal size for all work arrays. Each of these values is returned in the first entry of the corresponding work array, and no error message is issued by PXERBLA.

RWORK   *PCTRRFS only* (local workspace/local output) REAL array, dimension (LRWORK) On exit, RWORK(1) returns the minimal and optimal LRWORK.

LRWORK   *PCTRRFS only* (local or global input) INTEGER The dimension of the array RWORK. LRWORK is local input and must be at least
LRWORK ≥ LOCr(N + mod(IB−1, MB_B)).
If LRWORK = −1, then LRWORK is global input and a workspace query is assumed; the routine only calculates the minimum and optimal size for all work arrays. Each of these values is returned in the first entry of the corresponding work array, and no error message is issued by PXERBLA.

INFO   (global output) INTEGER
= 0:   successful exit
< 0:   If the i-th argument is an array and the j-entry had an illegal value, then INFO = −(i*100+j), if the i-th argument is a scalar and had an illegal value, then INFO = −i.

PSTRTRI/PCTRTRI

```
SUBROUTINE PSTRTRI( UPLO, DIAG, N, A, IA, JA, DESCA, INFO )
CHARACTER      DIAG, UPLO
INTEGER        IA, INFO, JA, N
               DESCA( * )
REAL           A( * )
```

```
      SUBROUTINE PCTRTRI( UPLO, DIAG, N, A, IA, JA, DESCA, INFO )
      CHARACTER        DIAG, UPLO
      INTEGER          IA, INFO, JA, N
      INTEGER          DESCA( * )
      COMPLEX          A( * )
```

Purpose

PSTRTRI/PCTRTRI computes the inverse of a real/complex upper or lower trian-gular distributed matrix sub(A) = A(IA:IA+N−1,JA:JA+N−1).

Arguments

UPLO   (global input) CHARACTER*1
        Specifies whether the distributed matrix sub(A) is upper or lower triangular:
        = 'U':   Upper triangular;
        = 'L':   Lower triangular.

DIAG   (global input) CHARACTER*1
        Specifies whether or not the distributed matrix sub(A) is unit trian-gular:
        = 'N':   Non-unit triangular;
        = 'U':   Unit triangular.

N   (global input) INTEGER
        The number of rows and columns to be operated on, i.e., the order of the distributed submatrix sub(A). N ≥ 0.

A   (local input/local output) REAL/COMPLEX pointer into the local memory to an array of dimension (LLD_A,LOCc(JA+N−1)).
        This array contains the local pieces of the triangular matrix sub(A). If UPLO = 'U', the leading n-by-n upper triangular part of the matrix sub(A) contains the upper triangular matrix to be inverted, and the strictly lower triangular part of sub(A) is not referenced.
        If UPLO = 'L', the leading n-by-n lower triangular part of the matrix sub(A) contains the lower triangular matrix, and the strictly upper triangular part of sub(A) is not referenced.
        On exit, the (triangular) inverse of the original matrix.

IA   (global input) INTEGER
        The row index in the global array A indicating the first row of sub(A).

JA   (global input) INTEGER
        The column index in the global array A indicating the first column of sub(A).

DESCA   (global and local input) INTEGER array, dimension (DLEN_).
        The array descriptor for the distributed matrix A.

INFO   (global output) INTEGER
        = 0: successful exit
        < 0: If the ith argument is an array and the j-entry had an illegal value, then INFO = −(i*100+j), if the ith argument is a scalar and had an illegal value, then INFO = −i.

PSTRTRS/PCTRTRS

```
      SUBROUTINE PSTRTRS( UPLO, TRANS, DIAG, N, NRHS, A, IA, JA, DESCA,
     $                    B, IB, JB, DESCB, INFO )
      CHARACTER         DIAG, TRANS, UPLO
      INTEGER           IA, IB, INFO, JA, JB, N, NRHS
      INTEGER           DESCA( * ), DESCB( * )
      REAL              A( * ), B( * )

      SUBROUTINE PCTRTRS( UPLO, TRANS, DIAG, N, NRHS, A, IA, JA, DESCA,
     $                    B, IB, JB, DESCB, INFO )
      CHARACTER         DIAG, TRANS, UPLO
      INTEGER           IA, IB, INFO, JA, JB, N, NRHS
      INTEGER           DESCA( * ), DESCB( * )
      COMPLEX           A( * ), B( * )
```

Purpose

PSTRTRS/PCTRTRS solves a triangular system of the form sub(A)*X = sub(B), sub(A)^T*X = sub(B), or sub(A)^H*X = sub(B), where sub(A) denotes A(IA:IA+N−1,JA:JA+N−1) and is a triangular distributed matrix of order n, and B(IB:IB+N−1,JB:JB+NRHS−1) is an n-by-nrhs distributed matrix denoted by sub(B). A check is made to verify that sub(A) is nonsingular.

Arguments

UPLO   (global input) CHARACTER*1
        = 'U':   sub(A) is upper triangular;
        = 'L':   sub(A) is lower triangular.

TRANS   (global input) CHARACTER*1
        Specifies the form of the system of equations:
        = 'N':   sub(A)*X = sub(B)   (No transpose)
        = 'T':   sub(A)^T*X = sub(B)   (Transpose)
        = 'C':   sub(A)^H*X = sub(B)   (Conjugate transpose)

DIAG   (global input) CHARACTER*1
        = 'N':   sub(A) is non-unit triangular;
        = 'U':   sub(A) is unit triangular.

N   (global input) INTEGER
        The number of rows and columns to be operated on i.e, the order of the distributed submatrix sub(A). N ≥ 0.

NRHS   (global input) INTEGER
        The number of right hand sides, i.e., the number of columns of the

PSTZRZF/PCTZRZF

```
SUBROUTINE PSTZRZF( M, N, A, IA, JA, DESCA, TAU, WORK, LWORK,
$                   INFO )
    INTEGER          IA, INFO, JA, LWORK, M, N
    INTEGER          DESCA( * )
    REAL             A( * ), TAU( * ), WORK( * )
SUBROUTINE PCTZRZF( M, N, A, IA, JA, DESCA, TAU, WORK, LWORK,
$                   INFO )
    INTEGER          IA, INFO, JA, LWORK, M, N
    INTEGER          DESCA( * )
    COMPLEX          A( * ), TAU( * ), WORK( * )
```

Purpose

PSTZRZF/PCTZRZF reduces the m-by-n (m $\leq$ n) real/complex upper trapezoidal matrix sub(A) = A(IA:IA+M−1,JA:JA+N−1) to upper triangular form by means of orthogonal/unitary transformations.

The upper trapezoidal matrix sub(A) is factorized as

$$sub(A) \;=\; (\; R \quad 0 \;) * Z,$$

where Z is an n-by-n orthogonal/unitary matrix and R is an m-by-m upper triangular matrix.

Arguments

M   (global input) INTEGER
         The number of rows to be operated on, i.e., the number of rows of the distributed submatrix sub(A). M $\geq$ 0.

N   (global input) INTEGER
         The number of columns to be operated on, i.e., the number of columns of the distributed submatrix sub(A). N $\geq$ M.

A   (local input/local output) REAL/COMPLEX pointer into the local memory to an array of dimension (LLD_A, LOCc(JA+N−1)).
         On entry, the local pieces of the m-by-n distributed matrix sub(A) which is to be factored.
         On exit, the leading m-by-m upper triangular part of sub(A) contains the upper triangular matrix R, and elements m+1 to n of the first m rows of sub(A), with the array TAU, represent the orthogonal/unitary matrix Z as a product of M elementary reflectors.

IA   (global input) INTEGER
         The row index in the global array A indicating the first row of sub(A).

JA   (global input) INTEGER
         The column index in the global array A indicating the first column of sub(A).

A   (local input) REAL/COMPLEX pointer into the local memory to an array of dimension (LLD_A,LOCc(JA+N−1)).
         This array contains the local pieces of the distributed triangular matrix sub(A).
         If UPLO = 'U', the leading n-by-n upper triangular part of sub(A) contains the upper triangular matrix, and the strictly lower triangular part of sub(A) is not referenced.
         If UPLO = 'L', the leading n-by-n lower triangular part of sub(A) contains the lower triangular matrix, and the strictly upper triangular part of sub(A) is not referenced.
         If DIAG = 'U', the diagonal elements of sub(A) are also not referenced and are assumed to be 1.

IA   (global input) INTEGER
         The row index in the global array A indicating the first row of sub(A).

JA   (global input) INTEGER
         The column index in the global array A indicating the first column of sub(A).

DESCA   (global and local input) INTEGER array, dimension (DLEN_).
         The array descriptor for the distributed matrix A.

B   (local input/local output) REAL/COMPLEX pointer into the local memory to an array of dimension (LLD_B,LOCc(JB+NRHS−1)).
         On entry, this array contains the local pieces of the right hand side distributed matrix sub(B).
         On exit, if INFO = 0, sub(B) is overwritten by the solution matrix X.

IB   (global input) INTEGER
         The row index in the global array B indicating the first row of sub(B).

JB   (global input) INTEGER
         The column index in the global array B indicating the first column of sub(B).

DESCB   (global and local input) INTEGER array, dimension (DLEN_).
         The array descriptor for the distributed matrix B.

INFO   (global output) INTEGER
         = 0: successful exit
         < 0: If the ith argument is an array and the j-entry had an illegal value, then INFO = −(i*100+j), if the ith argument is a scalar and had an illegal value, then INFO = −i.
         > 0: If INFO = i, the ith diagonal element of sub(A) is zero, indicating that the submatrix is singular and the solutions X have not been computed.

DESCA   (global and local input) INTEGER array, dimension (DLEN_).
         The array descriptor for the distributed matrix A.

TAU   (local output) REAL/COMPLEX array, dimension LOCr(IA+M-1).
         This array contains the scalar factors of the elementary reflectors. TAU
         is tied to the distributed matrix A.

WORK   (local workspace/local output) REAL/COMPLEX array, dimension
         (LWORK)
         On exit, if INFO = 0, WORK(1) returns the minimal and optimal
         WORK.

LWORK   (local or global input) INTEGER
         The dimension of the array WORK.
         LWORK is local input and must be at least
         LWORK $\geq$ MB_A $*$ (Mp0 + Nq0 + MB_A), where
         IROFF = mod(IA$-$1, MB_A), ICOFF = mod(JA$-$1, NB_A),
         IAROW = INDXG2P(IA, MB_A, MYROW, RSRC_A, NPROW),
         IACOL = INDXG2P(JA, NB_A, MYCOL, CSRC_A, NPCOL),
         Mp0 = NUMROC(M+IROFF, MB_A, MYROW, IAROW, NPROW),
         Nq0 = NUMROC(N+ICOFF, NB_A, MYCOL, IACOL, NPCOL),
         and NUMROC, INDXG2P are ScaLAPACK tool functions; MYROW,
         MYCOL, NPROW and NPCOL can be determined by calling the sub-
         routine BLACS_GRIDINFO.
         If LWORK = -1, then LWORK is global input and a workspace query
         is assumed; the routine only calculates the minimum and optimal size
         for all work arrays. Each of these values is returned in the first en-
         try of the corresponding work array, and no error message is issued by
         PXERBLA.

INFO   (global output) INTEGER
         = 0:   successful exit
         < 0:   If the ith argument is an array and the j-entry had an illegal
                value, then INFO = $-(i*100+j)$, if the ith argument is a scalar
                and had an illegal value, then INFO = $-i$.

Bibliography

[1] M. ABOELAZE, N. CHRISOCHOIDES, AND E. HOUSTIS, *The Parallelization of Level 2 and 3 BLAS Operations on Distributed Memory Machines*, Tech. Rep. CSD-TR-91-007, Purdue University, West Lafayette, IN, 1991.

[2] R. AGARWAL, F. GUSTAVSON, AND M. ZUBAIR, *Improving Performance of Linear Algebra Algorithms for Dense Matrices Using Algorithmic Prefetching*, IBM J. Res. Dev., 38 (1994), pp. 265–275.

[3] E. ANDERSON, Z. BAI, C. BISCHOF, J. DEMMEL, J. DONGARRA, J. DU CROZ, A. GREEN-BAUM, S. HAMMARLING, A. MCKENNEY, S. OSTROUCHOV, AND D. SORENSEN, *LAPACK Users' Guide*, Society for Industrial and Applied Mathematics, Philadelphia, PA, second ed., 1995.

[4] E. ANDERSON, Z. BAI, C. BISCHOF, J. DEMMEL, J. DONGARRA, J. DU CROZ, A. GREEN-BAUM, S. HAMMARLING, A. MCKENNEY, AND D. SORENSEN, *LAPACK: A portable linear algebra library for high-performance computers*, Computer Science Dept. Technical Report CS-90-105, University of Tennessee, Knoxville, TN, May 1990. (Also LAPACK Working Note #20).

[5] E. ANDERSON, Z. BAI, AND J. DONGARRA, *Generalized QR factorization and its applications*, Linear Algebra and Its Applications, 162-164 (1992), pp. 243–273. (Also LAPACK Working Note #31).

[6] I. ANGUS, G. FOX, J. KIM, AND D. WALKER, *Solving Problems on Concurrent Processors: Software for Concurrent Processors*, vol. 2, Prentice Hall, Englewood Cliffs, N.J, 1990.

[7] ANSI/IEEE, *IEEE Standard for Binary Floating Point Arithmetic*, New York, Std 754-1985 ed., 1985.

[8] ——, *IEEE Standard for Radix Independent Floating Point Arithmetic*, New York, Std 854-1987 ed., 1987.

[9] M. ARIOLI, J. W. DEMMEL, AND I. S. DUFF, *Solving sparse linear systems with sparse backward error*, SIAM J. Matrix Anal. Appl., 10 (1989), pp. 165–190.

[10] C. ASHCRAFT, *The Distributed Solution of Linear Systems Using the Torus-wrap Data mapping*, Tech. Rep. ECA-TR-147, Boeing Computer Services, Seattle, WA, 1990.

[11] Z. BAI AND J. DEMMEL, *Design of a parallel nonsymmetric eigenroutine toolbox, Part I*, in Proceedings of the Sixth SIAM Conference on Parallel Processing for Scientific Computing, SIAM, 1993, pp. 391–398.

[12] Z. BAI AND J. DEMMEL, *Using the matrix sign function to compute invariant subspaces*, SIAM J. Matrix Anal. Appl, x (1997), p. xxx. to appear.

[13] Z. BAI, J. DEMMEL, J. DONGARRA, A. PETITET, H. ROBINSON, AND K. STANLEY, *The spectral decomposition of nonsymmetric matrices on distributed memory computers*, Computer Science Dept. Technical Report CS-95-273, University of Tennessee, Knoxville, TN, 1995. (Also LAPACK Working Note No. 91), To appear in SIAM J. Sci. Stat. Comput.

[14] Z. BAI AND J. W. DEMMEL, *Design of a parallel nonsymmetric eigenroutine toolbox, Part I*, in Proceedings of the Sixth SIAM Conference on Parallel Processing for Scientific Computing, R. F. *et al.* Sincovec, ed., Philadelphia, PA, 1993, Society for Industrial and Applied Mathematics, pp. 391–398. Long version available as Computer Science Report CSD-92-718, University of California, Berkeley, 1992.

[15] J. BARLOW AND J. DEMMEL, *Computing accurate eigensystems of scaled diagonally dominant matrices*, SIAM J. Num. Anal., 27 (1990), pp. 762–791. (Also LAPACK Working Note #7).

[16] J. BILMES, K. ASANOVIC, J. DEMMEL, D. LAM, AND C. CHIN, *Optimizing matrix multiply using PHiPAC: A portable, high-performance, ANSI C coding methodology*, Computer Science Dept. Technical Report CS-96-326, University of Tennessee, Knoxville, TN, 1996. (Also LAPACK Working Note #111).

[17] R. H. BISSELING AND J. G. G. VAN DE VORST, *Parallel LU decomposition on a transputer network*, in Lecture Notes in Computer Science, Number 384, G. A. van Zee and J. G. G. van de Vorst, eds., Springer-Verlag, 1989, pp. 61–77.

[18] L. S. BLACKFORD, J. CHOI, A. CLEARY, J. DEMMEL, I. DHILLON, J. J. DONGARRA, S. HAMMARLING, G. HENRY, A. PETITET, K. STANLEY, D. W. WALKER, AND R. C. WHALEY, *ScaLAPACK: A portable linear algebra library for distributed memory computers - design issues and performance*, in Proceedings of Supercomputing '96, Sponsored by ACM SIGARCH and IEEE Computer Society, 1996. (ACM Order Number: 415962, IEEE Computer Society Press Order Number: RS00126. http://www.supercomp.org/sc96/proceedings/).

[19] L. S. BLACKFORD, A. CLEARY, J. DEMMEL, I. DHILLON, J. DONGARRA, S. HAMMARLING, A. PETITET, H. REN, K. STANLEY, AND R. C. WHALEY, *Practical experience in the dangers of heterogeneous computing*, Computer Science Dept. Technical Report CS-96-330, University of Tennessee, Knoxville, TN, July 1996. (Also LAPACK Working Note #112), to appear ACM Trans. Math. Softw., 1997.

[20] R. BRENT, *The LINPACK Benchmark on the AP 1000*, in Frontiers, 1992, McLean, VA, 1992, pp. 128–135.

[21] R. BRENT AND P. STRAZDINS, *Implementation of BLAS Level 3 and LINPACK Benchmark on the AP1000*, Fujitsu Scientific and Technical Journal, 5 (1993), pp. 61–70.

[22] S. BROWNE, J. DONGARRA, S. GREEN, E. GROSSE, K. MOORE, T. ROWAN, AND R. WADE, *Netlib services and resources (rev. 1)*, Computer Science Dept. Technical Report CS-94-222, University of Tennessee, Knoxville, TN, 1994.

[23] S. BROWNE, J. DONGARRA, E. GROSSE, AND T. ROWAN, *The netlib mathematical software repository*, D-Lib Magazine (www.dlib.org), (1995).

[24] J. CHOI, J. DEMMEL, I. DHILLON, J. DONGARRA, S. OSTROUCHOV, A. PETITET, K. STANLEY, D. WALKER, AND R. C. WHALEY, *Installation guide for ScaLAPACK*, Computer Science Dept. Technical Report CS-95-280, University of Tennessee, Knoxville, TN, March 1995. (Also LAPACK Working Note #93).

[25] ——, *ScaLAPACK: A portable linear algebra library for distributed memory computers - design issues and performance*, Computer Science Dept. Technical Report CS-95-283, University of Tennessee, Knoxville, TN, March 1995. (Also LAPACK Working Note #95).

[26] J. CHOI, J. DONGARRA, S. OSTROUCHOV, A. PETITET, D. WALKER, AND R. C. WHALEY, *A proposal for a set of parallel basic linear algebra subprograms*, Computer Science Dept. Technical Report CS-95-292, University of Tennessee, Knoxville, TN, May 1995. (Also LAPACK Working Note #100).

[27] J. CHOI, J. DONGARRA, R. POZO, AND D. WALKER, *ScaLAPACK: A scalable linear algebra library for distributed memory concurrent computers*, in Proceedings of the Fourth Symposium on the Frontiers of Massively Parallel Computation, McLean, Virginia, 1992, IEEE Computer Society Press, pp. 120–127. (Also LAPACK Working Note #55).

[28] J. CHOI, J. DONGARRA, AND D. WALKER, *The design of a parallel dense linear algebra software library: Reduction to Hessenberg, tridiagonal and bidiagonal form*, Numerical Algorithms, 10 (1995), pp. 379–399. (Also LAPACK Working Note #92).

[29] J. CHOI, J. DONGARRA, AND D. WALKER, *PB-BLAS: A Set of Parallel Block Basic Linear Algebra Subroutines*, Concurrency: Practice and Experience, 8 (1996), pp. 517–535.

[30] A. CHTCHELKANOVA, J. GUNNELS, G. MORROW, J. OVERFELT, AND R. VAN DE GEIJN, *Parallel Implementation of BLAS: General Techniques for Level 3 BLAS*, Tech. Rep. TR95-49, Department of Computer Sciences, UT-Austin, 1995. Submitted to Concurrency: Practice and Experience.

[31] E. CHU AND A. GEORGE, *QR Factorization of a Dense Matrix on a Hypercube Multiprocessor*, SIAM Journal on Scientific and Statistical Computing, 11 (1990), pp. 990–1028.

[32] A. CLEARY AND J. DONGARRA, *Implementation in scalapack of divide-and-conquer algorithms for banded and tridiagonal linear systems*, Computer Science Dept. Technical Report CS-97-358, University of Tennessee, Knoxville, TN, April 1997. (Also LAPACK Working Note #125).

[33] M. COSNARD, Y. ROBERT, P. QUINTON, AND M. TCHUENTE, eds., *Parallel Algorithms and Architectures*, North-Holland, 1986.

[34] D. E. CULLER, A. ARPACI-DUSSEAU, R. ARPACI-DUSSEAU, B. CHUN, S. LUMETTA, A. MAINWARING, R. MARTIN, C. YOSHIKAWA, AND F. WONG, *Parallel computing on the Berkeley NOW*. To appear in JSPP'97 (9th Joint Symposium on Parallel Processing), Kobe, Japan, 1997.

[35] M. DAYDE, I. DUFF, AND A. PETITET, *A Parallel Block Implementation of Level 3 BLAS for MIMD Vector Processors*, ACM Trans. Math. Softw., 20 (1994), pp. 178–193.

[36] B. DE MOOR AND P. VAN DOOREN, *Generalization of the singular value and QR decompositions*, SIAM J. Matrix Anal. Appl., 13 (1992), pp. 993–1014.

[37] J. DEMMEL, *Underflow and the reliability of numerical software*, SIAM J. Sci. Stat. Comput., 5 (1984), pp. 887–919.

[38] ——, *Applied Numerical Linear Algebra*, SIAM, 1996. to appear.

[39] J. DEMMEL, S. EISENSTAT, J. GILBERT, X. LI, AND J. W. H. LIU, *A supernodal approach to sparse partial pivoting*, Technical Report UCB//CSD-95-883, UC Berkeley Computer Science Division, September 1995. to appear in SIAM J. Mat. Anal. Appl.

[40] J. DEMMEL AND K. STANLEY, *The performance of finding eigenvalues and eigenvectors of dense symmetric matrices on distributed memory computers*, Computer Science Dept. Technical Report CS-94-254, University of Tennessee, Knoxville, TN, September 1994. (Also LAPACK Working Note #86).

[41] J. W. DEMMEL, J. R. GILBERT, AND X. S. LI, *An asynchronous parallel supernodal algorithm for sparse Gaussian elimination*, February 1997. Submitted to SIAM J. Matrix Anal. Appl., special issue on Sparse and Structured Matrix Computations and Their Applications (Also LAPACK Working Note 124).

[42] J. W. DEMMEL AND X. LI, *Faster numerical algorithms via exception handling*, IEEE Trans. Comp., 43 (1994), pp. 983–992. (Also LAPACK Working Note #59).

[43] I. S. DHILLON, *Current inverse iteration software can fail*, (1997). Submitted for publication.

[44] ——, *A Stable $O(n^2)$ Algorithm for the Symmetric Tridiagonal Eigenproblem*, PhD thesis, University of California, Berkeley, CA, May 1997.

[45] I. S. DHILLON AND B. PARLETT, *Orthogonal eigenvectors without Gram-Schmidt*, (1997). draft.

[46] J. DONGARRA AND T. DUNIGAN, *Message-passing performance of various computers*, Tech. Rep. ORNL/TM-13006, Oak Ridge National Laboratory, Oak Ridge, TN, 1996. Submitted and accepted to Concurrency: Practice and Experience.

[47] J. DONGARRA, S. HAMMARLING, AND D. WALKER, *Key Concepts for Parallel Out-Of-Core LU Factorization*, Society for Industrial and Applied Mathematics, Philadelphia, PA, 1996. (Also LAPACK Working Note #110).

[48] J. DONGARRA, G. HENRY, AND D. WATKINS, *A distributed memory implementation of the nonsymmetric QR algorithm*, in Proceedings of the Eighth SIAM Conference on Parallel Processing for Scientific Computing, Philadelphia, PA, 1997, Society for Industrial and Applied Mathematics.

[49] J. DONGARRA, C. RANDRIAMARO, L. PRYLLI, AND B. TOURANCHEAU, *Array redistribution in ScaLAPACK using PVM*, in EuroPVM users' group, Hermes, 1995.

[50] J. DONGARRA AND R. VAN DE GEIJN, *Two dimensional basic linear algebra communication subprograms*, Computer Science Dept. Technical Report CS-91-138, University of Tennessee, Knoxville, TN, 1991. (Also LAPACK Working Note #37).

[51] J. DONGARRA, R. VAN DE GEIJN, AND D. WALKER, *Scalability issues in the design of a library for dense linear algebra*, Journal of Parallel and Distributed Computing, 22 (1994), pp. 523–537. (Also LAPACK Working Note #43).

[52] J. DONGARRA, R. VAN DE GEIJN, AND R. C. WHALEY, *Two dimensional basic linear algebra communication subprograms*, in Environments and Tools for Parallel Scientific Computing, Advances in Parallel Computing, J. Dongarra and B. Tourancheau, eds., vol. 6, Elsevier Science Publishers B.V., 1993, pp. 31–40.

[53] J. DONGARRA AND D. WALKER, *Software libraries for linear algebra computations on high performance computers*, SIAM Review, 37 (1995), pp. 151–180.

[54] J. DONGARRA AND R. C. WHALEY, *A user's guide to the BLACS v1.1*, Computer Science Dept. Technical Report CS-95-281, University of Tennessee, Knoxville, TN, 1995. (Also LAPACK Working Note #94).

[55] J. J. DONGARRA AND E. F. D'AZEVEDO, *The design and implementation of the parallel out-of-core ScaLAPACK LU, QR, and Cholesky factorization routines*, Department of Computer Science Technical Report CS-97-347, University of Tennessee, Knoxville, TN, 1997. (Also LAPACK Working Note 118).

[56] J. J. DONGARRA, J. DU CROZ, I. S. DUFF, AND S. HAMMARLING, *Algorithm 679: A set of Level 3 Basic Linear Algebra Subprograms*, ACM Trans. Math. Soft., 16 (1990), pp. 18–28.

[57] ——, *A set of Level 3 Basic Linear Algebra Subprograms*, ACM Trans. Math. Soft., 16 (1990), pp. 1–17.

[58] J. J. DONGARRA, J. DU CROZ, S. HAMMARLING, AND R. J. HANSON, *Algorithm 656: An extended set of FORTRAN Basic Linear Algebra Subroutines*, ACM Trans. Math. Soft., 14 (1988), pp. 18–32.

[59] ——, *An extended set of FORTRAN basic linear algebra subroutines*, ACM Trans. Math. Soft., 14 (1988), pp. 1–17.

[60] J. J. DONGARRA AND E. GROSSE, *Distribution of mathematical software via electronic mail*, Communications of the ACM, 30 (1987), pp. 403–407.

[61] J. J. DONGARRA, R. VAN DE GEIJN, AND D. W. WALKER, *A look at scalable dense linear algebra libraries*, in Proceedings of the Scalable High-Performance Computing Conference, IEEE, ed., IEEE Publishers, 1992, pp. 372–379.

[62] J. DU CROZ AND N. J. HIGHAM, *Stability of methods for matrix inversion*, IMA J. Numer. Anal., 12 (1992), pp. 1–19. (Also LAPACK Working Note #27).

[63] R. FALGOUT, A. SKJELLUM, S. SMITH, AND C. STILL, *The Multicomputer Toolbox Approach to Concurrent BLAS and LACS*, in Proceedings of the Scalable High Performance Computing Conference SHPCC-92, IEEE Computer Society Press, 1992.

[64] M. P. I. FORUM, *MPI: A message passing interface standard*, International Journal of Supercomputer Applications and High Performance Computing, 8 (1994), pp. 3–4. Special issue on MPI. Also available electronically, the URL is `ftp://www.netlib.org/mpi/mpi-report.ps`

[65] G. FOX, M. JOHNSON, G. LYZENGA, S. OTTO, J. SALMON, AND D. WALKER, *Solving Problems on Concurrent Processors, Volume 1*, Prentice-Hall, Englewood Cliffs, NJ, 1988.

[66] G. FOX, R. WILLIAMS, AND P. MESSINA, *Parallel Computing Works!*, Morgan Kaufmann Publishers, Inc., San Francisco, CA, 1994.

[67] T. L. FREEMAN AND C. PHILLIPS, *Parallel Numerical Algorithms*, Prentice-Hall, Hemel Hempstead, Hertfordshire, UK, 1992.

[68] A. GEIST, A. BEGUELIN, J. DONGARRA, W. JIANG, R. MANCHEK, AND V. SUNDERAM, *PVM: Parallel Virtual Machine. A Users' Guide and Tutorial for Networked Parallel Computing*, MIT Press, Cambridge, MA, 1994.

[69] G. GEIST AND C. ROMINE, *LU factorization algorithms on distributed memory multiprocessor architectures*, SIAM J. Sci. Stat. Comput., 9 (1988), pp. 639–649.

[70] G. GOLUB AND C. VAN LOAN, *Matrix Computations*, Johns-Hopkins, Baltimore, second ed., 1989.

[71] G. GOLUB AND C. F. VAN LOAN, *Matrix Computations*, Johns Hopkins University Press, Baltimore, MD, third ed., 1996.

[72] W. W. HAGER, *Condition estimators*, SIAM J. Sci. Stat. Comput., 5 (1984), pp. 311–316.

[73] S. HAMMARLING, *The numerical solution of the general Gauss-Markov linear model*, in Mathematics in Signal Processing, T. S. *et al.*. Durani, ed., Clarendon Press, Oxford, UK, 1986.

[74] R. HANSON, F. KROGH, AND C. LAWSON, *A proposal for standard linear algebra subprograms*, ACM SIGNUM Newsl., 8 (1973).

[75] P. HATCHER AND M. QUINN, *Data-Parallel Programming On MIMD Computers*, The MIT Press, Cambridge, Massachusetts, 1991.

[76] B. HENDRICKSON AND D. WOMBLE, *The torus–wrap mapping for dense matrix calculations on massively parallel computers*, SIAM J. Sci. Stat. Comput., 15 (1994), pp. 1201–1226.

[77] G. HENRY, *Improving Data Re-Use in Eigenvalue-Related Computations*, PhD thesis, Cornell University, Ithaca, NY, January 1994.

[78] G. HENRY AND R. VAN DE GEIJN, *Parallelizing the QR algorithm for the unsymmetric algebraic eigenvalue problem: Myths and reality*, SIAM J. Sci. Comput., 17 (1996), pp. 870–883. (Also LAPACK Working Note 79).

[79] G. HENRY, D. WATKINS, AND J. DONGARRA, *A parallel implementation of the nonsymmetric QR algorithm for distributed memory architectures*, Computer Science Dept. Technical Report CS-97-352, University of Tennessee, Knoxville, TN, March 1997. (Also LAPACK Working Note # 121).

[80] N. J. HIGHAM, *A survey of condition number estimation for triangular matrices*, SIAM Review, 29 (1987), pp. 575–596.

[81] ——, *FORTRAN codes for estimating the one-norm of a real or complex matrix, with applications to condition estimation*, ACM Trans. Math. Softw., 14 (1988), pp. 381–396.

[82] ——, *Experience with a matrix norm estimator*, SIAM J. Sci. Stat. Comput., 11 (1990), pp. 804–809.

[83] ——, *Perturbation theory and backward error for $AX - XB = C$*, BIT, 33 (1993), pp. 124–136.

[84] ——, *Accuracy and Stability of Numerical Algorithms*, Society for Industrial and Applied Mathematics, Philadelphia, PA, 1996.

[85] S. HUSS-LEDERMAN, E. JACOBSON, A. TSAO, AND G. ZHANG, *Matrix Multiplication on the Intel Touchstone DELTA*, Concurrency: Practice and Experience, 6 (1994), pp. 571–594.

[86] S. HUSS-LEDERMAN, A. TSAO, AND G. ZHANG, *A parallel implementation of the invariant subspace decomposition algorithm for dense symmetric matrices*, in Proceedings of the Sixth SIAM Conference on Parallel Processing for Scientific Computing, SIAM, 1993, pp. 367–374.

[87] K. HWANG, *Advanced Computer Architecture: Parallelism, Scalability, Programmability*, McGraw-Hill, 1993.

[88] IBM CORPORATION, *IBM RS6000*, 1996. (URL = http://www.rs6000.ibm.com/).

[89] INTEL CORPORATION, *Intel Supercomputer Technical Publications Home Page*, 1995. (URL = http://www.ssd.intel.com/pubs.html).

[90] B. KÅGSTRÖM, P. LING, AND C. V. LOAN, *GEMM-based level 3 BLAS: High-performance model implementations and performance evaluation benchmark*, Tech. Rep. UMINF 95-18, Department of Computing Science, Umeå University, 1995. Submitted to ACM Trans. Math. Softw.

[91] C. KOEBEL, D. LOVEMAN, R. SCHREIBER, G. STEELE, AND M. ZOSEL, *The High Performance Fortran Handbook*, MIT Press, Cambridge, Massachusetts, 1994.

[92] V. KUMAR, A. GRAMA, A. GUPTA, AND G. KARYPIS, *Introduction to Parallel Computing – Design and Analysis of Algorithms*, The Benjamin/Cummings Publishing Company, Inc., Redwood City, CA, 1994.

[93] C. L. LAWSON, R. J. HANSON, D. KINCAID, AND F. T. KROGH, *Basic linear algebra subprograms for Fortran usage*, ACM Trans. Math. Soft., 5 (1979), pp. 308–323.

[94] R. LEHOUCQ, *The computation of elementary unitary matrices*, Computer Science Dept. Technical Report CS-94-233, University of Tennessee, Knoxville, TN, 1994. (Also LAPACK Working Note 72).

[95] T. LEWIS AND H. EL-REWINI, *Introduction to Parallel Computing*, Prentice-Hall, Inc., Englewood Cliffs, NJ, 1992.

[96] X. LI, *Sparse Gaussian Elimination on High Performance Computers*, PhD thesis, Computer Science Division, Department of Electrical Engineering and Computer Science, University of California, Berkeley, CA, September 1996.

[97] W. LICHTENSTEIN AND S. L. JOHNSSON, *Block-cyclic dense linear algebra*, SIAM J. Sci. Stat. Comput., 14 (1993), pp. 1259–1288.

[98] A. MAINWARING AND D. E. CULLER, *Active message applications programming interface and communication subsystem organization*, Tech. Rep. UCB CSD-96-918, University of California at Berkeley, Berkeley, CA, October 1996.

[99] P. PACHECO, *Parallel Programming with MPI*, Morgan Kaufmann Publishers, Inc., San Francisco, CA, 1997.

[100] C. PAIGE, *Some aspects of generalized QR factorization*, in Reliable Numerical Computations, M. Cox and S. Hammarling, eds., Clarendon Press, 1990.

[101] B. PARLETT, *The Symmetric Eigenvalue Problem*, Prentice-Hall, Englewood Cliffs, NJ, 1980.

[102] ——, *The construction of orthogonal eigenvectors for tight clusters by use of submatrices*, Center for Pure and Applied Mathematics PAM-664, University of California, Berkeley, CA, January 1996. submitted to SIMAX.

[103] B. PARLETT AND I. DHILLON, *On Fernando's method to find the most redundant equation in a tridiagonal system*, Linear Algebra and Its Applications, (1996). to appear.

[104] A. PETITET, *Algorithmic Redistribution Methods for Block Cyclic Decompositions*, PhD thesis, University of Tennessee, Knoxville, TN, 1996.

[105] E. POLLICINI, A. A., *Using Toolpack Software Tools*, 1989.

[106] L. PRYLLI AND B. TOURANCHEAU, *Efficient block cyclic data redistribution*, in EUROPAR'96, vol. 1 of Lecture Notes in Computer Science, Springer-Verlag, 1996, pp. 155–165.

[107] ——, *Efficient block cyclic array redistribution*, Journal of Parallel and Distributed Computing, (1997). To appear.

[108] R. SCHREIBER AND C. F. VAN LOAN, *A storage efficient WY representation for products of Householder transformations*, SIAM J. Sci. Stat. Comput., 10 (1989), pp. 53–57.

[109] B. SMITH, W. GROPP, AND L. CURFMAN MCINNES, *PETSc 2.0 users manual*, Technical Report ANL-95/11, Argonne National Laboratory, Argonne, IL, 1995. (Available by anonymous ftp from `ftp.mcs.anl.gov`).

[110] M. SNIR, S. W. OTTO, S. HUSS-LEDERMAN, D. W. WALKER, AND J. J. DONGARRA, *MPI: The Complete Reference*, MIT Press, Cambridge, MA, 1996.

[111] SUNSOFT, *The XDR Protocol Specification. Appendix A of "Network Interfaces Programmer's Guide"*, SunSoft, 1993.

[112] E. VAN DE VELDE, *Concurrent Scientific Computing*, no. 16 in Texts in Applied Mathematics, Springer-Verlag, 1994.

[113] R. C. WHALEY, *Basic linear algebra communication subprograms: Analysis and implementation across multiple parallel architectures*, Computer Science Dept. Technical Report CS-94-234, University of Tennessee, Knoxville, TN, May 1994. (Also LAPACK Working Note 73).

[114] J. H. WILKINSON, *The Algebraic Eigenvalue Problem*, Oxford University Press, Oxford, UK, 1965.

Index by Keyword

Index by Routine Name